*Prophets Facing Backward*

# Prophets Facing Backward

*Postmodern Critiques of Science and Hindu Nationalism in India*

*Meera Nanda*

*Rutgers University Press*

New Brunswick, New Jersey, and London

Library of Congress Cataloging-in-Publication Data
Nanda, Meera.
Prophets facing backward : postmodern critiques of science and Hindu
nationalism in India / Meera Nanda.
p. cm.
Includes bibliographical references and index.
ISBN 0-8135-3357-0 (alk. paper) — ISBN 0-8135-3358-9 (pbk. : alk. paper)
1. Science—History—India—20th century.  2. Nationalism—India.
3. Science—Philosophy—History—20th century.  I. Title.
Q127. I4N353 2003
509.54'09'04 dc21                                    2003006165

British Cataloging-in-Publication data for this book is available from the British
Library.

"Materialist Feminism: A Reader in Class, Difference and Women's Lives" by
Rosemary Hennessy and Chrys Inghram, 364-394, copyright 1997, is reproduced by permission of Routledge, Inc. part of the Taylor & Francis Group.

The publication program of Rutgers University Press is supported by the Board
of Governors of Rutgers, The State University of New Jersey.

Manufactured in the United States of America

*For Ravi*

# Contents

# List of Abbreviations

ASM    Alternative Science Movement
BJP    Bharatiya Janata Party (Indian People's Party)
BKU    Bharatiya Kisan Union (Indian Farmers' Union)
BKS    Bharatiya Kisan Sangh (Indian Farmers' Association)
KRRS   Karnataka Rajya Ratitha Sangh
KSSP   Kerela Sastra Sahitya Parishad (Kerela Science and Literary Council)
NSM    New Social Movement
PSM    People's Science Movement
RSS    Rashtriya Swayamsevak Sangh (National Volunteers Association)
SJM    Swadeshi Jagran Manch (Indigenous Awareness Forum)
SS     Shetkari Sanghathan (Farmers' Association)
VHP    Vishva Hindu Parishad (World Hindu Council)

# Preface

This is a book I *had* to write. I had no choice in the matter.

For many years, I have seen modern science being "put in its place." For many years, I have heard all the problems of the world we live in being laid at the doorstep of the Enlightenment. For many years, I have felt as if I'm drowning in platitudes about cultivating the "alternative sciences" of women, non-Western peoples, and other "victims" of the modern age.

I could never join the chorus, but nor could I shut out its jarring notes. Circumstances of biography had brought me into the humanities in the American academia at the height of its postmodernist fervor. The very fundamentals of what constitutes knowledge, how it is constructed and its impact on society were being questioned. It was simply not possible to remain neutral to these questions and just do your work—more so, if these debates were challenging your own deeply felt beliefs.

A microbiologist by training (Ph.D., Indian Institute of Technology, New Delhi, India), I came to America to study philosophy of science. For a variety of reasons, I dropped out after a couple of semesters at the Department of History and Philosophy of Science at Indiana University. At that time, continuing with my chosen profession of science writing and journalism made more sense. But I soon tired of skimming the surface of things. My itch for books, ideas, and scholarship returned. I sought out the academia one more time. In 1993, I enrolled for another doctorate degree at the Department of Science and Technology Studies at the Rensselaer Polytechnic Institute (RPI) in upstate New York, considered one of the pioneer programs in the field.

In science studies, I came face-to-face with a radical challenge to—even a denunciation of—all that I believed in. I found myself in a discipline whose very founding axioms I disagreed with. Using various nuances, science studies teach that modern science as we know it, down to its very content and criteria for justification is a construction of the dominant social interests of Western society. As a construct of power, modern science serves as a legitimator of Western and patriarchal power around the world. Many of my own compatriots from India had contributed to this literature, arguing that non-Western societies needed a "decolonization of the mind" that would come about by developing sciences that encoded Indian values which could still be found among the non-modern women and men in the farms and forests. Even worse, those enunciating these theories were people who professed to speak as feminists, radical democrats, and social-ists—political ideals of the left that I share. The difference, of course, was that while they saw themselves as belonging to the left of the political spectrum be-cause of their *suspicion* of modern "Western" science among other things, I saw myself as a leftist because of my *admiration* of modern science among other things.

Having grown up in a provincial town in northern India, I considered my edu-cation in science a source of personal enlightenment. Natural science, especially molecular biology, had given me a whole different perspective on the underlying cosmology of the religious and cultural traditions I was raised in. Science gave me good reasons to say a principled "No!" to many of my inherited beliefs about God, nature, women, duties and rights, purity and pollution, social status, and my relationship with my fellow citizens. I had discovered my individuality, and found the courage to assert the right to fulfill my own destiny, because I learned to de-mand good reasons for the demands that were put on me. When I came to New Delhi to do my doctorate, I was full of great idealism, and a great love for biology. When I finally gave up a career in research for science writing (I joined the staff of *The Indian Express* as a science correspondent after I completed my doctor-ate), it was because I thought science was too important to be confined to labo-ratories, and imprisoned in the pages of refereed journals. I was convinced that modern science had a role to play in religious reformation and cultural revolution in Indian society. Without knowing it then, I was speaking the language of the Enlightenment.

I soon found out that science studies had no room for the likes of me. En-lightenment was seen as the agent of colonialism, and modern science as a dis-course of patriarchy and other dominant Western interests. Salvation was to be found in debunking the universalist pretensions of science and encouraging al-ternative ways of knowing that would end the hubris of the West. Someone like me could only be pitied—which I often was—as a "colonized mind," dazzled by the superficial charms of the West.

The problem, I discovered, lay not only in the political ambience which over-whelmingly favored a debunking, radical skeptical stance toward science and modernity—that I could have put up with, even at the cost of intellectual loneli-ness. But the real problem lay in the theoretical rubric of social constructivism that justified the kind of relativist disdain for science that I was encountering. Starting with the Strong Programme, with its various cultural studies and femi-nist offshoots, combined with postcolonial theory and post-development cri-tiques, science was assumed to be "symmetrical" with all other local knowledges, all the way down to the content and cognitive values used for assessing and evaluating the evidence of experiments. In all cases alike, the ultimate source of justification was to be found in the prevailing social relations, cultural metaphors, and metaphysics. I saw this approach as denying the progress and universality of modern science. I also refused to make peace with the extreme indulgence, in the name of respecting cultural differences, toward patent falsehoods in other cultures' ontologies, and palpable irrationalities in their ways of relating to nature. What was being celebrated as "difference" by postmodernists, was more often than not, a source of mental bondage and authoritarianism in non-Western cultures.

This book started as my dissertation in which I decided to answer the constructivist science critics. I had two goals in my dissertation, both of which appear rather limited when I look at them from where I have ended up in this book. My first goal was to show why I could not consent to the dogmas of social constructivism, including feminist and postcolonial epistemologies. I tried to show that these critics did not do justice to science, either as an intellectual inquiry, or as a cultural weapon for the Enlightenment, especially in non-Western societies. My second goal was to examine the actual track record of postmodernist, "alternative science movements" in India. In my analysis of India's ecofeminist opposition to the Green Revolution, for example, I showed that the critics of science were eventually embracing the same traditions that have kept Indian women tied to home and hearth for centuries. I also tried to find a compromise between the relativism of social constructivism and an old-fashioned realism in the pragmatic realism of the classical American pragmatism of Charles Pierce and John Dewey. Interestingly, I was able to find an Indian homologue of a Deweyan secular-humanism in the teachings of the Buddha, as interpreted by India's most well-known dalit (untouchable) public intellectual, Bhimrao Ramji Ambedkar. Indian Enlightenment, I argued in my dissertation (as I do again in this book), will be built upon these indigenous resources of reason, naturalism, and humanism, updated and refreshed by modern science, and will be led by the victims of home-grown, religiously sanctioned oppressions. Skeptical reason, institutionalized by modern science, *is* the standpoint epistemology of the oppressed, I argued. My dissertation, in other words, was a confrontation with the founding assumptions of the entire paradigm of post-Kuhnian, anti-Enlightenment science studies, including feminist and postcolonial epistemologies.

It so happened that just as I was beginning my second stint in academia, dark clouds of Hindu nationalism were beginning to gather on India's horizon. In 1992, the long-simmering storm broke, and brought violence and mayhem in its wake. Mobs of frenzied Hindus destroyed an ancient mosque in Ayodhya with the intention of building a temple to their god Rāma at the very spot where the mosque had stood, and where they claim God Rāma was born eons ago. Soon afterwards, a coalition led by Bharatiya Janata Party, or BJP, a party with solid Hindu nationalist credentials, took control of the central government in New Delhi. Indulged and protected by the Bharatiya Janata Party, a culture of cruelty and intolerance in the name of the "Hindu nation" has become mainstream in India today. I consider Hindu nationalism to be one of the most dangerous developments in India's 55 years of independence.

It might come as a surprise to those who are unfamiliar with the importance of natural law in Hinduism, and with the history of Hindu nationalism, that science has always been at the center of Hindu nationalist revivalism. Hindu nationalists are obsessed with science, in the same way and for the same reasons as the "creation scientists" are obsessed with science. They display a desperate urge to "prove" that modern science verifies the metaphysical assumptions of the classical, Vedic Hinduism, and conversely, the sacred books of the Vedas and the Upanishads are simply "science by another name." Rather than take the contradictions between science and the Vedic conception of nature seriously, Hindu nationalists deny that there is any contradiction at all. This view of Vedic science is being pushed in schools, colleges, and the mass media to create a generation of Indians who think in Hindu supremacist terms.

In this book, I have broadened the scope of my inquiry to include the right-wing anti-modernist—or more accurately, reactionary modernist—movement of Hindu nationalism in my examination of how postmodern critiques of science are circulating in a modernizing society like India. The reality in India is that today, there is not one but two populist movements that stand shoulder-to-shoulder against secular modernity: they are the postmodernist new social movements on the left, and the Hindu nationalist movements on the right. I wanted to see what the causes of their origin and growth were. How did the postmodernist vision of a post-Enlightenment "alternative modernity" differ from that of the idea of "Hindu modernity?" Was the defense of epistemological parity between all sciences demanded by social constructivist theories any different from the much celebrated Hindu notion that different ways of knowledge are different only in name, and lead to the same truth?

Thus, I shifted my attention from social constructivist/postmodernist theories flowering in science studies and allied disciplines to what was going on on the ground in India. I plunged into the voluminous Hindu nationalist literature on "Hindu science" and "Hindu modernity." It is the result of my inquiry into the shaved ground between the postmodern and Hindu conceptions of science and modernity that I report in this book.

My inquiry has led me to conclude that what is going on in India today can best be described as "reactionary modernism," a development not very dissimilar from the one described by Jeffrey Herf in his 1984 study of the Weimar Republic and the Third Reich. Nuclear-tipped India is aggressively modernizing its technology, and equally aggressively re-traditionalizing its culture.

Now, Hindu ideologues might occasionally display Taliban-like sensibilities (demolishing a mosque, for instance, or destroying paintings that they find offensive, or inciting violence against films and books), but they are not crude revivalists. Their brand of revivalism is very sophisticated and often indistinguishable from the left-wing, postmodernist defense of "difference." It is not as if the Vedic science proponents are reading Michel Foucault, David Bloor, or Sandra Harding, although some of them do cite Thomas Kuhn and Paul Feyerabend. But the defenders of the faith simply assume that science and the project of modernity have been discredited "even in the West," and that they need not, any longer, treat them as phenomena of universal and critical import.

Re-traditionalization in India means not a rejection of the modern ideals of democracy or scientific reason, or even secularism: India has the dubious distinction of being perhaps the only society in the world where religious fanatics claim to be the "true" defenders of "genuine secularism!" As I learnt more about the mode of Hindu-style re-traditionalization of modern ideas, especially of science and secularism, I came to see that at the heart of Hindu supremacist ideology lies a very postmodernist assumption, namely, each society has its own distinctive norms of reasonableness, logic, rules of evidence, and conception of truth and there is no non-arbitrary, culture-independent way to choose among these various alternatives. What is more, I found a very postmodernist political assumption motivating the defenders of the faith in India, namely, each culture not only "naturally" prefers its own norms, but each culture has an obligation to be true to itself and cultivate a "modernity" in keeping with the "ways of knowing" ingrained in its own culture. The only difference was that while postmodernist intellectuals hoped for a new, more humane, feminist, and less reductionist science to emerge from the traditions of the oppressed and the neglected peoples in non-modern cultures, Hindu nationalists were claiming that such a humane, ecological, and non-reductionist science was already present in the worldview of the Vedas and the Vedānta, the dominant tradition of Hinduism. I found that the arguments that the postmodern and postcolonial left use to develop standpoint epistemologies of the colonized and the oppressed were being used by Hindu chauvinists, the enemies of the left, to present Hinduism itself as a paradigm of "alternative science" that will lead to the "decolonization of the mind." To be truly scientific, Hindu nationalists demand a return to the most orthodox elements of Hinduism. These same elements, incidentally, have a long history of accommodating the most irrational practices from astrology and magic to irrational taboos involving purity and pollution. These same elements have, for centuries, also justified depriving the vast majority of Indian people (women of all castes, the lower castes, untouchables, and tribal people) of their full humanity.

Furthermore, I came to realize that technological modernization without secularization and the acceptance of liberal Enlightenment values is dangerous, as it puts the force of modern technology behind fundamentally authoritarian and hierarchical social values. These values, in the case of India, are being mobilized by Hindu demagogues against religious minorities, especially the Muslims and the Christians, who are being stigmatized as enemies of "Hindu India." While the left-inclined anti-Enlightenment movements have not been able to stem the tide of new technology, they have been successful in silencing the modernist, Enlightenment-style thought in India. By denouncing science and secularization, the left has unwittingly contributed to the ugly phenomenon of reactionary modernism.

To my mind, the fact that this defense of orthodoxy could find support in avant-garde movements of the left in the new global academe was nothing less than a scandal and needed to be exposed. This is what this book aims to do.

The book opens with an extended introduction which lays down the thesis of reactionary modernity, and shows its connections to religious nationalism and fascism in its Nazi incarnation. The rest of the book is divided into three parts. The first part (chapters 2 to 4) is a study of the rise of Hindu nationalism and its defense of Vedic science. In the second chapter I argue that the lack of secularization of the civil society is a major cause of India taking the turn toward a reactionary-modernist Hindu nationalism. The other two chapters in this part examine questions related to Vedic science and how it is defended by its proponents as a legitimate, viable science, not just in antiquity but in the contemporary world.

The second part of the book (chapters 5 to 7) takes a closer look at the theoretical aspects of science studies. Chapter 5 reveals the relativist logic—what I call "epistemic charity"—of the Strong Programme of the sociology of scientific knowledge, cultural studies, feminist epistemology, and postcolonial theory. The next chapter looks at theories of hybridity, "cyborg" subjectivity, or borderland epistemology. I argue that the state of hybridity, much celebrated as an act of "resistance" and "agency" against the colonization of the mind, actually serves as the substratum for reactionary modernism. The seventh chapter is one of my favorites. In this chapter I try to introduce the reader to the "Deweyan Buddha" of India's dalits (untouchables), as interpreted by Ambedkar, India's foremost dalit intellectual for whom I have the utmost love and respect. This chapter is meant to offer a framework for a middle path and a workable epistemology for the Indian Enlightenment.

The third and final part of the book (chapters 8 to 10) looks at how postmodern critiques of science have influenced the agendas of new social movements in India. Chapter 8 revisits the "scientific temper" debate that took place in the early 1980s in India, a kind of pre-Sokal Science War I, which affected me deeply when it happened. The next chapter takes a critical look at ecofeminist ideas, while the final chapter looks at how ecofeminist ideas have been picked up by conservative farmers' movements that are supplying votes and recruits for the Hindu nationalist movements. The book concludes with a plea for another kind of revival in India—a revival of the spirit of the secular Enlightenment.

All through this work, I have keenly felt my parents' influence. My late father's uncompromising stand for his beliefs and his strict, almost monastic, self-discipline served me well as a model to live up to. My mother, forever the rationalist and the secularist, influenced me greatly by simply being herself. With no great fuss, and no fanfare, she has always stood up against ideas that offend reason.

My greatest debt, however, is to Ravi Rajamani, my partner and friend, and Jaya, our daughter. Their unfailing love and patience has sustained me through the many highs and lows of this project.

Meera Nanda
Thanksgiving Day, 2002
West Hartford, Connecticut

*Prophets Facing Backward*

# One

# Prophets Facing Backward

## *Betrayal of the Clerks*

---

*Saints should always be judged guilty until proven innocent.*
—*George Orwell,* Reflections on Gandhi

In the crash of the falling World Trade Center towers in New York City on September 11, 2001, one could hear, loud and clear, the intimation of an old specter rising: fascism. Signs of growing state authoritarianism are everywhere, even in advanced democracies. But fascism, wearing a clerical garb, and speaking the language of religious fundamentalism, is a real possibility in those largely non-Western societies where democratic institutions are young, or not fully established. While it was a radical interpretation of Islam that animated the suicidal killers of September 11, other religions are no less immune to being hijacked by fascistic movements.

In this book we will examine the rise of Hindu nationalism, a violent and chauvinistic movement that displays all the marks of fascism, including state-encouraged violence against religious minorities, like the pogrom that is going on against the Muslims in the state of Gujarat, even as I write these words. Like all fascists, Hindu nationalists set their political compass by the vision of a mythic golden age. These are the prophets who, even as they march forward, keep their faces turned backward toward an imagined past of Hindu glory.

The "clerks" whose betrayal will concern us in this book are men and women of secular learning, intellectuals who uphold left-wing *political* ideals, but who have lost all confidence in the classic left-wing *cultural* ideals of scientific reason, modernity, and the Enlightenment. These postmodern intellectuals and activists, in other words, display a passion for radical social transformation, alongside an equally passionate rage against the Enlightenment's promise of progressive social change through a rational critique of superstition, ideologies, and flawed reasoning. They have come to see the claims of universalism and objectivity of science as so many excuses for Eurocentrism and the "mental colonialism" of non-Western people. Their vision of a socially egalitarian world calls for, above all, an epistemological egalitarianism which respects the rights of non-Western and other non-dominant social groups to develop "their own" sciences, reflecting their culturally distinctive reason.

1

This book will tell the story of how these intellectuals, in their despair over the world they find themselves in, have helped deliver the people they profess to love—the non-Western masses, the presumed victims of "Western science" and modernity—to the growing forces of hatred, fascism, and religious fanaticism. We will look at how the Hindu nationalist dreams of a "Hindu modernity" have found a respectable home in the theories of "alternative epistemology" and "local knowledge" popular in the social constructivist, feminist, and Third-Worldist trends in the academia. How the vanguard of radical postmodernist thought in the latter half of the twentieth century has served as a bridge to reactionary modernist movements darkening the horizons in the twenty-first century is the theme of this book.

This book has one more task: to reaffirm the historic association of science with the goals of the Enlightenment, secularism, and democracy, especially in non-Western societies. I stand with Alan Sokal, that inimitable critic of all varieties of "fashionable nonsense,"[1] in affirming that "rational thought and fearless analysis of objective reality (both natural and social) are incisive tools for combating mystifications promoted by the powerful" (Sokal 1996, 64). With the threat of clerical fascism staring us in the face, it is time to recover the ground lost to the feel-good but dangerous relativism of postmodernism.

In this chapter I will try to draw a map of the terrain we will be covering in the rest of the book. I also try to provide clear definitions of the terms that we will be using to read the map. My goal is to place the rise of Hindu nationalism within the context of clerical fascism on the one hand, and the postmodern/postcolonial suspicion of modernity, especially modern science, on the other.

### *"Alternative Modernity" as Reactionary Modernism*

The attack on the World Trade Center symbolized the marriage between modern technology and religio-political passions that is becoming quite commonplace around the world. Radical Islamic groups routinely carry out their self-proclaimed *jihad* against the secular world using the tools—from the internet and television to bombs and other military hardware—made available by modern science, the driving force behind the secular world, whose worldview they oppose.

Or to take a potentially far more dangerous example, India and Pakistan, both increasingly riven by ethnic and religious politics internally and against each other, have armed themselves with nuclear weapons. In both countries, one finds that aggressive *technological modernization* is serving to further an equally aggressive *cultural re-traditionalization*, visible in the growing influence of religious nationalist ideas on the institutions of civil society and the state. These societies display a schizophrenia in which the modernization of the material-technological environment is embraced with great enthusiasm, while the modernization or secularization of cultural categories through which to understand these material developments is resisted as a sign of "Westernization."

To some extent, this disjunction between technological modernization and cultural conservatism is a normal part of modernization. As Daniel Bell argued in his classic *The Cultural Contradictions of Capitalism,* the sphere of production of goods (techno-economic) and the sphere of production of meaning (culture) "respond to different norms, have different rhythms of change, and are regulated by different, even contrary principles" (1996, 10). Industrialization of the techno-economic sphere does carry over a more functional, instrumental rationality into other spheres of social life. But the cultural realm is not moved solely by the drive for utility, or for class interests, for that matter. On the contrary, its affective and existential dimensions actively *resist* the utilitarian drive. For these reasons, it is not surprising to find an overdeveloped technological infrastructure (often producing more guns than butter, alas), coexisting with underdeveloped civic cultures which, for all their other virtues, cannot adequately ground a popular commitment to a liberal and secular democracy.

But what we are witnessing today in India and many other parts of the world is not this normal lag which can be expected to narrow with time. What we are witnessing is an aggressive rejection of the values of cultural modernity as such by intellectuals and opinion makers. "Modernization without Westernization" is the rallying cry of those who reject all secular ideologies of the modern age as Western and therefore alien. Whether grudgingly or enthusiastically, intellectuals in these societies accept technological modernization as an imperative for their continued survival. But they do not see secularization as a universal cultural imperative.

The lack of legitimacy for secular ideals among intellectuals has political consequences, for the simple reason that cultural modernization requires the active work of intellectuals. Intellectuals translate the science that goes into making the tools of the modern age into a new worldview and a new common sense. It is the job of intellectuals to deploy the metaphysics and the ethos of modern science to free their intellectual heritage from outdated and indefensible accretions and to translate the vital core of their heritage into a new vocabulary that can accommodate the disenchanted, rationally understandable view of the world. In societies where the dominant religions and the alliance of dominant classes are more hospitable to science (as in early modern Britain and America, for example), the creation of the new secular culture takes place from within the hegemonic institutions, such as the churches, schools, and the instruments of the state. In societies like that of seventeenth-century France, or twentieth-century India, where dominant social interests are more conservative, the pressure to create a new liberal culture has to come from Grub Street. Since intellectuals enjoy a relative autonomy from the dominant institutions of their societies, they are expected to challenge the common sense of their times. In either case, for a liberal, democratic culture to emerge alongside industrial and economic development, there is no getting away from the active cultural work of "educating democracy," as Alexis de Tocqueville called it. "Educating democracy" involves, in de Tocqueville's words (1988, 13), ". . . putting new life into its beliefs, purifying its mores . . . changing the laws, ideas, customs and mores needed to make democratic revolution profitable."

The situation in India could not be more different. With the demise of the Nehruvian consensus, starting around the Emergency imposed by Indira Gandhi in 1975, secular and left-inclined intellectuals, especially those associated with Gandhian initiatives in ecology, science, technology, and development-related matters, began to display a deep sense of disillusionment with the rationalist, secular, and liberal elements of modernity. Before the hegemony of deeply hierarchical religious traditions over social life could be loosened and a space for the individual could be carved out, these populist left-identified social movements began to turn to these same religious traditions to agitate for indigenous, Gemeinschaft-oriented, non-Western models of development. Unlike the old (mostly Marxist) movements which focused on redistribution of the gains of modernization, India's new social movements agitated for an alternative model of development altogether which was more in keeping with ordinary people's traditions and values.

Over the next two decades, these movements for alternative science and development began to connect with the anti-Enlightenment, postmodernist strains that had been growing in Western universities since the mid-1960s. Suspicion of all metanarratives of modernity was the distinguishing mark of postmodern intellectuals. There was a natural meeting of the minds between postmodernist and Third Worldist indigenist movements. This confluence led to new developments in postcolonial theory, feminism, and cultural theory, all united in a conception of culturally distinctive, alternative ways of being modern. (A more detailed treatment of postmodernism in science studies follows later in the chapter.)

In the closing decade of the twentieth century, these movements were joined by the movement for *Hindutva*, a semi-Sanskrit word for "Hindu-ness," which made alternative modernity, grounded in a highly selective version of Hindu culture and Hindu values, central to its political program. The Hindutva movement is an ultranationalist and chauvinistic movement that seeks to modernize India by recovering the supposedly pristine Vedic-Hindu roots of Indian culture. To that end, it is committed to acquiring the most modern technology and putting it in the service of a religious-nationalist resurgence. The Hindutva movement sees itself as the "Sangh Parivar," literally, the family of the Sangh, which stands for Rashtriya Swayamsevak Sangh (RSS), which roughly translates into National Volunteers Association. The Rashtriya Swayamsevak Sangh is a paramilitary organization formed in 1925 with the express purpose of cultivating and defending a unified Hindu community, and fostering a fervent patriotism for "Mother India," literally worshipped as an avatar of the goddess. The political party that has ruled India for close to a decade now, the Bharatiya Janata Party, evolved as the political front of the Rashtriya Swayamsevak Sangh. The Rashtriya Swayamsevak Sangh has spun off a large array of organizations active in all domains of social life. The Vishva Hindu Parishad (VHP), or the World Hindu Council, is the cultural arm of the Rashtriya Swayamsevak Sangh. The Vishva Hindu Parishad is active in the Hindu diasporas around the world and has also spun off separate organizations for women, the Durga Vahini ([the goddess] Durga's Brigade) and for youth, the Bajrang Dal ("Bajrang's Team," Bajrang being the monkey god, also known as Hanuman). In recent years the Rashtriya Swayamsevak Sangh family has grown to include new members agitating for indigenous science (Swadeshi Vigyan Movement) and a self-sufficient economy (Swadeshi Jagran Manch).[2]

The Hindutva ideology is simultaneously revivalist and deeply chauvinistic; the way it equates India with Hindu-ness automatically excludes Muslims and Christians as not fully Indians. As described by Vinayak D. Savarkar, the early twentieth-century theorist of Hindu nationalism, the authentic Hindu *volk* is a cultural-religious entity, an integrated body, nourished by the vital fluid of Hindutva.[3] Thus, only those communities whose religions are native to India, those who hold India to be not just their motherland but also their holy land, and whose religious worldviews developed in the ambience of Hinduism—for example, Sikhs, Buddhists, Jains—are a part of this Hindu body politic. Muslims and Christians, who have lived in India for generations and have known no other country, but whose religions are not native to the Indian soil, are forever excluded from the true Indian nation. To become "truly Indian," Muslims and Christians are asked to pledge allegiance to Hindu culture—by adopting Hindu names, praying to Hindu gods in schools and public places, learning Sanskrit, the sacred language of the Brahmins, refraining from beef-eating, and generally respecting Hindu sensibilities. The Hindu community is being consolidated by unifying diverse castes and language groups against the "aliens within," those who presumably, have their sacred places in Mecca, Rome, or Palestine.

Hindutva, described by some as "social racism," has already become a major tenet of social policy in India. With full blessings of the ruling party and in alliance with small and big Indian business houses, schools, temples, and public spaces openly carry representations of the Indian landmass as a Hindu mothergoddess—"Bharat Māta"—with rivers, mountains, cities associated with Hindu sacred history (McKean 1996; Sarkar 1996). The country is being sacralized. Those who do not pray to the gods that supposedly inhabit the landscape of the Hindu sacred land (*punyabhumi*) are increasingly made to feel like aliens in their own country; Hindus alone are to be counted among the children of Mother India.

A complicated politics of resentment and cultural redefinition has come to be adopted by all three intellectual-political currents—i.e. postmodernist, postcolonial, and Hindutva—against the core values of democratic individualism and skeptical Enlightenment. These values are simultaneously reviled, *and* redefined in indigenous concepts. They are reviled as foreign impositions, Western ideologies masquerading as universal ideals. At the same time, and often in the same breath, "deeper," "alternative," more "authentic" variants of scientific reason and human freedom are claimed to be found in selected versions of local cultures and religions.

It is true that different cultures respond to the global spread of capitalism, industrialization, and technological modernization in different culture-specific ways. But in order for any of these so-called "multiple modernities" to qualify as *modernity*—and not mere change—it has to share the common core of modernity which, as Max Weber pointed out, is made up of a secularized, disenchanted cosmos in which "there are no mysterious incalculable forces that come into play, but [where] one can, in principle, master all things by calculation" (Weber 1946, 139). Whatever else modernity may be, on a Weberian reading of history, modernity emerges only where and when the unquestioned legitimacy of a divinely preordained natural and social order begins to decline, and when human agency is set free from traditional cosmo-political legitimations of political authority.[4]

The proponents of alternative ways of being modern, whether from the Hindutva right or the postcolonial/postmodern left, deny that disenchantment or rationalization is a necessary feature of modernity. They see it as a specifically Western-Christian response to the rise of science, technology, and capitalism. Other cultures, it is claimed, can and should try to become modern while retaining their sacred cosmologies, and their religiously sanctioned traditional values. Indeed, they go even further and claim that a disenchanted, naturalist understanding of nature provided by modern science itself is a Western cultural construction. India, with its distinctive non-dualist understanding of nature, God, and human beings should cultivate a science that expresses these civilizational values, just as modern science expresses the values of the Judeo-Christian civilization of the West. Only modern science is disenchanted. Fully enchanted "alternative sciences" are possible which do not break the bond between God and nature.

Such redefinitions amount to rejecting modern scientific reason and secularism as culturally inappropriate for non-Western societies. Rather than see non-Western cultures as participating in a shared global history, providing competing but potentially convergent responses to universally shared human needs and aspirations, movements for "alternative modernity" see different cultures as providing radically different meanings to radically different experiences of modernization. Like the multiculturalists in the United States who tend to be anti-assimilationists, movements for alternative modernities, whether from the religious right or from the postmodernist left, refuse to assimilate non-Western cultures to any universal pattern. Like multiculturalism again, these movements tend to equate universal values with Eurocentrism, and aspire to find the unique historical destiny of non-Western societies.

This culturalist turn from both the right and the left has bred a reactionary variety of modernity in India. For at least three decades now, modernists and secular intellectuals have been on the defensive. Rather than engage with the continuing hold of religious traditions on popular imagination, they have been forced to fend off accusations of "mental colonialism" and elitism coming from postmodernists and Hindu nationalists alike. But despite strong indigenist and small-is-beautiful trends among them, the culturalist movements have been largely ineffectual against the allures of modern technology and industry. Since India opened up its economy to global industry and capital in the early 1990s, technological modernization has picked up pace. The result is the growth of a highly unbalanced public culture that has most of the technological accoutrements of modern life, without a popular acceptance of the norms of a liberal-democratic culture.

Following Jeffrey Herf's well-known study of a similar phenomenon in Nazi Germany, I will refer to this kind of modernity without liberalism as "reactionary modernism." Reactionary modernism, is very simply, "embrace of modern technology by those who reject Enlightenment reason" (Herf 1984, 1). I believe that the social conditions that led to this phenomenon in the Weimar Republic and the Third Reich—namely, "capitalist industrialization without a successful bourgeois revolution [and] weak traditions of political liberalism and the Enlightenment" (ibid., 6)—obtain in many parts of the developing world, including India. In these conditions, the dangers of fascistic nightmares cannot be ignored.

Herf describes how German intellectuals succeeded in "incorporating modern technology into the cultural system of modern German nationalism, *without diminishing the latter's romantic and anti-rational aspects*" (ibid., 2, emphasis added). Unlike the romantics who simply urged turning their backs on modern technology, reactionary modernists were able to combine an affirmative stance toward technological progress with dreams of the past, creating the infamous "steel-like romanticism" that Joseph Goebbels rhapsodized about.

I describe an Indian version of "steely romanticism"—the confluence of "dharma and the bomb"—in the next chapter. What is frightening about this phenomenon is how all aspects of the development of the nuclear bomb—including even the physics behind it—are being reconciled with a supernaturalistic, enchanted Vedāntic cosmology dressed up as "Vedic science" on the one hand (chapter 3), and with the age-old Hindu corporatism dressed up as a more "genuine" secularism and democracy on the other (chapter 2). I will show that reactionary modernism in India, like its counterpart in Germany, simultaneously denounces "Western" modernity with its emphasis on reason and secularism, *and* reinterprets modern ideas into a nativist idiom derived from traditional, upper-caste Hinduism. Far from being compatible with the core values of modernity, the Hindu cultural idiom, in actual reality, supports a highly irrational, illiberal, and hierarchical pattern of social life.

Herf describes how prominent German intellectuals, including Oswald Spengler, Carl Schmidt, Ernest Junger, and Martin Heidegger

> succeeded in removing technology from the world of Enlightenment reason, that is, of *Zivilisation*, and placed it into the language of German nationalism, that is, of *Kultur*. They claimed that technology could be described with the jargon of authenticity, that is slogans celebrating immediacy, experience, the self, soul, feeling, blood, permanence, will, instinct and finally the race, rather than what they viewed as lifeless abstractions of intellect, analysis, mind, concepts, money and Jews. By identifying technology with form, production, use value, creative German or Aryan labor . . . they incorporated modern technology into the anti-capitalistic yearnings that National Socialism exploited. (ibid., 224)

I believe that this mechanism of translating *Zivilisation* into *Kultur*, or the universal claims of reason into the jargon of cultural authenticity, can explain a whole range of phenomena associated with reactionary modernist culture. While Herf concentrates on how modern technology was Aryanized by the Nazis, in this book I will offer an analysis of how modern science is being Hinduized (and simultaneously Aryanized) by the ideologues of Hindutva. Modern science is being absorbed into an elite Brahminical-Vedantic form of Hinduism, without admitting any contradictions between the two, and thus, *without allowing any challenge to the latter's anti-naturalistic, anti-rational, and anti-democratic aspects*. The major mechanism of the Hinduization of science is an insistence upon finding an equivalence between the rationality of modern science and Vedic knowledge. The philosophical arguments for this epistemic equivalence, I will show, are endorsed by postmodernist critics of modern science.

Just as Hindutva confers a sacred status on the nation, it also sacralizes science, or conversely, scienticizes Hindu sacred knowledge. The sacred texts of the Hindus, it is claimed by Hindu nationalist intellectuals, are works of science which presage all the modern scientific theories, while containing a metaphysics and an epistemology which are affirmed by the most advanced findings of physics, biology, and ecology. Claims of scientificity give the Vedic lore a cover from rationalist-naturalist critique, and simultaneously confer upon Vedic Hinduism the prestige of being "modern" without becoming Western. But this much-touted "Hindu modernity" fails to cultivate a culture of modernity which requires, as mentioned above, a disenchanted nature and skeptical reason as new sources of authority in the institutions of the public sphere.

### Hindu Nationalism and Fascism: Shared Traits

We will examine the subversion of the spirit of modernity in the confluence of postmodernism, neo-Hinduism/neo-Gandhism and Hindu nationalism in the rest of the book. But at this point, I want to return to the admittedly controversial theme of clerical fascism that I started with. By clerical fascism I mean "a confluence of fascism and religious fundamentalism" (Laqueur 1996, 147). What shared characteristics bring about this confluence? Why is clerical fascism a useful concept in understanding the phenomenon of Hindu nationalism?

A note on the terminological distinction between fundamentalism and religious nationalism may be helpful here. Even though the Hindutva movement bears a family resemblance with religious fundamentalist movements, I prefer to refer to it as Hindu nationalism. Religious fundamentalism and religious nationalism are both species of the genus called "New Religious Politics" (NRP), a term introduced by Nikki Keddie in her 1998 essay. She characterizes New Religious Politics as socially conservative, populist movements that aim to gain political power in order to provide a new moral basis for the nation state derived from suitably reinterpreted religious sources (Keddie 1998, 697). While religious fundamentalist movements share all these features, they have connotations of dogmatic religious literalism that religious nationalists find objectionable. Moreover, given its Protestant Christian origins, non-Christian movements see the term "fundamentalism" as inapplicable to them. The nationalist dimension of Hindutva makes it a variant of new religio-political movements, while setting it apart from purely fundamentalist movements. As Mark Jurgensmeyer has correctly pointed out, unlike fundamentalist movements that might be motivated solely by religious beliefs, Hindu nationalists turn to Hindu sources, suitably reinterpreted, to find "prescriptions for their nation's political and social destiny" (1993, 6). In the process, Hindu nationalists sacralize the political discourse by conducting politics through Hindu myths, rituals, and symbols that confer a sacred status to secular entities like the nation, the state, society, and science. The secular becomes religious.

This fits the self-description of Hindu activists who see themselves as nationalists above all. There are two main components of their nationalism. The first is the demand for an explicit recognition of India as a Hindu nation by all Indians, Hindus, and non-Hindu Indians alike. They argue that because India is, and always has been, a Hindu nation, Hindu-ness or Hindutva must be explicitly acknowledged and institutionalized. Hindu values should be the guiding star of public and private life for the nation as a whole. This view is intimately connected to the second element of the Hindutva agenda, namely, "decolonization of the Hindu mind." Hindu nationalists, according to Koenraad Elst (2001, 10) "perceive themselves as the cultural chapter of India's decolonization. [They see themselves as] trying to free the Indians from the colonial conditions at the mental and cultural level, to complete the process of political and economic decolonization."

Hindu nationalism bears close similarities with clerical fascism. Fascism is one of the most overused words. I do not use this word lightly. I do not even use it, as other Indian scholars have, to draw attention to the chilling similarities between the demonization of Muslims and Christians in India and that of the Jews in Nazi Germany. Undoubtedly, the dynamic of state-condoned and often state-sponsored violence against the Muslim minority in India—the kind of violence witnessed in Gujarat recently[5]—bears a strong resemblance with the Nazi pogroms against the Jews. There is also a historical record of open sympathy between Hindu nationalist outfits like the Rashtriya Swayamsewak Sangh and the Hindu Mahasabha and Italian and German fascism.[6] I agree with these parallels. But obviously, similarity of symptoms does not implicate the same pathogen. In order to establish that, one would have to dig deeper, which I hope to do in this section.

I have two reasons for describing Hindu nationalism as clerical fascism.

First, I believe that religious nationalism in general bears functional similarities with fascism. *Religious nationalism and fascism are both peculiar responses to the forces unleashed by the introduction of modern industrial capitalism in societies with weak and/or discredited liberal traditions.*

Second, I believe that there is a substantial overlap between the metaphysics of Hindu nationalism and Nazism. While other religious fundamentalists may only share the motives and organizational forms of fascist movements, Hindu nationalism shares—and glorifies as "Aryan" and "scientific"—the same mystical, anti-rational, and holistic ideas that the Nazis borrowed from Hinduism. Both Nazism and Hindutva share a view of the world marked by resistance to the idea of transcendence, an insistence upon a lack of separation between human beings, nature, and the divine, and the sacredness of the natural and social orders. In themselves, these ideas do not necessarily promote the kind of illiberal ultranationalism one associates with fascism. On the contrary, they can even serve as a source of critical reflection on excesses of individualist and acquisitive societies. But in India, where the traditional caste and class elite have had a millennia-long elective affinity with these ideas, their resurgence is a cause of worry.

Let me expand upon these two issues.

Whatever else they may be, religious fundamentalism/religious nationalism and fascism are *reactionary modernist movements led by prophets facing backward.* Both seek to create a new society, a new Weltanschauung, a "third way" to replace the liberal and Marxist versions of modernity. The "new," however, is to be inspired by models derived from the nation's past, its own "authentic" repository of traditions and indeed, its eternal "soul." Even as the nation marches forward toward a more materially prosperous and technologically advanced future, it keeps its face turned backward to the past as a source of direction, inspiration, and resolve. Both fascism and new religio-political movements offer their societies an un-Faustian bargain: that a people can be made modern *and* retain their primordial cultural "essence," "soul," or "*svabhava*" (Sanskrit for "one's own nature"). Thus, they actively seek technical and military power as a way to harness the material benefits of modern civilization with an express purpose to rejuvenate national values and traditions. They offer a way of keeping the changes brought about by global industrial capitalism and the accompanying cultural transformations within the moral compass of "tradition"—a highly selective and contrived tradition, but which nevertheless resonates with popular religious pieties and customs.

The prophets facing backward display the zeal of the Jacobins, directed not against the ancien régime but against the ideals of the Enlightenment (Eisenstadt 1999). The participants in new religio-political movements, in other words, "fight back" against the core values of modern, secular liberalism which they see as eroding their way of life (Marty and Appleby 1992, 17). Unlike the average believers in a secular age who privatize their faith, separate it from the secular affairs of social life, or hunker down into enclaves where they can maintain the plausibility of their beliefs, religious nationalists refuse to privatize their piety. They fight for retrieving and embellishing selected values from their sacred tradition, and installing them as new standards for the conduct of civic life. As Marty and Appleby (ibid., 183) correctly point out: "Fundamentalists are reactive . . . but they are not mere protest movements. They are innovative world-builders who act as well as react, who see a world that fails to meet their standards and who then organize and marshal resources in order to create an alternative world for their followers to inhabit and vivify."[7]

To fulfill this aim of constructing alternative ways of living with the modern world, religious nationalists/fundamentalists use all means available to them, including modern political processes of electioneering, political organizing, massmarketing, and even modern weapons of mass destruction. But—and this is crucial—however innovative they may be, religio-political movements are "careful to demonstrate the continuity between their programs and teachings and the received wisdom of their religious heritage . . . a crucial element of their rhetoric . . . is that their innovations are based upon the authority of the sacred past" (Almond et al. 1995, 402). This allows them to disarm new ideas by encapsulating them into their sacred traditions, without allowing the new to question the old. (We will see extended examples of this in the case of the Hindu understanding of "secularism" and "Vedic science" in the next three chapters.)

Like religious fundamentalism/nationalism fascism, too, is a fighting creed. Fascists exhort "the people" to rise up and fight back the perceived decline and decadence, which they attribute to the corrosive effects of rationalism, materialism, and liberalism. Like religio-political movements, fascism, too, seeks salvation in the rebirth of a harmonious and organic collectivity that presumably existed before the onslaught of modernity. There is a general consensus among the scholars of fascism that the "fascist minimum" is best described as a promise of "national rebirth" after a period of perceived decadence. More technically, fascism is a "form of political ideology whose mythic core is . . . a palingenetic form of populist ultranationalism" (Griffin 1996, 26).[8] As Griffin explains, "palingenesis [palin: anew, again and genesis: birth] refers to a new start or regeneration after a phase of crisis and decline." The core myth of fascism is the idea of the rebirth of the nation through populist mass reawakening, leading *not* to the restoration of the past, but to a "substantially new order of society . . . inspired by historical precedents or the myth of a past golden age" (ibid., 33, 36). The reborn nation is a "higher," "integral," or "organic" collectivity, in which all sections of the society are perfectly integrated without the insidious competition between classes, castes, and genders. The actual fact of class/caste and gender conflict is covered up with the rhetoric of organic unity.

Rebirth of the values and traditions of the nation imagined as a premodern, organic community, bound by a sacred history, constitutes the meeting point of religious nationalist/fundamentalist and fascist movements. Obviously, all varieties of religious fundamentalisms (e.g., Islamic fundamentalism) do not demand that religious ideals be limited by national boundaries. Yet, there are instances—as in Hindu revivalism—when the very soil of the nation gets sacralized. In such cases, religious revivalism coincides with ultranationalism. Likewise, all varieties of fascism do not demand that the nation be imagined in religious terms. But given the fact that religion is a potent source of non-instrumental and communitarian values that can counter the threat of materialism and individualism that fascism sees as the source of decadence, it can serve as a mobilizing ideology for ultranationalist movements fighting for national rebirth. Indeed, shameful instances of active collaboration between the Church and the fascist regimes are, sadly, legion, including the Catholic Church's concordant with the governments of fascist Italy in 1929 and Nazi Germany in 1933.[9]

Why is it that religious revivalist movements grow in scale and intensity in some countries, but not in others at any given time? While the religious right remains a constant threat in advanced capitalist societies, it is, by and large, confined to the margins. The situation is very different in modernizing societies—as Germany and Italy were in the late nineteenth and early twentieth centuries, and where India is today. These societies are caught in a moment in the development of capitalism and democracy when the older modes of economic reproduction and social legitimation are in decay and the new ones have not fully replaced them. The breakdown of traditional protections for small-property owners under the sway of global capitalism and industrialization creates a mass of provincial propertied classes, who are simultaneously hostile to corporate capitalism (national or global) and to any form of socialism. Moreover, the arrival of democracy and universal suffrage gives these disaffected masses a political power, based upon their great numbers, which is disproportionate to their economic power. In the conceptual vocabulary of Barrington Moore Jr., the masses—really peasants or uprooted peasants transplanted to the cities—arrive at the historical scene of democracy without a democratic political culture.[10] The vision of a revivified, precapitalist, premodern nation, imagined in religious terms, serves as the mobilizing ideology of these socio-economic strata. In these groups, traditional conservatism finally finds a critical mass and takes on the characteristic of a mass movement that is simultaneously anti-capitalist/anti-big business, and anti-socialist/anti-trade unions.

In India, where mass democracy arrived before a secular and democratic civil society was fully consolidated under the cultural hegemony of the urban, industrial classes, the strains of the above kind of transition are especially acute. The Green Revolution has created a large mass of upwardly mobile, middle to small capitalist farmers who have made tremendous economic gains, but who are also threatened by the breakdown of traditional caste relations which they depended upon for super-exploitation of the landless, mostly untouchable and tribal farm workers. Moreover, these capitalist farmers are segmented in terms of their caste, regional, and linguistic affiliations. The language of religious nationalism cuts across these divisions and serves to unify them as they look for fresh material opportunities, while avoiding the cultural dislocation and existential angst that come with rapid change. The ideology of Hindutva, which promises the rebirth of the nation through the revitalization of a highly selective version of Hindu tradition, carries a tremendous emotional appeal for the millions experiencing rapid and unsettling changes brought about by linkages with global capital and technology: it speaks to their concerns in a way that resonates with their taken-for-granted assumptions about the world. (We will examine these issues at greater length in chapters 6 and 10.)

This unified Hindu identity, however, has a dark side, full of suspicion and anger at the "enemy" within—Muslims and Christians whose religions did not grow from the soil of India. The Muslims, especially, are targeted as the cause of India's historic and continuing backwardness, while Christians are seen as fishing for souls through their missionary work. The resurgence of the Hindu nation is made conditional on "Indianizing" the Muslims and Christians by making them accept the supremacy of Hindu culture. Indianization, as a Hindutva sympathizer described it, is the "*final solution for the Muslim problem in India*" (Elst 2001, 57, emphasis in the original).

While it eschews the biological racism of the Nazis, the social racism of Hindu nationalists is no less dangerous. More so, because Hindu social exclusion is backed by a philosophical worldview which proclaims Hindu beliefs and practices to be in accord with the laws of nature. And this brings us to the second issue of metaphysics. While other religious fundamentalists go back to their sacred teachings revealed by God, Hindutva proclaims its sacred sources—the Vedas—to be "nothing but" laws of nature, revealed by "Vedic sciences." As we will examine in greater detail in chapters 3 and 4 devoted to the question of "Vedic sciences," Hindu nationalists treat the entire family of Abrahamic religions (Judaism, Christianity, and Islam) with undisguised contempt. Islam is equated with stagnation and militarism, Judaism is seen as a defeated religion, overtaken by its rebellious offspring, while Christianity is declared to be superannuated and falsified by modern science, a once-great religion whose time has gone. The basic claim is this: only the Hindu conception of God as immanent in nature is in accord with the findings of modern science. Because the Hindu God is in nature and not separate from it, Hinduism alone can bring about a synthesis of reason and spirituality, progress, and ecological sustainability. The Hindu *punyabhumi* is no ordinary sacred land, it is the "cradle of civilization" itself, the birthplace of all science, discovered in a flash of insight by the Aryan ancestors of the Hindus.[11]

This notion that the Judeo-Christian-Islamic tradition is irrational, and a source of alienation from nature, has a dangerous ancestry. The ultimate goal of the Nazis was not "just" to liquidate the Jewish people, but to purge Christianity of the Judaic conception of God. What is often forgotten is that Nazism was a response, among other things, to rapid industrialization, urbanization, and a consequent feeling of alienation from nature. In a manner chillingly reminiscent of our own deep ecologists and Hindu nationalists, well-known Nazi ideologues, including Alfred Rosenberg, Heinrich Himmler, Adolf Hitler himself, and lesser lights like Savitri Devi and the theosophist, Helena Blavatsky, ascribed the alienation from nature to the Judaic dualism between a transcendent God and life-less nature (see Pois 1985; Mosse 1964; Goodrick-Clarke 1998 for these issues). The Nazis sought a genuine religion of nature that would allow Germans to live in harmony with nature.

In their quest for a non-dualist, immanent conception of God, the Nazis turned to the Vedic monism of ancient India, the supposed home of the "Nordic Aryans" before they got corrupted by "racial pollution" caused by interbreeding with the non-Aryan natives (the "swarthy Sudras," Alfred Rosenberg calls them). Rosenberg's *Myth of the Twentieth Century*, which was considered the manifesto of National Socialist ideology, turns to *varna* (the fourfold division of society, the basis of the caste system) and atman doctrines of the Vedic Hindus to find a new myth "suitable for the twentieth century" that can replace the distant, law-giver Judaic God. He praises "Aryan Hindus" for their "passionate yearning for the abolition of dualism" which allowed them to live according to the laws of the ordering principle of life—atman: "As a born master, the Indian felt his individual soul expand into the Atman which pervaded the entire universe and lived within his own breast as his innermost soul" (Rosenberg 1982, 8). The caste system was a part of the Aryan wisdom of organizing the society in accordance with the laws of nature. It was the forgetting of these laws of caste, according to Rosenberg and other Nazi Indophiles, that accounted for the degeneration of the great Aryan Hindus.

The celebration of Hindu non-dualism as a mark of Aryan superiority has not gone away. The neo-Nazi writings of the Italian fascist, Julius Evola, and German left philosopher-turned-neo-fascist, Rudolf Bahro and the writings of some deep ecologists continue to draw upon Vedic monism as an alternative conception of the relation between God and nature (Sheehan 1981; Biehl 1995).

Hindu nationalists returned the compliment with great public adulation of Hitler and Mussolini. The Hindu nationalist Rashtriya Swayamsewak Sangh, the parent organization of the Bharatiya Janata Party, India's current ruling party, showed deep fascination with both Mussolini and Hitler, but more so with Hitler, as his Aryan ideology was seen as Hindu in content and spirit (Jaffrelot 1996; Casolari 2000). It is well-known that Hitler was seen by many in India as an incarnation of Vishnu, a Western Aryan counterpart of the Hindu gods, Rāma and Krishna, come to restore the fallen world to its pristine goodness. Reverential references to Hitler as "savior of the world," "incarnation of God," and the "leader of Sanskrit-knowing Germans" were widespread in India in the 1930s (Bharati 1980). The Hindu Mahasabha (Hindu Great Assembly), a radical nationalist group established in 1915, explained the basis of the affinity with Hitler thus: "Germany's solemn idea of the revival of Aryan culture, the glorification of the Swastika, her patronage of Vedic learning, and the ardent championship of the tradition of Indo-Germanic civilization are welcomed by the religious and sensible Hindus of India with a jubilant hope. . . . Germany's crusade against the enemies of Aryan culture will bring all the Aryan nations of the world to their senses and awaken the Indian Hindus for the restoration of their lost glory" (quoted here from Goodrick-Clarke 1998, 66).

Of course, Hindu doctrines cannot be condemned for the sins of the Nazis. Neither can one surmise that fascism is immanent in India's culture; this same India, with the same mix of religions, has allowed a working democracy to survive and grow for the last 50 and more years. Vedic monism, in conjunction with other heterodox dualist and theistic trends, has provided, and continues to provide, great spiritual solace and a design for ethical living for millions of Hindus and non-Hindus around the world. Hinduism cannot be held responsible for the manner in which some of its ideas were raided by Aryan supremacists.

But what is deeply troubling is that Hindu nationalists are putting the exact same spin on Vedic monism and varna as the Nazis did, and the neo-Nazis still do. Rather than interpret monism as a *mystical* pantheism, which is what it was meant to be, proponents of "Vedic science" insist upon treating monism as a *scientific* doctrine, based upon a uniquely Hindu conception of rationality, and congruent, supposedly, with the most advanced theories of quantum physics, cognitive sciences, and ecology (chapters 3 and 4).

That fascist movements would find holist /non-dualist philosophical ideas useful is not surprising at all in view of the fascist minimum of populist ultranationalism. Holist views of nature and society in which the collective is held to be larger than the individual, the organism more than the sum of its parts, are eminently suited for illiberal and totalitarian philosophies. Such philosophies can mobilize individuals to sacrifice their freedom for the sake of the collective good, and to even accept the indignities meted out to them in the name of their duty and destiny. The genius of all fascisms is to try to contain the social energies released by modern industrial capitalism within the framework of an organicist collective.

What makes Hindu nationalism especially susceptible to fascistic tendencies is one simple fact: India has the dubious distinction of having gone the farthest in translating the holist, corporatist vision into reality. The dominant Hindu conception of a good society has always sanctified hierarchy over equality, duty over rights, and an undifferentiated, mystical unity over individuation and separation. It is no surprise at all that in a time of rapid social change, Hindu nationalists should see the breakdown of community life as a threat to Hinduism itself. It is no surprise that Hindu nationalists would try to reinterpret modern freedoms and ideas into a traditional idiom of "integral humanism" characteristic of the caste society. The ideals of non-individualistic and hierarchical society have the blessings of the dominant religious traditions in India, and consequently, enjoy a deep emotional resonance among ordinary people.

Two clarifications are in order before we move on to consider how the postmodernist and postcolonial despair over modernity has inadvertently lent legitimacy to the worldview of Hindu nationalism.

First, while I locate the meeting point of fascism and fundamentalism in their shared attempt to recreate the lost glory of the nation, it is not my intention to castigate all attempts to look back in order to move forward. Far from it. I fully admit that *all* prophets look back. Here, I believe Marx was right on the mark when he wrote in his *Eighteenth Brumaire of Louis Bonaparte*, that

Men make their own history but under circumstances directly found, given and transmitted from the past. The tradition of all the dead generations weighs like a nightmare on the brain of the living. And just when they seem engaged in revolutionizing themselves . . . in creating something entirely new, precisely in such epochs of revolutionary crisis, they anxiously conjure up the spirits of the past . . . and borrow from them names, battle slogans and costumes in order to present the new scene of world history in this time-honored disguise . . . (Marx 1978, 595)

Invocation of the past—"conjuring up the dead"—is how new struggles are legitimated. There is a very good reason why this should be so. Ancestral traditions and religious mores are the substratum from which the most human motivations and values spring. Secularism and scientific rationality in the West, after all, did not start out as secularism and science, but as "fundamentalist" ways of becoming religious. Even the iconoclasts of the Enlightenment wrapped up Newton in the toga of Cicero and Lucretius. It is perfectly understandable why non-Western societies, facing truly monumental changes, would reach back into their own past.

The problem with Hindu nationalists looking back to India's ancient heritage is not *whether* they should look back at all: they have no choice but to. The problem is *which* traditions are they looking back to? What elements of Indian heritage do they seek to reawaken? All cultures contain a multiplicity of traditions, often at odds with each other. With its numerous gods and a host of philosophies about God, nature, and the relation between them, Hinduism is well-known for its multiple traditions which vary not only with caste, but also with the gender and age group.

The question one has to ask is if the return to Vedic non-dualism or monism—admittedly, the dominant tradition, favored by the Brahmins and quite often equated with all of Hinduism—can really help anchor the new worldview of science and secularism in Indian culture. I believe firmly, for reasons that will become clearer as we explore the history of Indian science, that revitalizing the monist tradition as a model and source of science of nature will amount to opening the door to the worst kind of superstitions and magical thinking, of which we already have an excess in India. Moreover, given the castism that Vedic monism and holism have supported for centuries, returning to this philosophy of Brahminical orthodoxy goes entirely against the spirit of a liberal and secular democracy that India has constitutionally committed itself to. Revitalizing Vedic monism will amount to a reactionary retrieval of the past.

Finally, while I see clear signs of creeping fascism in India, the country has not yet capitulated. There is great danger, no doubt. But there is also hope. More than half-a-century of electoral democracy with universal adult suffrage has mobilized Indian people on a crosscutting array of secular interests which will hopefully keep the religious-nationalist passions in check. With all their flaws and shortcomings, institutions of judiciary, the mass media, and civil rights watch-dogs function with reasonable freedom in India. I will admit that it is difficult to remain hopeful in the face of the recent massacre of innocents in Gujarat. In any case, at the time of writing (mid-2002), clerical fascism is still more of a threat than a reality in India.

I now turn to the intellectual trends that have unintentionally aided the project of Hindutva. We will expand upon the themes introduced in the following sections in the rest of the book.

### The Total Critique of Modernity: Deconstruction of Science

The intellectuals—from philosophers and priests to journalists and political activists—have a job to do. Among other things, they are supposed to tell us: What is wrong with the world we live in? What is preventing the satisfaction of our collective longings? This act of interpreting the malady can create new objects of longing. If we think of class stratification as the obstacle to a better society, we long for a society without classes, for example. Naming and explaining the source of our discontent creates new objects of longing which, in turn, can engage the imagination of the people and propel them to social action.[12]

Over the last three decades, starting around the 1960s in the West and a decade or so later in many parts of the Third World, influential secular and progressive intellectuals have pointed the finger at one common enemy of human emancipation—the modern age itself. Bringing a radical conservative sensibility to traditional left-wing concerns with alienation, patriarchy, imperialism, and ecology, self-described postmodernists have criticized the modern world of industrial capitalism and liberal humanism not for failing to live up to its own ideals, but for upholding and cherishing these ideals in the first place. In judging the modern world against a melange of Utopian ideals derived in equal measure from the world-that-we-have-lost (the world of traditions, metaphysics, "situated knowledges") and a world-not-yet-born (the world of radical difference and hybridity), postmodernist intellectuals have launched a "total critique" of the modern world (Ferry and Renaut 1985; Antonio 2000).

What are the ideals of modernity that the postmodernists indict? Modernity and postmodernity are both notoriously difficult to peg down. Modernity, as I understand it, is another word for the humanism of the Enlightenment. The generic "enlightenment" is a spirit, a temper of questioning inherited dogmas, summarized by Kant in his famous motto, *sapere aude*—"dare to know," take the risk of discovery, exercise the right of unfettered criticism and accept the loneliness of autonomy. This spirit of questioning is generic and universal, and has flowered many times in history, in many different countries (India in the time of the Buddha, for example, or Athens of Socrates and the great rationalist tradition of the Mutazili school in Islam). Enlightenment in this sense is an ongoing process which is still incomplete, and will forever remain incomplete. What is referred to as "the Enlightenment," on the other hand, is the process involving a series of debates and controversies spanning all of Europe and beyond in the "long eighteenth century." In the words of Jonathan Israel, "The Enlightenment marks the most dramatic step towards secularization and rationalization in Europe's history . . . the Enlightenment not only attacked and severed the roots of traditional European culture in the sacred, magic, kingship and hierarchy, secularizing all institutions and ideas, but (intellectually and to a degree in practice) effectively demolished all legitimations of monarchy, aristocracy, women's subordination, ecclesiastical authority and slavery, replacing them with the principles of universality, equality and democracy" (2001, vi).

Although no single cause can be isolated, this revolutionary movement was rooted in the rise of Cartesianism in the mid-seventeenth century and the subsequent rise of the mechanical philosophy of Galileo and Newton. Modern science was central to the challenge the Enlightenment posed to arbitrary authority, be it the authority of priests or of kings. Science presented new standards for arriving at the truth that relied on experiment, trial and error, and a belief in the mathematical regularity of the universe. Inspired by the new philosophy, the Enlightenment philosophers and public intellectuals posed a serious challenge to all abstractions for which no physical reality seemed to exist. The project of the Enlightenment, in other words, was a "revolt against superstition" propelled by a confidence in secular-scientific knowledge to challenge all inherited dogmas backed by political power (see Jacob 2001 for a succinct summary).

The rise of postmodernism has totally discredited the necessity of, and even the possibility of, questioning the inherited metaphysical systems, which for centuries have shackled human imagination and social freedoms in those parts of the world which have not yet had their modern-day enlightenments. I will broadly include as postmodernists those intellectuals who have lost faith in the promise of modernity and the (European) Enlightenment. They argue that modern science and modern secular cultures/institutions have lost their liberatory potential and have turned into sources of the subjugation and mental colonialism of non-Western people, women, and cultural minorities. This sentiment gets translated into an opposition to the project of development and modernization in what used to be called the Third World. To varying degrees, postmodernist intellectuals attribute renewed relevance to all that modernity has set aside. Thus, one of their major preoccupations is the preservation and cultivation of "local knowledges" embedded in traditional cosmologies, religions, and traditional practices of agriculture; medicine, etc. These local knowledges, postmodernists insist, are legitimate sciences in their own right. They cannot and should not be judged by the standards of rationality set by modern science. To do that amounts to Western hubris.[13]

To understand the radical nature of postmodernism, let me unpack the idea of modernity to show the centrality of reflexivity—epitomized by modern science—to it. To paraphrase Jürgen Habermas, the condition of modernity is characterized by these four features: *(a)* individualism, holding simply that each person is entitled to his or her subjective freedom; *(b)* the right to criticism, granting that nothing needs to be taken for granted; *(c)* autonomy of action; and *(d)* philosophy of reflection, enabling human beings to understand the world and themselves without metaphysical guarantees of religion (Habermas 1995, 17). At the heart of these modern characteristics lies the expanded role of reason in a society's self-understanding. Modernity, Habermas insists, "can no longer borrow the criteria by which they take their orientation from the models supplied by another epoch; it has to create its normativity out of itself" (ibid., 7). This does not mean that modern societies renounce all traditions, or that modern women and men start with clean slates—that is impossible. What this means is that modern people do not sanction a practice simply because it is a part of their traditions. Rather, as Anthony Giddens puts it, tradition itself has to be "justified in the light of knowledge which is not itself authenticated by traditions" (1990, 38).

*Postmodernism challenges the very possibility of knowledge that is not, in the final instance, authenticated by a local cultural tradition.* As modern science is the very paradigm of universal knowledge that claims to transcend cultural differences and local traditions, it has become the ultimate target for deconstruction. Postmodernist critics challenge the core assumptions that constitute the self-understanding of modern science as an epistemologically progressive and universal enterprise (see Sokal 1996; Brown 2001; Kitcher 1998; and Klee 1997 for a fuller development of these assumptions):

1. There is a world in which there are objects, processes, and properties that are independent of us and our beliefs about them.
2. The aim of science is to give a reliable, albeit imperfect and tentative description and explanation of these objects, processes, and properties.
3. Science has learned how to learn about nature. We have developed a variety of tools and techniques (observations, logic, and statistical inferences) for learning how things are.
4. Such methods may occasionally lead us astray. But science has made remarkable progress so far. This progressive character is manifested in increased powers of prediction and intervention in nature.
5. These increased powers of prediction and intervention give us the right to claim that the kind of entities described in scientific research exist independently of our theorizing about them and that many of our descriptions are approximately true.

6. There is only one science. Scientific truths are true for all societies. Criteria of justification of scientific claims cut across national and cultural differences.

Each one of these assumptions has been challenged. The prevailing view in social and cultural studies of science is that scientific norms of rationality, objectivity, and truth are cultural conventions; different cultures with different metaphysical assumptions about nature and the workings of the human mind will find different facts about nature equally convincing, on equally rational grounds. There is, in other words, nothing special and universal about the methods and findings of modern science. Andrew Ross, a cultural critic of science, summarizes the social constructivist view in these words: "Western technoscience is a highly local form of knowledge and is unlikely to have a world monopoly on good scientific ideas. Scientific knowledge is not given by the natural world, but is produced or constructed through social interactions between/among scientists and their instruments, and these interactions are mediated by the conceptual apparatuses created in order to frame and interpret the results" (1996, 10–11).

In other words, change the social environment and the conceptual categories and you will have different "facts" which will be equally "scientific"—and there will be no reason to prefer the facts arrived at by what we know as modern science over any other science of other cultures. What we take as scientific truth varies from place to place, from culture to culture. That we believe modern science is actually taking us closer to the truth is merely a convention that has developed out of the contingencies of Western dominance over the rest of the world's cultures. (We will examine this thesis in a more nuanced manner in chapter 5.)

Critiques of science and the scientific worldview are as old as science itself. But prior to the current wave of postmodernism, relativization of scientific reason to culture was limited to the right-wing fringes, including proto-fascist nationalists like Oswald Spengler and Aryan science theorists like Alfred Rosenberg. Most academic philosophers and sociologists of knowledge in the early middle part of the twentieth century, notably, Karl Mannheim, Robert Merton, and Peter Berger, exempted the content and/or validity of science from social relations. Even the Marxist critics of science, notably J.B.S. Haldane, John Bernal, and Joseph Needham who saw modern science as part of the ideology and practical needs of capitalism, were primarily critical of the capitalist economic system which they saw as standing in the way of a fuller, more complete development of science, while applauding the progressive role of the already existing ("bourgeois") science in challenging the premodern, feudal worldview. These stalwarts of the old left could still appreciate the contradictory Janus-face of science as "both a battle cry for freedom and a rationale for oppression and domination, a weapon for enlightenment against superstition and a chauvinist and racist dismissal of knowledge of third-world cultures, both a condition of our existence and a target of our politics . . ." (Levins 1986, 5).

Things began to change in the 1960s and the 1970s. The liberatory face of science was lost from view. The radical social movements that emerged in the United States and Europe out of the anti-Vietnam, anti-imperialist, ecology, and feminist movements of this time were divided over the science question. Initially, the radical science movements offered a mostly *political* critique of science that did not challenge the *epistemology* of science. Movements like the Union of Concerned Scientists, Science for the People, and Scientists Institute for Public Information mostly focused their attention on the abuses of science by the military-industrial complex, rather than on the biases in the content of science. Accordingly, they exhorted scientists to deploy the already existing sciences to serve socially progressive ends (Moore 1996).

But the Third Worldist, feminist, and other identitarian trends in the new social movements began a more radical epistemological critique of science which gradually overtook the political critique. Theorists of critical or Western Marxism, especially the ideas of Herbert Marcuse, Eric Fromm, and others associated with the Frankfurt School, who were (correctly) opposed to Leninist theories of "scientific socialism," gained a wider hearing among the new social movements. The Western Marxist critiques of scientificity and positivism of the Soviet brand of Marxism which claimed to find a law-like movement in history, spread to the very idea of scientific law, or scientific fact. Whereas the earlier Marxist movements had seen ideology as a distorted reflection of the world, critical Marxists came to see science serving as ideology, and ideology as containing elements of scientific truth. By insisting upon the "totality" and continuity of idea, interests, and events, Western Marxists demanded a radical contextualization of science, and decried as "positivist" any claim of autonomy or the self-grounding of science (Gouldner 1980).

In this politically charged atmosphere, the appearance of Thomas Kuhn's 1962 classic *The Structure of Scientific Revolutions* brought about a paradigm change, both in the academic sociology of science and in the political movements critical of science. As is well-known, Kuhn argued that scientific practice always takes place under a paradigm. The ruling paradigm decides what kind of problems scientists study, what criteria they use to evaluate a solution, and what experimental procedures they find acceptable. All aspects of what scientists do are so filtered through the paradigm they are in that no paradigm-neutral, objective evaluation of scientific "facts" in two different paradigms is possible. Scientific theories in different paradigms are simply incommensurable, and there is no way to judge a more accurate representation of reality. Indeed, there is no "reality" which is not relative to a paradigm. Although in his later writings Kuhn did assert the continuity of cognitive values across paradigms which can be used to assess progress in science, his view of incommensurability opened the door to radical relativism and anti-realism.

Many sociologists of science welcomed Kuhn's work as finally enabling them to challenge the taboo against a sociological interpretation of the *content* of science. (Prior to Kuhn, social factors were only invoked to explain failed theories. Social interests were seen as a source of error, while truth was supposed to be true regardless of social interests that would be served if one rather than another theory turned out to be true.) If rational acceptance itself depends upon the paradigm, which on Kuhn's account, includes the training, the socialization, and the metaphysical beliefs of scientists, then, sociologists reasoned, there is a legitimate place for social and cultural factors in the evaluation and justification of scientific facts. Two well-known programs, the Strong Programme promoted by David Bloor and Barry Barnes at the University of Edinburgh, and the Empirical Program of Relativism, championed by Harry Collins at the University of Bath, both starting around the early 1970s, led the way to a new post-Mertonian paradigm in the sociology of science which is generally referred to as "social constructivism." Later, these programs were joined by the actor-network school developed by French sociologists, Bruno Latour and Michael Callon, and by ethnographers of laboratory work, led chiefly by Bruno Latour, Steve Woolgar, and Karin Knorr Cetina. Together with their more politically radical allies in feminist and cultural studies, these programs came to form the core of the relatively new interdisciplinary area of study called STS, which stands for Science, Technology, and Society Studies, or Science and Technology Studies. In the beginning, Science, Technology, and Society Studies was almost exclusively a British practice. But now there are influential practitioners throughout North America, as well as in France, Germany, Netherlands, Scandinavia, Israel, and Australia. Among postcolonial societies, India boasts a host of active programs and notable scholars belonging to what has been dubbed the "Delhi School of Science Studies." Science, Technology, and Society Studies has all the trappings of an autonomous academic discipline, including journals, societies, professional societies, textbooks, and so on (see Hess 1997; Shapin 1995 for sympathetic surveys).

Although epistemologically radical, the Strong Programme and related social studies of scientific knowledge were not motivated by any politically radical agendas. Indeed, the more politically engaged science critics, especially those with Marxist sympathies, were critical of the Strong Programme's insistence on "symmetry" or even-handedness between true and false scientific claims, seeing both as explicable by social interests. As Ulllica Segerstråle (2000, chapter 17) points out, the motivation of these programs was primarily epistemological, that is, to correct what they saw as blind spots of Mertonian sociology and the positivist philosophy of science. They were interested in carving out a new intellectual niche for sociology in science, rather than criticizing science per se.

But this social constructivist paradigm provided the opening for those political critics of science who wished to argue for a simultaneous transformation of society and science. Various feminist, postcolonial, and standpoint epistemologists embraced the social constructivist agenda in order to argue for different logics and different background assumptions that would, presumably, lead to "feminist," or "Indian," or "Islamic" sciences. The idea of modern science as the lingua franca of the modern world was abandoned.

### Indian Intellectuals and the Siren Songs of Postmodernism

Whatever its later sins, modernity has already served, in the words of Karl Jaspers, as a "second axial-age" in the West[14]—that is, the rise of science and liberal democracies together have allowed a sufficient reshaping of collective thought to satisfy the demands for greater individual autonomy and a greater room for public reason. Science, in other words, has been a part of the dominant culture in the West for a considerable length of time. Rationalization and secularization of culture have already come to dominate the mainstream of Western societies. One can at least understand—even though disagree with—why there could be a popular disaffection with science which is now seen as allied with the ruling interests.

But how is one to understand the appeal of the postmodernist critique of science and modernity in still-modernizing societies where modernity is far from having ushered in a second axial age? In conditions where the minimum conditions of modernity—industrialization, urbanization, rationalization—are yet to be met, one can legitimately wonder, how can postmodernity have any appeal?

Yet, there is no denying the appeal. Indeed, the paradox is that while postmodernism in the West is largely an academic phenomenon, postmodern themes in many parts of the so-called Third World are at the cutting edge of new social movements—mobilizing large masses of people from a cross section of classes—for alternative development, environmental protection, anti-imperialism/anti-globalization, and even feminism. Nowhere is the championship of postmodernist, anti-Enlightenment themes more widespread than in India, the land aptly described by V. S. Naipaul as the land of "a million mutinies." What is more, intellectuals of Indian origin have taken a lead in the development of postmodern and/or postcolonial critiques of science and modernity in the Western academia. Anyone interested in postmodern social theory cannot escape the often verbose writings of Gayatri Spivak, Homi Bhabha, or the subaltern historians including Ranjit Guha, Gyan Prakash, and Partha Chatterjee. I can attest from personal experience in science studies that the writings of ecofeminist Vandana Shiva and the proponents of "alternative science" like Ashis Nandy, Shiv Visvanathan, Claude Alvares are treated as key texts.

In the non-Western parts of the world, modernization brings tremendous dislocations, breaking up traditional community structures without leaving anything promising in their place, at least in the short term. Flush with unreasonably high expectations of a socialistic revolution on the one hand, and with nationalistic, anti-colonial passions on the other, non-Western intellectuals have remained ambivalent about modernization. Even as they are drawn to the ideas of equality, liberty, and fraternity, they have remained acutely conscious of the Western origin of these ideas. The West has become an intimate enemy which fascinates the postcolonial intellectuals, even as they resent it for alienating them from the masses and their cultures in their own societies. Postcolonialism, in the words of Gayatri Spivak (1993, 280), is a "deconstructive philosophical position [that allows intellectuals to] say an impossible 'no' to a structure [they] critique, yet inhabit intimately."

The question is *why* intellectuals should work so hard just to say "an impossible no" to the intellectual heritage of the West which, by their own admission, is indispensable to their own intellectual and political lives and to the political culture of their societies. The answer lies in a very peculiar schizophrenia one commonly finds among a class of radical intellectuals whose radicalism is cut off from the real struggles of ordinary people.

As Edward Shils (1961) diagnosed with great insight early on, intellectual schizophrenia is especially acute in India, a country with a truly unique and long-standing intellectual tradition of its own, and also the country with the longest experience of colonialism. Shils found that *in actual fact*, Indian intellectuals were quite deeply rooted in the dominant Hindu worldview and traditions, including its more irrational and authoritarian elements like astrology, arranged marriages, and caste prejudices. But they *saw* themselves as tragic, homeless souls, uprooted and out-of-touch from the "real" India. Shils gives reasons for this gap between reality and self-perception, chief among them being: English education, overcompensation for the heritage of Brahminical exclusivity without the courage to break out of the Brahminical worldview, the lack of a real civic-culture and the absence of a socially progressive indigenous bourgeoisie with whom modern intellectuals could relate.

This angst that Shils described has only grown as Indian intellectuals have become more intimately drawn into a cosmopolitan, English-speaking culture readily accessible in India's big, and even not-so-big, cities. Recently, Vikram Chandra (2000), an Indian novelist writing in English described the pervasive sense of angst among Indian intellectuals about failing to see "the 'real India' of the slums, faraway villages . . . and the jungles of tribals. . . . Real India that is completely unique, incomprehensible and approachable only through great and prolonged suffering and unveils herself only to the most virtuous." From the fear of the mongrel nature of their own selves, Chandra observes, educated, middle-class Indians have created this "golem-demon of the all-devouring West" whose recognition they desperately seek, but whose corrupting power they hold responsible for their alienation from the real India, supposedly left untouched by modernity. To ward off the demon God of the West, the leftists among the intellectuals pray to the "Lost Valley," while the rightists hanker after "Mount Restoration." Chandra concludes that the "Lost Valley and Mount Restoration are exactly alike" and the "Lefties and the Righties [are] joining hands in praying to the God of Authenticity."

There is one consequence of this schizophrenia that is relevant to our study: a populist style of thinking. As Shils wisely describes it (1969, 46), populism is a belief in the creativity and superior moral worth of the ordinary people, of the un-educated and the un-intellectuals. Populist thinking is how Indian intellectuals deal with their anxiety about whether they have allowed themselves to be corrupted by an admired foreign culture. Unconsciously or otherwise, they seek to compensate for this sense of guilt by insisting upon finding virtue in the masses: "to identify with the people, to praise the cultures of the ordinary people as richer, truer, wiser and more relevant than the foreign culture in which they had themselves been educated, has been a way out of this distress . . . 'the people' in whom 'real India' lives have become their sacred objects" (Shils 1961, 73).

The predilection of Indian intellectuals toward populism stems from the lasting influence of Mohandas Karamchand Gandhi. India gained its freedom without a peasant revolution from below (as in China), or a fascist revolution from above (as in Japan). But then, neither did India experience a successful bourgeois revolution (as in Britain) that could lay the foundations of a liberal civil society. Post-independence India took on a modern liberal constitution without experiencing any substantial break with its past cultural traditions. This relatively conservative, non-revolutionary "modernity" was the result of the unusual alliance the Indian National Congress, under the leadership of Gandhi, succeeded in building between the emerging industrialist, capitalist interests in the cities and the peasant masses (not the landed elites) in the countryside (Moore 1966; Vanaik 1990). Using the religious idiom of a moderate, inclusive Hinduism, Gandhi was able to build a coalition of the urban bourgeoisie and the masses of impoverished peasants in the countryside. Speaking the language of Hinduism and traditions, Gandhi was able to provide this strange amalgam of Westernized intellectuals, merchants, industrialists, and ordinary tillers of the soil with a model for a future India based upon an idealized version of "village republics" and "trusteeship of the rich." He was thus able to galvanize the rural masses against the British without challenging the existing power relations in the society, and simultaneously, commit the emerging urban middle classes to a preservationist, culturally conservative model of development. This alliance between the urban bourgeois and the rural masses has left a permanent tendency toward rural populism in India's development programs. The angst of Indian intellectuals toward their own modernist impulses comes from this history.

In the early decades after India's independence, the Gandhian agenda was tempered with Nehru's forward-looking, industry-oriented socialistic policies. As the fortunes of Nehru's brand of democratic socialism began to decline starting around the mid-1970s, Gandhian ideals, mixed with ideologies of "people's power," "direct democracy," and "total revolution" directed at the institutions of the Indian state and its Western collaborators came to dominate the intellectual and political space.[15] As the influence of Marxist and socialist ideas also declined with the fall of the Soviet Union, Gandhian populism became even more dominant. There was nothing else that was left standing intact.

The rise of postmodernism in the West has lent a new academic gloss to this neo-Gandhian cult of "the people." In the last three decades of the twentieth century, the disillusionment and the guilt of Western postmodern intellectuals made a common cause with the populism of Indian and other intellectuals from other postcolonial societies. The same forces of globalization that spread modern technology and capitalism all around the world also globalized the critics of the globalized modernity. Increasing levels of migration brought intellectuals from non-Western societies to major metropolitan universities in the West. Postmodern critiques already fashionable in these universities seemed to satisfy these intellectuals' nationalistic and populist urges to resist the West and at the same time, to affirm the traditions of the West's "victims," lumped together in one big mass without adequate consideration for internal class and cultural contradictions as "the Third World people," "the marginalized," "the subaltern," or simply the "other." According to Aijaz Ahmad (1992, 196) a Marxist critic of Edward Said's influential *Orientalism* (1978), many of the migrant intellectuals from Asia who entered the American and Western academia from the late 1960s onward, were from relatively privileged backgrounds back in their native countries. Postmodern condemnation of Western knowledge as having constructed the East as inferior seems to be tailor-made for them, Ahmad believes, as it gave these "upwardly mobile professionals . . . a narrative of oppression" and victim-hood which they needed to advance their careers, but which they lacked in their real lives. If entire civilizations had been victimized by the West, then even those who had not personally experienced much oppression in their lives could justifiably take a stand against it.

Many migrant intellectuals came as "Third World" intellectuals, but began to identify themselves as "postcolonial" intellectuals, that is, they too began to see modernity in their native lands through the prism of the postmodernist suspicion of the promise of modernity.[16] An uncanny chasm appeared between the post-colonial intellectuals and ordinary men and women in the so-called postcolonial societies. On the ground, ordinary women and men in the modernizing societies struggled as best as they could to affirm new ideals and new freedoms in the world they found themselves in. Sitting in their Ivory Towers, however, the self-described postcolonial intellectuals could only sneer at these new ideals as "mental colonialism" and lament the loss of the original civilizational direction of their native lands. Like their postmodernist colleagues in the West, postcolonial intellectuals, too, began to see in modernity only a singular, relentless, deterministic march of nihilistic disenchantment, instrumentalization, and domination.

The difference, of course, is that while the Western postmodernists could at least take the hegemony of modern, mostly liberal, ideas for granted, the postcolonial critics were condemning modernity even before it had a chance to take root in the lives of their societies. Whereas postmodernism in the West could serve as an internal self-correction of the excesses of capitalist modernity, postmodernism in modernizing societies like India serves to kill the promise of modernity even before it has struck roots.

### The Betrayal of the Clerks

Under the circumstances of an incomplete modernity that prevail in India, the postmodern-style total critique of modernity amounts to a grand betrayal of the intellectuals of their vocation. This betrayal is in part responsible for the growth of reactionary modernity that we are witnessing in India under the sway of Hindu nationalist parties. With self-consciously left-wing humanists embracing a nativist and anti-rationalist agenda made respectable by highfaluting postmodern theory, there is hardly any organized resistance left to the Hindu nationalists. This is not to deny that the left and secular intellectuals are carrying out a valiant struggle against the Hindu nationalist policies of cultural indoctrination and ethnic cleansing. But what is missing is the existence of a well-articulated secular worldview which has the power to mobilize popular opinion, and which is not afraid to challenge the purported "wisdom" of popular traditions. Even in those areas where the secular left has some political presence—as in issues related to the environment, deforestation, sustainable development, and women's rights—the left movements have mobilized the masses on a populist basis, invoking and celebrating a romanticized vision of traditions, a vision which comes fairly close to that invoked by the religious right. The new social movements of the secular, left-wing intellectuals in India run the risk of fighting a merely strategic war against the religious right, while losing the battle for the hearts and minds of the masses. But unless they win this battle, they cannot win the war.

It is the calling of intellectuals to speak out on behalf of universal ideals and principles. Intellectuals who could rein in political passions in the name of universal good have existed with varying numbers and prestige in all societies through history. But ever since the Dreyfus affair in late nineteenth-century France that gave birth to the social type we recognize as modern intellectuals, we expect our intellectuals to simultaneously defend the autonomy of their artistic, literary, and scientific endeavors from political and social passions and interests, and at the same time, to intervene in the social, political debates of their societies.[17] In the last two centuries, the establishment of nation states and spread of public education has created a new class of intellectuals and technical intelligentsia who, for the first time in history, are free from the supervision of the Church and the patronage of kings and lords. These largely secular and cosmopolitan intellectuals constitute a new "flawed universal class," as Alvin Gouldner (1979) calls them.

What gives this new class the right to speak for the entire society, indeed the entire world, and not just for their own class? According to Gouldner, the universalism of modern intellectuals is grounded in the "culture of critical discourse," a new style of justification of assertions which seeks to elicit un-coerced, rational consent, without invoking the speaker's societal position or authority. The growth of the culture of critical, reflexive discourse is the cumulative product of the Enlightenment and secularization in modern nation states. The intellectuals' relative autonomy gives them the moral authority to intervene on behalf of the universal principles that may otherwise be trampled upon in the heat of controversy. Because they could separate their ideas from the passions in the streets, the clerks could legitimately challenge the street from the position of critical outsiders who speak on behalf of a transcendent good. Only by purposefully declining the cosy comforts of patriotism, as Martha Nussbaum (1996) has argued passionately, could the intellectuals aspire to become "the citizens of the world."

Julien Benda warned in his classic *La Trahison des Clercs*, when the clerks betray their calling—when they begin to exalt the particular over the universal, the "realism of the multitude" over the moral good—then there is nothing left to prevent society's slide into tribalism, where nations and their customs become objects of "religious adoration." What is worse, when the clerks forsake their commitments to universal principles, they give nationalist passions a wholly new legitimacy; they allow what Benda called *"intellectual organization of political hatreds"* so that the most murderous nativisms begin to "claim to be founded on science." Benda adds prophetically, "we know what self-assurance, what rigidity, what inhumanity are given to these passions by this claim [of scientificity]" (1928, 21, 22, emphasis added).

Indeed, the "intellectual organization of political hatreds" is precisely what has come to pass in India. As we will examine in much greater detail later, Hindu nationalists are covering up their program for Hindu supremacy by claiming to be working in the interest of "science." Hindutva ideologues proclaim Hindu beliefs and practices to be in accord with the laws of nature, discovered through "direct realization" and "reason" by their "Aryan" ancestors. While other religious fundamentalists hold God as the source of their sacred books, Hindutva proclaims its sacred sources—the Vedas—to be "nothing but" the science of nature. On this account, Hindu supremacy is "nothing but" a defense of science and reason.

This need to shore up metaphysics as science is especially acute in Hinduism because Hinduism claims the order of nature itself as the basis of morality and ethics. For this reason perhaps Hindu nationalists are more extreme in their claims of Vedas as science. But to take on the aura of science, to present the mythos of the holy books as true logos, fully consonant with the best of modern science, is a prominent feature of *all* religious fundamentalist (Armstrong 2000) and fascist movements (Pois 1985). By presenting the mythos as logos, religious fundamentalist movements stand to gain the stamp of "objective truth" for their religious metaphysics, which after all, structures the ethical and moral universe of the believers. Erasing the difference between science and myth, furthermore, enables the religious nationalist movements to propagate myths as science in schools, the media, in public policies, and even law.

Benda's fears of "intellectual organization of political hatreds" are indeed coming true.

It will be my contention in this book that the political legitimacy and philosophical arguments for the reconciliation of science and myth have been prepared not by the Hindu right, but by self-styled "left" intellectuals who see all sciences as cultural constructions. *If science is inseparable from the cultural assumptions, if logos is a product of mythos, it becomes possible to argue that mythos itself is logos.* If there are no objective criteria to differentiate science from non-science, what is to prevent non-science from taking on the coloring of science?

In India, it is the neo-Gandhian, ecofeminist and postcolonial intellectuals who embraced the social constructivist view of science with great gusto. It is the postmodernist/postcolonial demand for "alternative science" and "Indic modernity" that is finding a practical—and murderous—expression in the project of Hindu science.

### The Crimes of Modernization?

One objection has to be faced and answered: What if both Hindu nationalism and postmodernist critics are responding to a real crisis? What if modernization has in fact unleashed such gross inequities, such horrendous cultural displacements, and such anguish among the vast majority of Indian people that a total critique of the modern age is justified? What if science has really become, as the phrase goes in India, "the reason of the state," giving the de-cultured, West-

ernized elites a carte blanche to treat the non-modern masses as so many pota-
toes in a sack? What if technological modernization and cultural modernity is
truly robbing people of all creativity, turning them into pathetic clones of their
erstwhile colonizers? If so, then aren't the postmodernist intellectuals and the
Hindu activists, in their own overlapping ways, justified in looking for locally
grown alternatives?

To draw a detailed balance sheet of India's 55 years of modernization is too big
a task. I will cite only two sets of evidence, one quantitative and the other qualitative, to argue that the despair over modernization is totally disproportionate to
the actual facts on the ground. The evidence will show that it is not the poor and
the culturally marginal classes/castes who are clamoring for indigenous sciences or authentic models of development. Rather, it is the upward mobile urban
middle classes, the newly enriched middle-caste agrarian classes who are the
chief beneficiaries of anti-modernist ideas. This enables them to enjoy the benefits of new technology, new consumer goods, and new economic opportunities
without losing control over their traditional subordinates, the women, the lower
castes, and the poor.

There is simply no denying that even after 55 years of freedom, India remains
a land of gross inequities and perfectly avoidable deprivations. As of November
30, 2001, for example, the Indian government held close to 60 million tons of food
grains in reserve, while there were more than 200 million persons who were unable to meet their basic caloric needs, and while 47 percent of all children suffered from chronic malnutrition. Or to take another example, while India has
emerged as the powerhouse of information and biotechnology, with a huge pool
of talented engineers and scientists, close to 300 million of India's people are still
illiterate, including 54 percent of all "untouchable" males and close to 80 percent
of all "untouchable" females. While India spends barely 1.5 percent of its gross
domestic product annually on primary education, its defense budget comes to
about 18 percent of its gross domestic product. (All figures are from *Human Development Report*. See United Nations Development Program 1998.)

Dismal statistics like these give ammunition to the critics. But the economic
data also tell another story—of slow but *substantial* improvement, not just of incomes but of human development indices as well. India started out with a massive pool of the poor and destitute. In 1973, 55 percent of all Indians lived below
the poverty line. The 2001 census finds about 25 percent of all Indians living below the poverty line.[18] There is a debate over whether the growth in the gross domestic product due to liberalization in 1991 has hastened the decline of poverty
or not. But even if one were to be suspicious of extravagant claims of a rapid
trickle down, it cannot be denied that economic reforms have at least not ad-
versely affected the trend of decline in poverty. Undoubtedly, 25 percent of In-

dia's people below the poverty line are 25 percent too many. But still, India has been able to make a dent in the chronic, abysmal poverty that has affected so many millions for such a long time.

This decline in poverty is accompanied by improvements in indices of human development. Nearly every human development index compiled by the United Nations Development Program from life expectancy, decline in infant mortality, and adult literacy has shown a reasonably healthy growth since the corresponding level in 1960.[19] The Census in 2001 shows a significant growth in literacy, 75 percent among males and 54 percent among females (see Parikh and Radhakrishna 2002 for more details). No doubt there is much still to be done. But the trends are definitely moving in the upward direction.

In one of the rare qualitative studies of this kind, the well-respected agronomist N. S. Jodha (1988) found some interesting results which challenge the conventional wisdom of modernization as a source of hardship and anomie. Jodha found that over a period of two decades, even those villagers in western India who had not seen an increase in real incomes reported a significant increase in well-being. The villagers felt their lives were getting better because they did not have to depend upon the patronage of their caste superiors, they no longer felt compelled to follow inherited occupations for they had more choices and greater access to modern amenities. Freedom from patronage, opportunities for individual choices, a belief in progress: all these are modern aspirations which these villagers had discovered for themselves. Many of these improvements, incidentally, were made possible thanks to state intervention, the same state that is treated by postmodernist critics as authoritarian and colonial in its mind-set.

More recent ethnographic studies, even those sympathetic to postmodernist theory (Gupta 1998, for example), have confirmed that there is hardly any nostalgia for the good old days among the poor. Even in such a quintessentially peasants' and women's movement as Chipko (the tree-huggers movement to prevent deforestation in northeastern India), ordinary rank-and-file were not agitating against development projects. They were agitating instead for better local control over and a better local share of the economic pie (Rangan 2000). It is the village elite in Gupta's study and the Gandhian leadership of the Chipko movement who speak the language of traditionalism.

Indeed, as many recent analyses of the class composition of Hindu nationalism have shown (see Desai 2002; Brass 2000, for example), indigenism is serving as the ideology of the rising middle classes among the rural landowners who have made tremendous gains through the introduction of new Green Revolution technology and the generous subsidies provided by the Indian state. Hindu nationalism is helping to unify the new bourgeoisie, which includes not just the traditional dominant castes but now increasingly also the traditional Shudra castes (one rung above the untouchables), who make up most of the neo-rich in the countryside. The new entrants to modern economy are increasingly getting linked to global economic institutions, especially since the beginning of liberalization in 1991. This has increased their level of insecurity, while at the same time, increasing their appetite for newer consumer products and newer lifestyles. It is these newly upwardly mobile classes/castes, caught between their traditional inhibitions and modern allures, that are largely responsive to Hindutva's calls for a "Hindu modernity."

Those intellectuals in India and in the Western academia, who have embraced a wholesale critique of the very principles of modernization, have to answer these class contradictions. While they claim to speak for the amorphous mass of the non-modern "people" they are adopting the standpoint of the traditional elites who feel threatened by the new cultural attitudes and the demands of their traditional subordinates.

### Conclusion

Secular modernity is facing a crisis of legitimation in India. Some of it is undoubtedly due to the stresses and strains of modernization: those who have reaped the economic gains now want to shore up their cultural dominance, while those who have been left out feel betrayed. But without leadership and intellectual guidance from elites with deep philosophical, instead of material, complaints against secular modernity, it is unlikely that popular discontent would have taken the religious nationalist turn it has taken.

*Part I*

# Hindu Nationalism and "Vedic Science"

*Two*

# Dharma and the Bomb
## *Reactionary Modernism in India*

---

*You will understand the Bhagavad-Gita better with your biceps.*
*—Swami Vivekananda, Hinduism*

Two events in the last decade mark the dramatic changes that have taken place in India's politics and society since it set out on its "tryst with destiny"[1] in 1947 as a secular and democratic nation state.

On December 6, 1992 a mosque in Ayodhya was razed to the ground by Hindu mobs to make way for a temple to a Hindu god, Rāma. Barely five years later, on May 11, 1998, India literally sent shock waves around the world when Indian scientists tested nuclear weapons in the western Indian desert.

A decade after the mosque was destroyed, Hindu radicals are still clamoring to build the Rāma temple in Ayodhya. And five years after South Asia turned nuclear, the borders between India and Pakistan remain more tense than ever.

Historians will look back at these events as major turning points in India's history. Chest-thumping Hindu nationalists have already begun to characterize them as heralding the dawn of the "Hindu century." I, on the other hand, see them as heralding India's turn to reactionary modernity.

These events mark a fork in the road. For most of the 55 years of independence, especially during the Nehru years, India has tried to combine at least the promise of a secular-liberal cultural modernity along with technological modernization. With the Rāma temple agitation and the building of the bomb, India is making the classic choice of all reactionary modernists: modern technology without the ideals of modernity, "Vedic sciences" without the ethos of science, biceps with the *Bhagavad Gita*, and bombs with dharma . . .

The era of dharma and the bomb will be an era when India will have nuclear bombs in its silos and the Vedas in schools. It will be an era when Toyota-driven "chariots" with Bollywood-style plastic idols will usurp the gods. It will be an era, in other words, when dharma—the traditional Hindu idea of cosmic and moral order—will be invoked to bring the masses into the public arena to give their consent to illiberal and ultranationalistic social policies at home, and nuclear-tipped belligerence in South Asia.

Welcome to Hindu modernity!

Because the two events—the politicization of dharma leading to the Rāma temple agitation, and the development of the bomb—are so symptomatic of India's turn toward reactionary modernity, this chapter will begin with a closer look at them.

A closer look at these events is meant to highlight a distinctive and rather peculiar feature of reactionary modernism that both of these events share, namely, an aggressive hyper-modernism. As we examine the texture of these events, it will become obvious that Hindu nationalism asserts itself not by rejecting the modern ideas of democracy, secularism, and scientific reason, but by aggressively restating them in a Hindu civilizational idiom. The champions of Hindu nationalism pretend to set themselves apart from their Islamic and Christian counterparts by claiming to be enlightened champions of democracy, secularism, science, all of which they claim to find in the perennial wisdom of the Vedas, Vedānta, and in the original, uncorrupted Vedic institution of four varnas or castes. When they use the modern word "secular," they mean the traditional hierarchical tolerance of the relativity of truths that prevailed in a caste society. When they use the word "science," they mean an enchanted, supernatural science based upon the idealistic metaphysics of classical Hinduism that treats the divine as constitutive of all of nature. In both cases, *the modern is simply subsumed under the traditional by declaring both to be equivalent in function and rationality.* The *break* that secularism and modern science made with what passed as "tolerance" and "science" in premodern societies is simply not recognized. If any difference is registered at all, it is only to declare the superiority of the Indian alternative. All the virtues of the modern West, and none of its vices, are claimed to be found in the wisdom of ancient Hindu sages.

It will be my contention that this seemingly hyper-modern attempt at the translation of modern secularism (this chapter) and scientific worldview (next two chapters) into the language of Hindu dharma actually subverts the task of creating a secular and humanistic worldview that can support the values of a tolerant and plural democracy. The seeming openness to modern ideas serves only to confirm the greatness of orthodox Hindu traditions which, conveniently whitewashed to hide their deeply inegalitarian history and irrational content, are held up as resources for an "authentic" Hindu modernity. To the Indian masses caught between the perilous promise of modernity and the psychological comfort of traditions, the Hindu nationalists offer a way to become "modern" by returning to "traditions." The unasked question is whether the Brahminical traditions of the Vedas and the Vedānta favored by Hindu nationalists can *actually* enable Indian people to develop a genuinely more rational and secular civil society.

More specifically, I will try to answer two questions in this chapter. One, how have the Hindutva demagogues succeeded in creating this monstrous hybrid of Hindu modernity that combines such opposites as modern secularism and the "tolerance" of a caste society? What assistance have the postmodern adulations of difference provided to the success of this widespread belief in the "modernity of traditions?" Two, how can one explain the success of Hindu demagogues inciting the mobs to violence against religious minorities, all in the name of Hindu ideals of "true tolerance" and "authentic secularism?" How does one explain the mass enthusiasms this kind of reactionary modernist rhetoric is generating across all sections of Hindus?

I will suggest that the answers to these questions lie in the historic weakness of the Enlightenment and secularization in India. India officially declared itself to be a secular state without paying due attention to the secularization of civil society. Postmodernist condemnation of science further weakened any challenge to the hold of religiosity on the public sphere.

A note to the readers: While I think this chapter is important in understanding how religious nationalism justifies itself in a secular, modernist vocabulary, the readers who are primarily interested in the debates about science may want to skip the latter parts where the intricacies of the Indian debates over secularism are dealt with. All the rest of the chapters in the book deal with issues of science and social movements more directly. For those interested in understanding the phenomenon of Hindu nationalism, however, this chapter should be of considerable interest.

### Two Scenarios: Dharma and the Bomb
#### Dharma

On September 25, 1990, a Toyota van made to look like an ancient chariot (*rath*) carrying L. K. Advani, a prominent Bharatiya Janata Party leader, currently the nation's deputy prime minister, set off on a 10,000 km, month-long procession (*yātra*) from the temple town of Somnath in western India to another famous temple town, Ayodhya in the northeast. Ayodhya is presumed to be the birthplace of Rāma, a mythological god-king. Hindu nationalists claim that in 1528, the Mughal emperor, Babar, had torn down a temple that marked Rāma's birthplace and replaced it with a mosque, the Babri Masjid. Advani's chariot journey was meant to fan popular passion in support of building a Rāma temple next to the mosque, with its sanctum sanctorum located in the heart of the mosque. Just two years later, on December 6, 1992, Hindu mobs razed the mosque to the ground.

Advani's procession came at a time when the communalization of Indian politics was already at an all-time high. Indira Gandhi had started playing the religion card in search for short-term electoral gains. The tradition was continued by her son, Rajiv Gandhi, who bent over backward to first mollify the most conservative elements of the Muslim community over the Shah Bano affair in 1985,[2] and then, to mollify the Hindus, pressured the judiciary to open the gates of the disputed mosque in Ayodhya to Hindu worshippers. In the meantime, caste divisions had also taken on a huge political presence when affirmative action was extended to the non-untouchable lower castes (the Shudras) in 1990, leading to massive street protests by upper-caste youths.[3]

In the middle of this highly charged atmosphere, L. K. Advani set out in his "chariot" to Ayodhya. A very modern, air-conditioned Toyota was decorated to look like a Bollywood version of a chariot from the *Ramayana* or *Mahabharata*, complete with traditional Hindu religious iconography of the sacred symbol "Om," lotuses, and lions. While the chariot itself carried no idols—only L. K. Advani and saffron-clad Hindu holy men and their associates, many dressed as Hanuman (the monkey god) and other characters from the *Ramayana*—the procession took on a ceremonial life of its own. Many along the route treated the "chariot" as a temple-on-wheels and honored it with incense, coconuts, and sandalwood, all the offerings proper for worshipping a divine object. Demonstrations of religious fervor and political militancy blended into each other. Women performed devotional songs and dances, young men, carrying bows and tridents and dressed as characters from the *Ramayana* paraded in the yātra, some even presenting Advani with bowls of their own blood (Jaffrelot 1996, 416–417; Davis 1996, 47). En route, the yātra was met with political rallies in which holy men exhorted the masses to support nationalist causes and candidates. The yātra left a trail of murderous riots in its wake, as it was followed by the convergence of tens of thousands of Hindu "volunteers," including a large number of women, who attacked the mosque and placed a Hindu flag on its dome. This was the beginning of the end. The mosque was destroyed two years later.

The religious frenzy created by this yātra paid rich electoral dividends. It destabilized the existing government, forcing new general elections in 1991. The Bharatiya Janata Party won handsomely, increasing its share of seats in the national parliament from 85 in 1989 to 120, and the share of votes from 11.36 percent in 1989 to 20.08 percent (Jaffrelot 1996, 439). Since then, it has become clear that appeals to religious bigotry bring more votes in India these days than appeals for peace.

Advani's infamous procession is only one example of the new genre of public spectacle that has honed the skill of meshing the most instrumentalist, murderous politics with the religious sentiments of the people. One of the most creative movements was the "Rāma shila puja" (Rāma temple brick worship ceremonies) in which Hindus from all over the world were asked to hold public ceremonies for consecrating the bricks meant for the Rāma temple in Ayodhya. Some 300,000 such ceremonies were held and 83 million rupees were raised (ibid., 385). There are many more examples, the most shameless being the "pride yātra" that the Bharatiya Janata Party and allied groups carried out in Gujarat in September

2002 on the heels of the worst carnage in that state against Muslims earlier in the year.

The seamless meshing of religiosity and political calculation makes these spectacles most dangerous. Eyewitness and media accounts show that elected officials, from ministers to the village-level leaders, openly participate in these pilgrimage-style political rallies. Priests and holy men openly campaign for votes for Hindutva candidates, local merchants extend their hospitality and the police, meanwhile, protect the Hindu mobs. This mixing of religion, big money, and the machinery of the state has become the major source for spreading the Hindutva message across the country.

### The bomb

In May 1998, the media around the world carried pictures that should have sent a chill down our collective spines. They showed crowds of ordinary, everyday men and women dancing in the streets of New Delhi to celebrate India's successful nuclear weapons tests. For these mobs, the technological hardware of the bomb was a symbol of their national greatness, their strength, and even their virility. It was a Hindu bomb against the Islamic bomb of Pakistan. Opinion polls showed over 90 percent approval of India's decision to go nuclear.

The ideologues of Hindu nationalism and many ordinary people on the streets claimed that the bomb was foretold in their sacred book, the *Bhagavad Gita*, in which God Krishna declares himself to be "the radiance of a thousand suns, the splendor of the Mighty One. . . . I have become Death, the destroyer of the worlds." While Robert Oppenheimer used the Hindu imagery after the first nuclear test in 1945 to express fear and awe at what science had wrought, the Hindu partisans see in this imagery a cultural and religious justification for their nuclear weapons. Indeed, some observers have gone so far as to claim that for many Indians, the nuclear tests were a *religious experience* in which they saw "the triumph of divine power . . . the workings of providence, grace, revelation and a history guided by an inexorable faith" (Harman 2000, 738).

There is plenty of evidence for a distinctively Hindu packaging of the bomb. Even though the Bharatiya Janata Party government eschewed religious rhetoric in its official pronouncements, it gave the other members of the Hindutva family, the Rashtriya Swayamsewak Sangh and the Vishwa Hindu Parishad, a free rein to claim the bomb for the glory of Hindu civilization and Vedic sciences. Shortly after the explosion, Vishwa Hindu Parishad ideologues inside and outside the government vowed to build a temple dedicated to Shakti (the goddess of energy) and *vigyan* (science) at the site of the explosion. The temple was to celebrate the *vigyan* of the Vedas that are supposed to contain all the science of nuclear fission and all the know-how for making bombs. Plans were made to take the "consecrated soil" from the explosion site around the country for mass prayers and celebrations. Mercifully, the fear of spreading radioactivity scuttled these plans. But the Hinduization of the bomb has continued in many other ways. There are reports that in festivals around the country, the idols of Ganesh were made with atomic orbits in place of the traditional halo around his elephant head. Other gods were cast as gun-toting soldiers. At an official level, the weapons and the missiles under construction are routinely given distinctly mythological names, from *Agni* (the fire god) to *Trishul* (trident, the symbol of God Shiva). The religious imagery was sufficiently pronounced to have alarmed a group of American scholars of Hinduism who issued a letter of concern to "protest the use of religious imagery to glorify and to legitimate nuclear exercises."[4]

Some in the peace movement have claimed that it is the Western media's biases that have led them to emphasize the exotic while ignoring those masses of Indian people who are opposed to the bomb. Seen in isolation, all the talk of Shakti temples and atomic gods may be nothing more than a sideshow, the fanatical fringe's ploy to whip up popular support. But the religious rhetoric cannot be seen in isolation, because it is taking place in the larger context of the state-sponsored introduction of Vedic science and religious values in public education and other institutions of the public sphere and the frantic attempts by Hindu nationalists to rewrite the history of the Indus valley civilization as the cradle of the "Aryan" civilization.

### Two Hypotheses: Religiosity and Authenticity as Political Forces

The developments described above raise two important questions that the rest of the chapter will try to answer. One, why has Hinduism emerged as a political force now, after more than 50 years of multi-religious, democratic, and secular polity? Two, how can one explain the "modernist" character of Hindu nationalism which, even as it fans tribal passions, claims to stand for "secularism" and "tolerance?"

Let us consider the first question first. Well-known thinkers/writers in the West (Peter Berger, Karen Armstrong, Mark Jurgensmeyer, for example) and in India (Ashis Nandy, T. N. Madan, the doyens of Indian social theory, and Koenraad Elst, speaking for the Hindu right) have singled out excessive secularism, and the imperialist hubris of scientific reason, as the chief causes of religious-political movements. Those who participate in these movements are seen as responding to assaults on their faith by liberal and secular elites talking the language of science.

I will propose a contrary explanation and argue that in India, it is not the excesses but the weakness of secularism, evident in the lack of secularization of the civil society, that is the major cause of the phenomenal rise of Hindu nationalism. Under the veneer of a neo-Hindu-inspired state ideology of "equal respect" of all religious orthodoxies, a high degree of popular religiosity, inegalitarian and superstitious in content, has grown unchecked. It is this religiosity which is now providing the popular base for Hindu nationalism. *It is the incompleteness of the project of Enlightenment, rather than an excess of it, that explains India's turn to reactionary modernism.*

In counting popular religiosity as a source of religious politics, rather than just a reaction to, or a rationalization for, "real" (read social-economic) causes, I am following the recent work by three of the most well-known scholars of new religio-political movements, namely, Martin Marty, R. Scott Appleby, and Nikki Keddie. Marty and Appleby distinguish religious-political movements by the fact that they specifically defend religion as *religion*, that is, as a source of ultimate concerns and "erosion-free" eternal values. Fundamentalists see themselves as fighting against secular elites who they see as reducing religion to false consciousness (Marty and Appleby 1992, 815). The participants' own reasons for action—namely, defense of religion—must be treated seriously.

Most of the time, this does not happen. Even otherwise astute analysts are reluctant to count religious faith as a motivating factor for those who join or support religio-political movements. Admittedly, subjective motivations are hard to assess. But even when participants openly admit, for example, that they have traveled long distances to Ayodhya to build the temple because they believe they are honoring God Rāma, analysts tend to downplay these "idealistic" reasons in favor of more "material" interests, such as women's desire to participate in the public sphere, the youth's hostility to caste divisions, or the elites' need to mobilize the entire village behind their own class interests (see Jaffrelot 1996, 424–436 for examples). There are countless other left-liberal commentators, far less astute than Jaffrelot,[5] who simply dissociate "real" Hinduism from the faith of those who join the Hindutva agitations. Then there are those Gandhian, subaltern historians, and postcolonial romantics who give full autonomy to mass religiosity when it comes to ascribing "tolerance," an indigenous feminist "resistance," and ecological consciousness to the peasant masses, but who deny that the religious faith of these same masses has any connection with the religious ideology of the nationalist parties. Henry Munson, a scholar of Islamic fundamentalism, has this caustic comment on the situation which expresses my own views very well: "[No self-respecting] liberal or leftist scholar would argue that the activists in environmental, feminist, gay rights, peace or liberation-theology movements were really trying to cope with stress engendered by rapid modernization." Yet, these same scholars seem to have no compunctions in dismissing the arguments of the fundamentalists as mere "symptoms of their inability to cope with the stress brought about by post-Enlightenment modernity" (Munson 1995, 163).

Nikki Keddie offers an explanatory framework which tries to restore to religious faith its agency to motivate and mobilize. She has argued that the levels of popular religiosity explain why some countries turn to religio-political movements (e.g., South Asia, United States) while others facing similar structural problems caused by modernization and globalization do not (e.g., East Asia, Western Europe): "Significant new religious political movements tend to occur only where in *recent decades* (whatever the distant past), religions with supernatural and theistic content are believed in, or strongly identified with, by a large proportion of the population. In addition . . . a high percentage of the population identifies with the basic tenets of religious tradition regarding its god or gods, its scriptural text and so forth. The only single word for this phenomenon is . . . religiosity" (Keddie 1998, 702, emphasis in the original).

While economic and cultural displacements caused by modernity and globalization prepare the ground for fundamentalism, the latter come to take root, grow, and come to power *only* in those societies where traditional religious symbols and doctrines have considerable emotional resonance. In India, internal stresses and strains of modernization, coupled with the collapse of the socialist alternative worldwide, have led to deep disarray. Yet, why is an aggressive, supernaturalist, and ritualistic form of Hinduism, rather than any other ideology, filling in the gap left behind by the collapse of the Congress-style secularist centrism? Could it be that this kind of religion-tinged politics resonates with the popular religiosity of the Hindu majority? For religious symbols to move the masses, these symbols must seem worth fighting for—and even killing for. Only those who define their identities and their moral universe through their gods will answer the Hindutva call to arms against those with other gods, and those without any gods (the much reviled godless "secular elites").

Obviously, popular religiosity is not without resources for inter-religious harmony. Those who define their identity through their gods, also learn to imitate the goodness and compassion of their gods. Indeed, many among those who thronged to Ayodhya were there for their love of Rāma, and not out of animosity against the Muslims, with whom they live as neighbors. Yet, unfortunately, Hindutva parties are drawing upon this religious piety. When Advani's chariot rolls into their village, with himself cast as a protector of faith, the response he evokes is a class apart from ordinary political mobilizations.

To suggest that mass religiosity is itself a part of the problem is to invite trouble. To avoid misunderstanding, let me anticipate the more common criticisms. Critics tend to paint any mention of religion as playing a negative political role as a sign of a colonial and Orientalist mind-set that assumes India to be inherently religious where "religion provides the only motive for action in public life," as Vinay Lal argues (1995, 167). Or they tend to read it as suggesting that "there is something seriously wrong with religions in non-Western world," as Mark Jurgensmeyer would have it (1993, 2). But I make no such assumptions. Those who participated in Advani's rath yātra, for example, displayed a whole variety of motivations from considerations of caste, dissatisfaction with the status quo, historic grievances against non-Hindus, and an inchoate search for a new identity. But the fact that these diverse motivations could be aroused and channeled into collective action using religious motifs shows the continuing potency of religion in the civil society. Even if some of the gods are newly fabricated—like the aggressive, militaristic version of God Rāma, Bharat Māta, the patron goddess of India, or the "atomic Ganesh"—they are carefully dressed up with the sanctity borrowed from the old and dearly loved gods. It is this continuing potency of religion, its power to call the masses to arms and *justify acts of violence and intolerance as righteous* that is of concern to me. This does not show that something is inherently wrong with non-Western religions. It only shows that these societies have not yet had the revolution in values that is needed to create a liberal, secular culture.

Two more caveats are in order. I do not read the persistence of high levels of religiosity in India as any kind of Indian essentialism. Contrary to some who insist that India can never become secular because secularization is a Protestant phenomenon, and because Hinduism, unlike the three monotheistic religions, is a "totalizing religion" that "claims all of life" (Madan 1998), I believe that Hinduism has only *temporarily* evaded the forces of Enlightenment and secularization. It is the historical contingencies of anti-colonial nationalism, and not a radically different form of religiosity, that have sheltered and nurtured a romanticized version of Hinduism in public life. Finally, there are those who, like John Hawley (2000), believe that secularists have no appreciation of the importance of religion as an essential part of human life. But secularism and secularization, as I understand them, are not arguments for elimination of religion, but for keeping it honest (next section). It is *because* religion is vitally important for modern societies, that it is worthy of a constant and critical engagement.

Moving on to the second issue: the purported "modernity" of Hinduism. One noticeable feature of the rath yātra and the bomb was that both were presented by Hindutva ideologues as signs of "national awakening," a veritable "Hindu Renaissance" (Gurumurthy 1993; Govindacharya 1993). Even at the height of the temple controversy, when the social fabric of India was being rent apart by religious passions, the Vishwa Hindu Parishad was publishing advertisements claiming that "a Hindu India is a Secular India" (Elst 2001, 171). L. K. Advani, the perpetrator of the yātra that did so much to fan popular passions, could still claim that the temple movement was an attempt to "purify the nation's public life and public discourse" that has become corrupted by Western ideas and divorced from India's own Hindu heritage (Bharatiya Janata Party 1993). Hindutva's most notorious demagogue, Sadhavi Rithambra, could proudly tell her audiences that "Hindu civilization has never been one of destruction. . . . Wherever you come across ruins, wherever you come across broken monuments, you will find the signature of Islam. Wherever you find creation, you discover the signature of the

Hindu"—this spoken from the ruins of a mosque brought down by Hindu mobs! (Quoted here from Kakar 1998, 159.)

Here we see reactionary modernism in the making. The idea of secularism and religious tolerance is being removed from the world of Enlightenment reason and fitted into the language of Hindu dharma. Secularism is the end product of the process of secularization, of rationalization, and de-mystification of religious metaphysics. In the Hindutva discourse—"Hindu India, Secular India"— secularism is being divorced from secularization and read into religious metaphysics itself. Not secularization, but a *de*-secularization—a "return" to India's Hindu-ness (when did India leave it behind in the first place?)—will ensure a secular society. And this, when blood had not even dried in the riot-torn streets in the aftermath of vandalism at Babri Masjid!

Contemporary Hindutva ideologues have inherited this penchant for combining contraries—secularism and Vedic Hinduism, for example—from the novel form of Hinduism, often referred to as "neo-Hinduism" or "neo-Vedānta," that emerged in nineteenth-century India, and has come to form the self-understanding of many a middle-class, English-educated Hindu. As a rough approximation, neo-Hinduism is the brand of Hinduism that is taught by Maharishi Mahesh Yogi, Deepak Chopra, and their clones in the countless yoga-meditation-vegetarian ashrams that dot the landscape of North America and Western European countries. Neo-Hinduism emerged out of the encounter between Hindu intellectuals and the new currents of Western science (especially the ideas of Newton, Darwin, and Spencer), Western political philosophy (especially the works of Comte, Mill, and Paine) and Christian theology. Prominent among the neo-Hindu thinkers are nearly all the leading lights of the so-called "Indian Renaissance," including Raja Ram Mohan Roy (1772–1833), Bankim Chandra Chattopadhyaya (1838–1894), Swami Vivekananda (1862–1902), Aurobindo Ghose (1872–1950), Mohandas K. Gandhi (1869–1947), and Sarvapalli Radhakrishnan (1888–1975), and to a lesser extent, Jawaharlal Nehru (1889–1964).

These reform-minded Indian intellectuals were persuaded of the importance of these new ideas for the reform of their own religion and society. But given the contingencies of colonialism, they were not honest to these values. Even when they themselves were convinced of their importance, they did not use these values to publicly interrogate their own traditions so that contradictory ideas and practices could be challenged and demystified. Rather, keen to assert their national pride against the colonizers, these intellectuals tended to subsume the new ideas into the unreformed tradition. Rather than agitate against those elements of the inherited tradition that negated the content and the spirit of the modern worldview, neo-Hindu intellectuals began to find homologies between the new worldview of science, liberalism, and even Christian ideas of monotheism, and the high-Brahminical Vedic literature, especially the philosophy of non-dualism contained in the Advaita Vedānta of the eighth-century philosopher, Shankara-charya. These monistic/non-dualist trends in Hindu philosophy were elevated to prominence by the Romantic anti-Enlightenment currents in nineteenth-century European thought. Thus neo-Hindu intellectuals ended up retrofitting Western ideas of reason and liberalism that they admired into the mystical elements of Hindu tradition that the Westerners admired. The illiberal and irrational role these mystical ideas had actually played throughout India's history was never confronted with the honesty and courage it required.[6]

The neo-Hindu view of the "modernity" of Brahminical orthodoxy has had a profound influence on the self-identity of Indian intellectuals, educators, and even at times the judiciary, to say nothing of the "god men" (and god women, too) who minister to the spiritual needs of the growing middle classes and the "non-resident Indians" (people of Indian origins living outside India). It is this neo-Hinduism that has now come to constitute the official ideology of the mainstream of the Hindu nationalist movement.

While traditional Hinduism is itself a construct, neo-Hinduism is a special kind of social construction: it satisfies the nationalistic pride in India's Hindu heritage, but it has no real hold on civil society. While countless neo-Hindu swamis can write books on the "scientificity of the Vedas" or the "tolerance" and "modernity" of Brahminical teachings, the reality is that the ordinary practicing Hindu believes in many superstitions, is far from tolerant, and is often modern only in his/her consumption habits. This gap between the rhetoric of "modernity of traditions" and actual reality has allowed a massive self-deception and doublespeak to become India's official cultural policy. Indians have come to proclaim as their heritage exalted ideals of modernity, which in fact their own traditions do not adequately support. The result has been an extremely shallow and fragile modernity which exhausts itself in the acquisition of technological baubles, without informing the life-world of ordinary people.

This gap between ideals and reality is nowhere more evident than in the professions of secular*ism* as an official policy, and the abysmal lack of secular*ization* of the civil society. This brings me back to my thesis that the lack of secularization of the civil society and the continued persistence of high levels of popular, supernatural religiosity in Indian society has contributed to the rise of Hindu nationalism. I am not suggesting, obviously, that the high level of popular religiosity is itself and necessarily reactionary. I am only suggesting that under the current circumstances of transition to modernity, unchecked and unreformed popular religiosity is serving as fertile soil for a nationalistic mobilization aimed against the members of Islamic and Christian faiths.

I agree with those who argue that religious identities are not "out there," always-already formed with distinct caste and class interests. Religious identities, as Thomas Hansen (2001) has persuasively argued, are organized and reproduced through performative practices, including religious rituals in the public arena. My contention is simply that an inchoate and unorganized popular religiosity structures the common sense of the masses which resonates strongly with Hindutva's performative rituals. All components of popular religiosity—the view of the world, rituals, and ethics—serve as rules of grammar that Hindutva parties share with the everyday Hindu believers of all castes, including those castes who have been left out of the fold of the twice-borns. The importance of recognizing the role of religiosity in the workings of Hindutva is that it opens us to the possibility that Hindutva may not be a distortion of "real" religion, designed to dupe and manipulate the "innocent" masses. It opens us to the possibility that Hindutva may actually have a significant amount of popular consent, and that the social engineering of Hindu nationalists may actually be tapping into the popular longings for a renewal. That is where the weakness of the Enlightenment and secularization in India shows up. The grammar of political-cultural discourse is still largely religious, the authority of gods and gurus in matters of public and personal life still undiminished. Secularism and science have been given a bad name before they could revise the rules of grammar.

In what follows, I will first make a theoretical detour and explain what I mean by secularism and secularization and why India cannot become a secular country without a prior secularization of consciousness. The rest of the chapter will explain how celebration of civilizational difference, the right of non-Western societies to live according to their own lights, so central to contemporary critiques of universalism, lies at the heart of the current crisis of reactionary modernism in India.

### India: Secularism without Secularization

I will use the term "secularism" to denote a variety of legal-political arrangements between the state and civil society, and secularization to describe the growth of a sensibility, a Weltanschauung. Secularism is a *formal*, juridical separation of the Church and the state. A state is secular that "guarantees individual and corporate freedom of religion, deals with the individual as a citizen irrespective of his religion, is not constitutionally connected to a particular religion, nor seeks either to promote or interfere with religion" (Smith 1996, 178).

Secularization is a *substantive* separation of the sacred and the profane spheres in the consciousness of men and women. A society is secularized to the degree that "religious institutions, actions, and consciousness, lose their social significance" (Wilson 1982, 149). Religious institutions do not disappear, neither do all individuals become atheists, but with secularization, religion progressively ceases to provide legitimations for more and more of the mundane, this-worldly aspects of social life in both the public sphere (e.g., science, economy, law) and the private sphere (e.g., marriage, divorce, gender and caste relations, food restrictions, etc.) (Bruce 1996). Organized religion no longer serves as the "sacred canopy," to borrow Peter Berger's evocative phrase, sheltering and nurturing all of life, but becomes simply one more sphere of social life, which individuals may or may not choose to participate in, and which may or may not succeed in influencing public policy. Religion loses its taken-for-grantedness and becomes a choice.

In today's globalized world, all societies, everywhere, are subjected to the same social-structural processes that work to break up the sacred canopy—especially industrial capitalism (which practically runs on instrumental rationality), the formation of the nation states, and the universal spread of modern science. That is, even those societies that did not experience the Protestant Reformation and the Scientific Revolution—the two main cultural forces that initiated the process of secularization in the West—are now caught up in the process of secularization. But the universalization of the *process* does not ensure the uniformity of *consequences*; some Protestant Christian societies (e.g., Great Britain) have experienced a steeper decline in the scope of religion than other equally technologically advanced capitalist societies with a dominant Protestant culture (e.g., United States). The difference is explained by the social context, including the history of state-Church relations, the relative strength of the Enlightenment-style movements, and the nature of the dominant religion itself, especially its openness to science and the presence or absence of doctrinal legitimacy for a separation between a transcendent god and a godless nature.[7] One cannot assume, in other words, that religion will suffer the same degree of decline wherever the winds of capitalist modernity blow. But, neither can one assume that the sacred canopy can remain standing anywhere, unscathed, in the face of the whirlwind of modernization.

India is a country that *most needs* a sharp decline in the scope of religion in civil society for it to turn its constitutional promise of secular democracy into a reality. But India is a country that is *least hospitable* to such a decline in the scope of religion. There are two main reasons for India's relative resistance to secularization. First, the historic failure of the Enlightenment-style movements, and second, Hindu metaphysics which does not allow any separation of the sacred from the mundane.

As discussed in the previous section, the rising middle-class intellectuals were initially attracted to Western-style liberal and socialist ideas, but their nationalism led them toward Hindu revival, leading to the growth of neo-Hindu thought. Politically, the urban professionals, traders, and the budding industrialists put their weight behind Gandhi's Hinduism-tinged populism. The only consistent voices for a rational critique of religious common sense came from the untouchable and lower-caste intellectuals, supported by occasional movements for cultural reform launched by the communist parties which failed to connect with the lower-caste movements. But by and large, these Enlightenment-style movements failed to win consistent support from organized left parties, and failed to have much impact on the mainstream of Indian society.

The philosophical worldview of Brahminical Hinduism creates its own unique roadblocks against secularization. In the name of "difference," the critics of modernity on the right and on the populist left have accepted the non-dualist, totalizing nature of Brahminical Hinduism as India's destiny, while damning the Enlightenment-style challenge to the metaphysical assumptions and ethical demands of Hinduism as "scientism" and "Orientalism." The trouble is that this holist metaphysics is far from hospitable to the civic equality that the Indian Constitution promises for all citizens, regardless of caste, gender, and creed. Those who insist upon bringing India's modernity in accord with its "civilizational uniqueness" will have no choice but to re-define the nature of the rights and liberties in this civilizational idiom which gives priority to differential duties and an organicist "harmony." The Bharatiya Janata Party and allies are openly committed to just such a philosophy, the philosophy of "Integral Humanism," which rejects the idea of a nation as a contract between individuals, and conceives society as an organism with different parts regulated by the overall needs of the entire organism (see Upadhyaya 1965). Elements of such a redefinition, argued with much erudition and theoretical sophistication, are visible in the work of theorists of "difference" among the Gandhian-postmodernist critics of modernity as well (notably, Parekh 1992).

But if India wants to realize *in practice*, not just in formal law, the liberal secular ideals promised in the Constitution—equality of all regardless of creed, caste, and gender—it has no choice but to internally reform Hinduism. This would necessarily mean, in the Indian context, a renewed respect for the scientific worldview and a strengthening of the Enlightenment-style critique of the content and influence of Hindu metaphysics. Because a naturalist-rationalist erosion of the supernatural is not forthcoming from within the Hindu network of priests, ritualists, theologians, as it was, say, within the liberal Protestant denominations in the West, it will have to be introduced from without. To quote Jose Casanova (1994, 30): "The Enlightenment and the critique of religion [become] independent carriers of processes of secularization wherever the established churches become obstacles to the modern process of functional differentiation. By contrast, wherever religion itself accepted, perhaps even furthered, the functional differentiation of the secular spheres form the religious sphere, the radical Enlightenment and its critique of religion became superfluous."

A scientific critique of metaphysics is uniquely important in the case of Hinduism because the dominant Vedic and Vedāntic traditions treat God as immanent in nature and therefore allow no demarcation between the supernatural and nature, the super-empirical and the scientific. Metaphysics serves as physics in

Hinduism, and is used to naturalize inegalitarian ethics (dharma). To clarify the distinction between the two, and to free morality from what Hindu metaphysics presents as the natural order, is the unfinished task of Indian Enlightenment (see Nanda 2002, for a more detailed argument for the Indian Enlightenment).

The relationship between metaphysics and politics is not one of strict determination. A culture's ontology is not its destiny. But neither is a culture's fundamental ontology as infinitely open to infinite "social constructions" as the anti-essentialist critics seem to assume. Technically speaking, a deeply religious people can still decide, as a matter of political expediency, to run their public affairs by secular rules. A secular society only requires that the state and the public institutions are not openly and officially associated with any religion and that their laws, policies, rules, and regulations are not justified in the light of religious doctrines. Deeply religious societies, in order to secure freedom from the state (e.g., the United States) and/or to secure inter-religious harmony (e.g., India) can decide to conduct their public affairs in a secular fashion. On the other hand, deeply secular, irreligious societies, as in Britain and in Western Europe, can continue to allow traditional churches to be officially recognized by the state.

Many in India, including those who are deeply opposed to Hindu nationalism, point to the United States as an example to argue that high levels of popular religiosity need not come in the way of a secular state in India. But the experience of the United States cannot be so easily generalized because the mainline churches in the United States themselves served as carriers of secularization. Until around the First World War, the mainline churches in America took modern science and the ideals of progress and democracy very seriously and helped the believing public to reconcile their faith with modernity by removing irrational elements from the former. Indeed, the rise of conservative Christian movements is a response to the success of the theological liberalism of American churches.[8]

In the absence of internal reforms in mainline institutions of Hinduism, a yawning gap has emerged between the constitutional norms of religious neutrality and civic equality expected of the state and public institutions, and the religious and hierarchical values by which most people live their lives. Hindu nationalists are attempting to close this gap by reinterpreting the constitutional norms to bring them in conformity with traditional religious values. The other way to close the gap is to try to change the social values to bring them closer to the secular and democratic norms adopted at the time of independence. The latter option is the option of secularization, which in India will require an active critique of religious reason by democratic means of education, demystification, and by making alternative, naturalistic explanations of natural and social phenomena more accessible. The rise of religious nationalism in India today is a result of the incompleteness of this project of Enlightenment.

## Indian Secularism: Hindutva and Postmodern Critics

The success of the rath yātra bears out the truth of the hypothesis we started out with—namely, religio-political movements tend to occur only where religions with supernatural and theistic content are believed in, or are strongly identified with, by a large proportion of the population. The high levels of religiosity ensure that even those who are motivated by religious piety end up getting drawn into religio-political movements. Indeed, opinion-poll surveys and anecdotal evidence show that in India, especially at the village level, "it was reverence for Rāma rather than anti-Muslim feelings which was most in evidence . . . the RSS-VHP-BJP used the instrumentalization of religious symbols for political purposes but many of its followers mobilized for religious ends" (Jaffrelot 1996, 476–477). While it is hard to gauge the motives, 43 percent of Hindus in India approved of the demolition of the mosque, only four percentage points less than those who disapproved, and nearly 60 percent of Hindus did not want the demolished mosque to be rebuilt by the government (*India Today*, January 15, 1993). More than 10 years later, 48 percent of the Hindus polled supported the Vishwa Hindu Parishad-Rashtriya Swayamsewak Sangh demand to build a temple at the site where the mosque had stood (*India Today*, February 4, 2002). Clearly, the symbols of Rāma and the temple have lost none of their potency for mass mobilization, even if this mass mobilization does not automatically end up as a vote for the Bharatiya Janata Party in elections.

That God Rāma, the goddess Bharat-Māta, or the "Atomic Ganesh" would bring the masses out in pilgrimage-style political processions is hardly a surprise. Consider the mytho-political role of *Ramayana*, the epic celebrating the life of God Rāma, which the nationalists have enlisted in their own cause as an exemplar of an ideal Hindu society. For nearly a thousand years, at least ever since India's contact with Islam around the eleventh century, *Ramayana* has served as a political text which simultaneously legitimizes theocracy by divinizing the rulers, and at the same time, fuels xenophobia by demonizing the outsider (Pollock 1993). The *Ramayana* provides the "cultural code in which proto-communalist relations could be activated and theocratic imagination could be justified" (ibid., 288). The mytho-political power of this epic has been kept alive by the mass media, especially films and television. In 1987–1988, the state-sponsored television ran a widely popular dramatization of the *Ramayana*, which came in handy for the ongoing Rāma temple agitation.

It is not surprising that the Hindu nationalists would choose the story told in the *Ramayana* as a mobilizing myth, or that their rhetoric would appeal to religious piety for God Rāma. The Bharatiya Janata Party, Vishwa Hindu Parishad, and its allies did not create this popular religiosity; the 55 odd years of modernization have led to a growth, not a diminution of, the appeal of traditional and assorted new-age gods, gurus, and rituals. Neither are the Hindu nationalists the first to exploit the popular religiosity for political purposes—the "secular" Congress party, under the reign of Indira Gandhi and Rajiv Gandhi, had already honed the art of playing the religion card to get votes.

But what is unique and dangerous about Hindu nationalism is the insistence that this popular religiosity contains within itself a model of the authentic Hindu secularism, "tolerance" and "pluralism" which is under threat from the state-imposed "pseudo secularism." The Hindu community is secular by definition, Hindutva supporters claim, because it has always been "tolerant" of other faiths. Hinduism also has its traditional methods of separating religion from the state by dividing up the two functions between the Brahmins and the Kshatriyas, respectively. As long as the modern state can institute policies that protect Hindu interests and Hindu culture, Hindu priests need not intervene in the affairs of the state. The Hindutva ideologues demand the Indian state drop its stance of neutrality toward all religions, openly embrace and cultivate the Hindu-ness of India through educational institutes, institute a ban on conversions and cow slaughter, allow tax breaks for Hindu ceremonies, etc.

All religio-political movements demand a greater and open role for religion in the affairs of the state. Hindutva is no exception. But it has its uniquely diabolical twist to the standard fundamentalist script: *making India Hindu, will make it secular.* Cultivating the Hindu worldview in schools and public institutions will make them more open to reason, science, and modernity. Genuine secularization demands an aggressive desecularization; decline of religious bigotry requires a religious revival. As the Orwellian Vishwa Hindu Parishad slogan would have it, "Hindu India, Secular India."

Ironically, this call for a return to Hindu traditions to find an authentic model of secularism and tolerance has found support from unexpected sources: the anti-Enlightenment critics of modernity who claim to speak from an emancipatory postmodernist perspective. I want to briefly examine the overlap between the right-wing and the left-wing critics of secularism/secularization.

The gist of the Hindutva complaint against a modern secularism and secularization is that a mistrust of religion conception of is un-Hindu and anti-secular. It is un-Hindu because Hinduism is a holist religion and does not allow a separation of the realms of the sacred and the profane, as Christianity and Islam permit. It is anti-secular because Hinduism is the source of secularism in India. Without the tolerance and pluralism of Hinduism, secularism will not be possible. What should be kept out of the public sphere are not *all* religions, but only those that are intolerant of tolerance. An authentic secularism in India will openly embrace Hinduism as the state religion, and clamp down on religions that smack of "Semitic intolerance," namely, Islam and Christianity.

For example, N. S. Rajaram (1995), a prominent Hindu nationalist, simply redefines the idea of secularism as "pluralism." No formal separation of state and the Church is needed in those religions which are supposedly, always-already tolerant of pluralism. In a totally spurious reading of the West, Rajaram claims that Christendom was a theocracy and needed secularism to create space for individual choice. Since Hindu India has never been a theocracy and has always respected individual "choice" in matters of faith, it has no use for the Western-style suspicion of religion or for any attempt to secularize it. This position completely and conveniently erases the role Hinduism has always played in regulating and disciplining all aspects of political life, from the legitimations of the rulers, the exclusion of the lower orders, to the total banishment of the untouchables and women from the public sphere. Yes, traditional Hindu society did give each sect and caste a "choice" to adhere to their own practices, but then, it also proportioned the status and their rights according to their proximity to Brahminical ideas and way of life. The "tolerance" and "pluralism" that Hindutva celebrates has, in fact, justified the cruel indifference of a caste society.

To take another, far more dangerous example, the Bharatiya Janata Party, India's current ruling party, is committed by a formal oath that each member of the Party is required to take, to uphold the philosophy of "Integral Humanism"[9] which explicitly declares Hindu dharma to be supreme over the Indian Constitution. According to this philosophy, the function of the state is to protect the nation, and because the Indian nation is held together by Hindu dharma, the Indian state exists to defend Hindu dharma: "the government of independence, by democracy, for dharma" (Upadhyaya 1965, 13). Hindu cosmic and moral law—dharma—in this philosophy is the supreme guide, and just like Brahmin sages could remove the kings who erred against dharma in ancient times, religious considerations can override all branches of the elected government.

In spite of their public profession of this radical philosophy that recommends a religious state, the Bharatiya Janata Party and allies claim that "Hindu India is Secular India." How is this accomplished? *Hindutva ideologues simply deny that Hindu dharma is a religion. It is instead elevated to natural law, which is right for all people, everywhere, at all times.* All other religions become derivatives—some good, some bad, but all derivatives—of the eternal natural laws discovered by Hindus. Like an indulgent parent, Hindu India will grant freedom, within limits, to other religions practiced within its territory insofar as they "are the same as" (i.e., do not contradict) the perennial truths contained in Hindu dharma. This is the vision of "religious tolerance" that animates Hindu nationalism. It is not tolerance toward others who may be different but are still your equals. The Hindutva tolerance toward monotheistic religions like Islam and Christianity is the tolerance of a parent toward the follies of immature teenagers.

The Hindutva contention that the separation of religion from the state is an inappropriate Western import has found support from unexpected quarters— the anti-Enlightenment, anti-secularist social theorists who work with postmodernist assumptions, if not the postmodernist jargon. In the late 1980s and early 1990s, two of India's foremost social theorists—Ashis Nandy and T. N. Madan—wrote furious tracts against secularism, which they reduced to an imported ideology of Westernized elites, out of touch with the simple faith of ordinary people. This position found support from other well-known intellectuals including the late M. N. Srinivas, the world-renowned anthropologist, Partha Chatterjee, a postcolonial theorist, Dipesh Chakrabarty, a subaltern historian, and Madhu Kishwar, India's best-known feminist. These attacks on the supposed crimes of the secularized elites have shifted the fulcrum of Indian politics so much to the right that even self-described left-liberals (see the writings of Bharucha 1998; Kakar 1998) decry any criticism of popular religiosity as elitist and "Orientalist," reducing secularism purely to a political stratagem of state-neutrality.

The gist of the famous "anti-secularist" position is that "Hinduism must be secularized within the spiritual framework [of Hinduism itself]" (Madan 1993, 675). To question the spiritual framework of Hinduism from a naturalist worldview of modern science amounts to a modernist, Westernized "minority trying to shape the majority after its own image" (Madan 1998, 298). The patron-saint of anti-secularists is Gandhi, who Madan and other anti-secularists pit against their arch-enemy, the secularist, Nehru. Gandhi stands for an indigenous secularism from within Hindu spirituality, while Nehru represents a Western-style secularism that would dare to question the worldview of Hindu spirituality.

Why must India secularize from within Hindu spirituality? Why is disenchantment and secularism in the sense of separation of the sacred from the profane not an option for India? On both these issues, the answers of India's best-known social theorists overlap with the most rabid Hindu nationalists.

The basic claim is twofold. One, the Hinduism of the masses is tolerant and pluralistic as it is, and does not need to mimic Western patterns. Two, Hinduism is holistic and therefore any separation between nature and supernatural, mortals and gods, public and private, faith and politics is foreign to it. Indeed, the claim is that it is *because* Hindus don't draw boundaries between the sacred and profane, or between gods and demons, that they are tolerant of others. Any attempt to disenchant Hindu religiosity—to remove gods from the public sphere, to remove the supernatural from the natural—will make it intolerant on the lines of dualist Semitic religions (Nandy 2001).

For all their vehement opposition to Orientalism for having essentialized India as all-that-the-West-is-not, one finds an insidious cultural essentialism at work in the anti-secularist tirade. Starting from the undeniable fact that the separation of Church and state has roots in the Judeo-Christian tradition of separating God from the Caesar and from nature, Madan and Nandy concur that secularization is a "gift of Christianity" (Madan's phrase), which obviously makes it inappropriate for the largely Hindu India. (Indeed, this is precisely what the proponents of Hindu nationalism had been saying all along.) But this reductionist understanding of secularization does not do justice to the Western experience which had at least three other components which transcend the Christian experience and have become universal in scope: namely, the formation of the modern state, the growth of modern capitalism, and the modern scientific revolution. Moreover, the anti-secularists, influenced as they are by the postmodernist critique of the very idea of modernity, have paid scant attention to those dissident minority traditions in India—the rationalist and neo-Buddhist movement among the dalits, for example—which have been struggling to remove the traditional legitimations for caste and gender oppression derived from a sacralized understanding of nature.

As in the Hindutva discourse, the inherent pluralism and holism of Hinduism are held up as adequate—nay, superior—resources for tolerance and inter-religious dialogue. Even after the bloodletting in Ayodhya, to say nothing of the innumerable anti-minority riots in which gods are directly invoked and even imitated in weaponry and attire, Nandy and his fellow anti-secularists continue their romance of Hindu gods as benign playmates who "invite us to live in a plural world" (Nandy 2001, 131). Even after all we know of the long history of upper-caste Hindu intolerance and violence against the lower castes, popular religiosity is declared to be "definitionally non-monolithic and operationally pluralistic," while any attempt to keep religion out of the public sphere is declared to be "ethno-phobic and frequently ethno-cidal" (Nandy 1998, 324). The only difference that separates the left-wing anti-secularists from the right-wing anti-secularists is where they presume to find these sources of purported tolerance. While it is the Brahminical texts of Vedānta that the right wing turns to, the left-wing Gandhians turn to a romanticized view of folk traditions and the everyday "innocent faith" of ordinary people.

The assumption is that faith itself can never motivate the faithful to commit acts of violence or hatred. In other words, those who came out in droves carrying consecrated bricks to build a temple and those who celebrated the destruction of the mosque, were not responding to what their faith in Rāma meant to them. Rather, they were "only" expressing their political grievances against the modern world which has marginalized them. That faith itself can serve an ideological function, that evil done in the name of religion is as much a manifestation of religion as the good that is done in the name of religion, is a possibility not even contemplated by the romantic anti-secularists who straddle the space between Gandhian anti-modernism and a full-blown postmodernism in India.

### How "Western" is Indian Secularism?

It is a safe bet that anyone who did not know much else about the Indian political system, but only read anti-secularist critiques from the left and from the right, would come out convinced that a grave injustice is being done, that India is in the grip of "mental colonialism," adopting a flawed Western model of godless secularism while ignoring rich resources of religious tolerance at home. It is not unlikely that such a reader would form an impression that the Indian state, especially under Nehru, was some kind of an atheist outfit, out to punish the faithful in order to eliminate all traces of religion in public life.

Nothing could be further from the truth. The joint right- and left-wing attacks on secularism as enshrined in the Indian Constitution amount to a sad joke. Both sides refuse to acknowledge the fact that far from mimicking the Western models, the kind of secularism that the Indian state has practiced in the last 55 years is a genuinely Indian invention, with strong inspiration from neo-Hinduism. Indeed, many of its limitations are a result of these romanticized views of Hinduism. It is not Westernization, but a flawed indigenization of the idea of secularization that is the source of the problem.

India is a very peculiar kind of secular society. It promises the state's neutrality in matters of religion not by separating the state, equally without preference, from all religions, but by allowing the state to interfere (presumably) equally, without preference, in all religions. This is the famous doctrine of *sarva dharma samabhava*, which translates into "equal respect for all religions." In other words, *there is no wall of separation between the state and the Church, but only an injunction to the state to show equal respect for all religions, as it celebrates, reforms, and otherwise interferes in their affairs.*

The Indian Constitution promises all citizens full freedom to believe in, practice, and propagate their religious faiths.[10] However, the same constitutional clause (Article 25) that confers the freedom of religion on citizens, also confers upon the state the right to "regulate and restrict" any *secular* activities (e.g., economic and political) associated with religious practices, if these activities conflict with public order, morality, and, more importantly, with the guarantee of civic equality promised as a fundamental right by the Constitution. (The Judiciary has been very careful not to interfere in the purely *religious/ceremonial* aspects of religious beliefs and practices.) Thus, the Constitution itself interferes radically with Hinduism by abolishing untouchablity, removing the differential caste rights and responsibilities, and granting fundamental rights of equal citizenship to all, regardless of caste, gender, and creed. The Constitution, moreover, opened all Hindu, Jain, and Sikh temples to all castes, and generally pointed in a reformatory direction in matters of gender justice by recommending the passage of a Uniform Civil Code that would regulate the personal law of all religious communities. Admittedly, the reform of Muslim, Christian, and other minority personal laws has not gotten the same attention as the reform of the Hindu caste and gender laws—at first, under Nehru's government, because of consideration for the minority sensitivities just after the partition, and later, under the regime of Rajiv Gandhi, for political expediency.

The state is not only permitted to interfere in order to reform religion, but also to promote and celebrate it in the public sphere. Thus the courts have interpreted the Constitution to permit the state to sponsor religious pilgrimages, extend patronage to places of worship, aid parochial schools. While the state cannot raise taxes and sponsor *any one* religion, it can, under the doctrine of *sarva dharma samabhava*, spend public money on *all* religions. Religious communities are promised autonomy from state intervention in their religious affairs, and at the same time, given state assistance as a part of the promotion of "Indian culture." In principle, both the autonomy and the patronage is supposed to be equal for all, but in practice, Hinduism, being the religion of the majority, and given the personal religiosity and/or biases of politicians, has come to serve as the de facto cultural policy of the Indian state.

With all this power to intervene, the Indian state is still seen as committed to disestablishment and secularization for two good reasons. One, it does not, at least in theory, establish Hinduism as the state religion by extending it any special privileges. Secondly, the Constitution separates the fundamental rights of citizenship from the dictates of Hinduism which relativized rights and obligations to caste and gender. Indeed, the establishment of democratic India, with equal rights of citizenship, itself was a secularizing process, as it separated citizenship from religious legitimations.[11]

The word "secularism" in India, then, means something very different from what it has come to mean in the West. The Indian variant carries no connotations of indifference to religion in the public sphere. While the Constitution enjoins the citizens to cultivate a "scientific temper"—roughly, a spirit of inquiry and a respect for scientific evidence—the Constitution was never intended, nor ever used, as a manifesto for atheism, or even for secular humanism. Rather, to quote Rajeev Dhavan, a scholar of Indian constitutional law, "India's secular state was designed to celebrate all faiths, and also enjoined to eliminate some especially invidious practices sanctioned by the religions in question" (2001, 319).

Where did this celebratory understanding of secularism come from? The fundamental basis of Indian secularism "equal respect for all religions," is Hindu in its inspiration, deeply nationalistic, but only very indirectly "Western." It would be great if popular Hinduism, as practiced and lived by ordinary everyday Hindus, was really so respectful of all religions. That would be a very precious resource for a genuinely open and tolerant society, and India would be the envy of the entire world. But, alas, Hinduism as containing—and therefore, respecting—the ultimate truth of all religions was more a Hindu nationalist assertion of self-pride, rather than a sociological fact of Hindu life.

The idea that the non-dualism of Vedānta contains the core of spiritual experience that all religions seek—of non-separateness, of oneness with the Creator and his creation—was first articulated by Swami Vivekananda, who was moved as much by a nationalistic will to assert the greatness of Hinduism, as by a reflection on the spiritual experiences of his guru, Ramakrishna. Because other religions were, supposedly, seeking the same experience of the godhead epitomized in Vedānta, a Hindu could respect them all as different expressions of the same truth contained in his own religious teachings. This is the underlying rationale for saying that Hinduism respects all religions as different paths to the truth. The only catch is that the truth that they all seek is supposed to be the same truth first discovered by the Hindus.

It has been suggested that the reason finding "toleration" in Vedānta became a preoccupation of Vivekananda and other neo-Hindu intellectuals was the fact that it was a political value favored by the British out of their own experience of secularization at home (van der Veer 1994). Vivekananda set out to beat the British at their own game by "proving" that the supposedly superior values of the colonialists were already present in Hinduism in a much more refined "spiritual" form than the crass "materialism" of the Westerners. The existence of multiple sects and practices in India were interpreted as proof of Hinduism's innate tolerance and "equal respect for all religions." This reading of Hinduism was embraced by Gandhi as his creed and thence entered the Indian Constitution as the basis of India's unique brand of secularism. It is this view of all religions as aspiring toward the truths of Hinduism that allowed the Indian lawmakers to interpret secularism as equal respect for all religions; by respecting other religions, they were respecting Hinduism. Is this tolerance or self-love?

Thus, Indian secularism is not Western in any straightforward manner. It grew out of a colonial response to Western influences: only in that sense is it "Western." But it built upon the existing religious resources of Hinduism and in that sense it is deeply Hindu. The question we need to ask is how well this celebration of "tolerance" in Hinduism reflects the lived reality in India.

### The Romance of "Tolerance"

The foregoing leads to an obvious objection: Why should anyone object to accepting non-Western traditions of pluralism and tolerance as a *modus vivendi* in which religion is not artificially kept out of the public sphere and all religions are treated with the same amount of respect? In other words, why *shouldn't* India turn to its own heritage and become secular in its own way that retains and respects the religiosity of all? Why should the European experience of Church-state separation be universally respected?

The answer is simple: The much-celebrated "tolerance" of Hindu traditions cannot serve as sufficient grounds for religious pluralism simply because it *does not exist*. Like secularism, tolerance and pluralism mean something entirely different in the Indian context; they are cultural corollaries of the caste system. Tolerance, as it is generally understood today, in the twenty-first century, implies the right of people to be different and not to be penalized or treated any differently for being different. While Indian intellectuals freely use the word with its modern connotations, they forget that what passes for "tolerance" in Hinduism is a hierarchical ordering of difference. To take a trivial example, if you eat beef, you must accept being classified as an untouchable, and on this condition your difference will be tolerated.[12] From the foods you eat, the families you marry into, to the gods you worship, difference is a measure of status, measured against the ideas and practices of the twice-born. Differences are not accepted as a right to be respected, but as so many criteria for assigning a position in the established social order. True, Hinduism has a multiplicity of gods, sacred texts, and traditions. But the multiplicity has never, in any meaningful way, translated into so many different and equally good ways to worship and to live. Far from it. Even such a staunch apologist for all things Hindu as Sarvepalli Radhakrishnan unapologetically admits the intrinsically hierarchical vision of Hindu "tolerance":

> Hinduism does not distinguish ideas of God as true and false, adopting one particular idea as the standard for the whole human race. It accepts the obvious fact that mankind seeks its goal of God at various levels, in various directions, and feels sympathy with every stage of the search . . .

> Hinduism accepts all religious notions as facts and arranges them in the order of their intrinsic significance. . . . The worshippers of the Absolute are the highest in rank, second to them are the worshippers of the personal God, then come the worshippers of incarnations like Rama, Krishna, Buddha; below them are those who worship ancestors, sages and deities, and lowest of all are the worshippers of petty forces and spirits . . . (1993, 18)

This is a pretty accurate description of Hindu "tolerance" borne out by the studies of popular Hinduism as practiced by ordinary believers. Indeed, even the gods and goddesses are graded in terms of purity and impurity, with the "vegetarian gods" superior to those who accept animal sacrifices (Fuller 1992). Difference is indeed "tolerated," but at the price of inequality.

At a doctrinal level, the neo-Hindu tolerance to other religions again resembles the hierarchical exclusivity of the caste system. The often repeated ecumenical mantra in which Indians take great pride, namely, "all religions are true" actually operates by encompassing other religions as inferior aspirants for the ultimate truth contained in the Vedānta (namely, the identity of the individual soul with the Absolute Soul). All religions are equal in worth insofar as they are so many attempts to get to the unity of the divine and the cosmos already contained in Vedānta. "Tolerance" of others then simply becomes a way of asserting one's own superiority. Paul Hacker, a well-known Indologist, who first critically examined Hindu ecumenism, has correctly suggested that what passes for tolerance is actually "inclusivism" which amounts to "claiming for, and including in one's own religion what really belongs to an alien sect" (1995, 244). Other religious traditions are not respected as legitimate expressions of different ways of conceiving God, nature, and humanity, but simply as different stages of maturity, on the way to realizing what Hindus already know as dharma.

It was Vivekananda, the nationalist revivalist who popularized the reading of Hinduism as a "universal religion . . . inclusive enough to include all the ideals . . . that already exist in the world . . . into the infinite arms of the religion of the Vedānta" (Halbfass 1988, 408). On this interpretation, Hinduism was not one religion among the many other religions of India or of the rest of the world, but the very essence of all religions. As S. Radhakrishnan (1993, 11), never tired of repeating, "Vedānta is not a religion, but religion itself in its most universal and deepest significance." This is the basis of Hinduism's "tolerance" which aggrandizes itself even as it claims to respect others.

### *"Religionization" of the Public Sphere*

How has this romance of tolerance as "equal respect for all religions" fared in the last 55 years? While the Indian Constitution has struggled valiantly and admirably to protect India's religious pluralism, it has also encouraged a religionization of the public sphere which is now serving as the breeding ground of religious nationalism.[13]

In the absence of a strict wall of separation, but only an injunction to be even-handed in its interventions, the Indian state has not hesitated to intervene in religious matters. Over time, this has led to an open mixing of politics and religion, with Hinduism considered as the undisputed natural religion of the state. Far from serving as a tool for Westernization, as understood by right and left anti-secularists, the *Indian state has been an agent of Sanskritization*, that is, it has democratized an upper-caste, neo-Vedāntist understanding of Hinduism through educational institutions, radio, television, and official state rituals (Srinivas 1966, 132). Traditional rituals, from astrology and *vastu*,[14] to faith healing and fire-ceremonies for rains, and traditional ritualists from astrologers to jet-setting godmen, have found new patrons among politicians and state agencies. Far from a decline, the supernatural has found a new lease of life, thanks to "secularism," Indian style.

In the 50-plus years after independence, a de facto alliance between politicians, business-houses, and religious establishments has emerged. Using religion to make political appeals started with Gandhi, was moderated under Nehru, returned with a vengeance with Indira Gandhi, was continued by her son, Rajiv Gandhi, and since then has taken an openly anti-Muslim and anti-Christian turn. From India's first president, Rajendra Prasad praying at the Somnath temple, to Indira Gandhi offering prayers at the Vishwa Hindu Parishad's Bharat Māta Temple (a temple dedicated to India as mother goddess), elected officials, in their official capacity, bowing and humbling themselves before idols, priests, and assorted godmen is a common sight. The use of state facilities, state-run broadcast media, and funds for religious ceremonies, complete with Sanskrit prayers and traditional rituals of Hindu worship is routine (see McKean 1996 for examples).

*Because Hinduism is not seen as a religion of the Hindus alone, but rather the tolerant mother of all that is true in all religions,* conspicuous religiosity by public figures has even got the stamp of approval from the Indian judiciary. In a 1996 decision, India's Supreme Court equated the invocation of Hinduism in election campaigns as merely a pan-Indian cultural expression and not a political use of religion. Indian courts themselves routinely take it upon themselves to act as interpreters of religion and have often justified their rulings in terms of "true" Hinduism or Islam, as they understand them. Contrary to the Hindutva complaint that the secular state has interfered with Hinduism while being protective of Islam, Indian courts and legislature have protected Hindu practices, including protecting the cow, certain parts of rituals that keep the untouchables out, and restrictions on religious conversion (Mahajan 1998).

Although quantitative data on popular religiosity are not available, modernization of the economy and technology has not lessened the scope of religion in legitimating secular spheres of public and private life. On the contrary, freedom of religion, coupled with the fact that conspicuous religiosity is politically expedient, has meant a lush growth of religiosity in all spheres of life. Going by numbers alone, there has been an explosion of temples and other places of worship. In Delhi, the number of registered religious buildings increased from 560 in 1980 to 2,000 in 1987 (*India Today,* June 15, 1987). Temples to deities popular with middle classes—notably Hanuman, the monkey god, and Santoshi Ma—have experienced an explosive growth in the number of worshippers (Lutgendorf 1997). Temples and places of pilgrimage are supported by business houses and industrialists and promoted by state agencies for tourism and culture. As it has become increasingly respectable to mix politics and religion, political leaders openly act as guests and sponsors, in their official capacities, of religious ceremonies. An increasing number of bureaucrats, police officers, judges, and army personnel have begun to openly affiliate themselves with religious organizations and Hindu nationalist parties.

Gurus are entering the boardrooms of Indian businesses, offering meditation, yoga, and prayer for stress-reduction, and "Vedic management" techniques that classify employees depending upon their inborn natures (*gunas*) (*India Today,* May 24, 1999). Temples are modernizing by setting up multimedia shows, "laboratories" of "spiritual sciences," and websites for online worship (*India Today,* February 16, 1998). For all the modernization, the content of religiosity remains instrumental and supernatural: miracles, supernatural interventions for material success, and faith healing are routinely expected from gods and gurus. (Other

Indian faiths, including Christianity and Islam are not free from miracles, either.) If the new middle classes indulge in this new-age-ish "sumptuary spirituality," to use a phrase borrowed from Lise McKean, the newly emerging capitalist farmers in the rural areas and small towns are "upgrading" their social standing by adopting the traditions of the upper castes as well as elements of the new spirituality. As they climb the social ladder, they only give a modern gloss to, but do not discard the hierarchical and patriarchal elements of traditions; caste violence and anti-women attitudes are commonplace. Nearly all medieval social relations—arranged marriages, dowry, child marriages, ill-treatment of widows, female infanticide and feticide, witchcraft, possession, and even sati (widow immolation)—have survived modernization, some to a greater and others to lesser extent.

If secularization is measured by a change in the social imaginary from the supra-empirical to the empirical, from the supernatural to nature, and from the customary and/or scriptural to faith in one's own reason, Indian society does not show much evidence of it. The fundamental assumptions about nature, human beings, and the good still derive from religious traditions that have not undergone anything resembling a reformation or enlightenment.

Out of this unreconstructed social imaginary, Hindutva parties are creating a new cultural identity for Hindus which is aggressively nationalistic and xenophobic. Adept at mobilizing religious symbols and sentiments, Hindu zealots find it easy to turn secular grievances and aspirations into a religious idiom or, as in the case of the Rāma temple, hitch innocent piety to political purposes. Here is a brief sampling of some of the ways in which the religious is being transmuted into the political by Hindu nationalists:

- Hindu holy men and women have entered the political realm. They openly ask for votes for the Bharatiya Janata Party in the name of gods, while accepting the reverence of the faithful coming to touch their feet. They also have been active in the re-conversion movement, inducing Christians and Muslims to "return" to the Hindu fold.
- The politicizing of religious festivals is common. In the last two decades, the annual festival of Ganesh (or Vināyak, as the elephant god is called in southern India) has been turned into a major public celebration, with enormous idols installed at public places, followed by processions with distinct anti-Muslim overtones (Fuller 2001).
- Unemployed and often illiterate untouchable youths, denied all other avenues of dignity and advancement, are being courted and induced to join as storm troopers against religious minorities (Anandhi 1995).
- Under the pretext of "freedom of religion" Hindutva parties have begun to arm their members with *trishul*s or tridents, the emblem of God Shiva. The *trishul*s are cleverly disguised knives; about six inches long and sharp enough to

kill. So far, four million *trishul*s have been distributed in the state of Rajasthan alone (Setlavad 2001).

These tactics are bearing fruit. The Rashtriya Swayamsewak Sangh which started with only 99 members in 1925, today has a membership exceeding 2.5 million in 30,000 branches nationwide, and growing, while the Bharatiya Janata Party boasts more than 10 million party members (*India Today*, December 15, 1996). These organizations have made other inroads into civil society by running over 5,000 schools, attracting children from middle-class, urban families. The presence of these groups on college campuses, trade unions, urban slums, and the countryside has also grown tremendously (*India Today*, January 15, 1993).

All these developments spell trouble for the future of India's democracy.

### Conclusion: The Dangers of Difference

The phenomenal rise of Hindu nationalism in India cannot be called a "revival"—for it was never dead in the first place. Neither is it a case of "de-secularization," "de-privatization," or "return" of the sacred in the public sphere—for the public sphere in India was never secularized, nor religion privatized, in the first place.

Secularism, India-style, was not premised upon the secularization of the society. It was premised upon the "inherent" ecumenism and tolerance of Hinduism. To be a Hindu was seen as being secular—that is, tolerant and broad-minded—almost by definition. Indeed, so "broad-minded" are Hindus that they simply see all religions native to India as part of Hinduism itself, regardless of whether Sikhs, Buddhists, and neo-Buddhists really want to be so seen! Such a self-aggrandizing interpretation of secularism was bound to turn Hinduism into the de facto religion of the Indian state, which prided itself on being a "secular democratic republic."

The surprise is not the rise of the Bharatiya Janata Party and its allied Hindu nationalists—it was a disaster waiting to happen. The surprise is how little principled resistance there has been to the ideal of a Hindu nation. It is true that the organized left parties, intellectuals, journalists, human rights activists, and many non-governmental organizations have marched in the streets, organized peace rallies, investigated and exposed the crimes against non-Hindu minorities. But the secular opponents of the Hindu right have failed to create and sustain a secular consciousness among ordinary people.

I have argued in this chapter that the postmodern Gandhian left has been as much a prisoner of cultural nationalism as the Hindu right wing. Secularization was not just neglected, but actively decried by both sides in favor of indigenism and cultural authenticity. The only beneficiary has been the Hindu right which is planting the seeds of hatred in the name of religion.

*Three*

# Vedic Science, Part One

*Legitimation of the*
*Hindu Nationalist Worldview*

> *The Rig-Veda is a book of particle physics.*
> —*Raja Ram Mohan Roy,* Vedic Physics

> *The attempt to efface the features of the struggles*
> *between science and religion is nothing but a hopeless*
> *attempt to defend religion.*
> —*Sadiq al-Azm,* A Criticism of Religious Thought

The Hindutva theoreticians and propagandists like to describe themselves as "intellectual Kshatriyas" [1]—Kshatriyas, of course, being the warriors in the fourfold varna order of India. These self-proclaimed guardians of "Hindu India" have taken it upon themselves to defend Hindu culture, its core values, and its worldview from liberals and Marxists at home, and from Eurocentric stereotypes abroad.

Modern natural science is the weapon of choice of these intellectual warriors. Hindu nationalists are obsessed with science. They are obsessed with science the way creation scientists are obsessed with science. They use the vocabulary of science to claim that the most sacred texts of Hinduism—the Vedas, the Upanishads which contain the essence of Vedic teachings (also called the Vedānta), and Advaita Vedānta—are, in fact, scientific treatises, expressing in a uniquely holistic and uniquely Hindu idiom, the findings of modern physics, biology, mathematics, and nearly all other branches of modern natural science. From the most orthodox priests, to the modern globe-trotting, English-speaking swamis, academics (many of them in American and Canadian universities), pamphleteers and journalists, down to ordinary men and women sympathetic to Hindu nationalist ideology, all proclaim that Hinduism is simply another name for scientific thinking. Some declare the entire Vedic literature as converging with the contents and methods of modern science. Others concentrate on defending such esoteric practices as Vedic astrology, *vastu shastra* (building structures in alignment with the cosmic "life-force"), Ayurveda (traditional Hindu medicine), transcendental meditation, faith healing, telepathy, and other miracles as scientific. We will call both of these claims together as "Vedic science," as their shared aim is to prove that the mythos of the Vedas contains within itself, and even surpasses, the logos of modern science.

Vedic science is supposed to lead to a better, a more whole natural science that will cure the reductionism and matter-spirit dualism of "Western" science. Vedic science apologists promise to "raise" the lower knowledge (*apāra vidya*) of "mere matter" provided by modern science by integrating it into the "higher knowledge" (*pāra vidya*) of the spirit disclosed by their own traditions. In the words of David Frawley (2001, 153–154), an American-Hindu astrologer and Ayurvedic doctor, "the Indic tradition largely accepts Western science as valid within its own sphere, but regards that sphere as limited. . . . The Indic tradition aims to spiritualize Western civilization by integrating its science into a deeper spiritual view which also accepts the spiritual sciences of Yoga and Vedānta . . ."

This and the next chapter will look at how Hindutva's intellectual warriors justify the scientificity of Vedic literature and practices. What kind of arguments do they offer as they go about eroding all boundaries between modern science and sacred Vedic lore? How do they understand the enterprise of science and its relationship with religion and the rest of the culture? This chapter will begin with a review of the contemporary uses—and dangers—of the Vedic science paradigm. This will be followed by two short theoretical detours; the first will establish the necessity of disenchantment or de-divinization of nature for the creation of modern consciousness, while the second will examine the ontology and epistemology of Vedānta. Next, the influence of Orientalist theories of an "Aryan" golden age on nineteenth-century Hindu revivalism will be examined. We will continue our investigation in the next chapter, where we will explore the various philosophical arguments put forth to dress the mystical idealism of the Vedas in scientific clothes.

The discussion will range across a diverse set of literatures. We will encounter arguments derived from Hindu theology, evidence culled from the history of science in ancient India, the history of the Indian Renaissance and Hindu revivalism. We will also examine the changing relationship between science and religion in the Judeo-Christian tradition. We will encounter some ugly ghosts rising from the ashes of "Aryan science." We will also have an occasion to see social constructivist theories of science being put to work by the champions of Vedic sciences.

Because we will range across so many literatures, it is useful to describe in advance where we are heading and what we can expect to find at the conclusion of our exploration. At the end of this and the next chapter, I hope to have established that the Hindu nationalist movement operates like a scavenger, scooping up scraps of liberalism, feminism, postmodernism, socialism, new age ecology, modern technology, and an unusually high dose of modern science. Having gathered all these bits, Hindu nationalists turn into bricoleurs combining them into patterns—pastiches or hybrids, as postmodernists would call them—that fit into orthodox Vedāntic cosmology which makes material nature an effect of divine consciousness, endowing natural objects with consciousness and agency. If bricolage is the preferred postmodern style, Hindu intellectuals are postmoderns par excellence.[2] Much before anyone had even heard of postmodernism, they have been hammering together a worldview which can accommodate anything and everything into Hinduism, as long as its civilizational "difference" (read superiority) is not challenged.

This neo-Hindu eclecticism shares the following three traits with postmodernism:

1. A radical relativization of rationality to respective cultures. Unlike the postmodernists, who reject all metanarratives, Hindu nationalists rush to claim the metanarratives of science for themselves. But like the postmodernists, they don't treat these metanarratives as genuine and universally valid *advances* in knowledge against which the validity and methods of local knowledges and social practices can be judged, revised, and even rejected. Modern science is treated as epistemologically equivalent to Vedic science, both expressing "the same truth" at different levels, in their own, culturally appropriate ways, so that the latter can be substituted for the former without any loss of objectivity. At the same time, whatever is acclaimed in the West as scientific at any given time is reinterpreted as a fulfillment of classical Hindu ontology and epistemology. The idea that advances in modern science can, and often do, *contradict and negate* elements of traditional sciences is simply not entertained. Indeed, the presumption of equality of all ways of knowing as simply different names for "the same truth," rules out even the theoretical possibility of such contradiction.

   Postmodern theorists stand shoulder-to-shoulder with Hindu nationalists in decrying as "positivist," Orientalist, and oppressive any attempt to judge non-Western knowledge systems against advances in modern science. I call this epistemological move of establishing equivalence and denying contradictions between modern science and local knowledges as "epistemic charity." I show how epistemic charity lies at the heart of social and cultural constructivist theories of science and their offshoots in feminist and postcolonial critiques of science (chapter 5).

2. A radical anti-Eurocentrism. Unlike the postmodernists who deny all cultural essentialisms, Hindu nationalists want to declare scientific thinking as the essence of Hinduism, and Hinduism as the essence of India. But like postmodernists, they insist upon the preservation of civilizational difference and authenticity. So even while they claim modern science as an affirmation of Hindu cosmology and epistemology, they reject those elements of the scien-

tific worldview, especially its naturalism and empiricism, which conflict with the mystical idealism of orthodox Hinduism. In terms they share with radical critics of modernity in the West, Hindu nationalists reject any tampering with the indigenous worldview and values as a mark of "mental colonialism." Hindu nationalism is the fulfillment of the postcolonial demand for creating "alternative modernities" based upon non-Western civilizational values.

3. A radical re-enchantment of nature and de-differentiation of society. Hindu nationalists present Hinduism as the ultimate answer to the postmodernist disenchantment with "dualisms" between scientific facts and cultural values, nature and culture, individual and society, etc. The *entire* postmodernist discourse of locating the crisis of modernity in the "dualisms" and "reductionism" of "Western ways of knowing" is *fully and without alterations* incorporated into the discourse of Hindu modernity and Hindu science. These epistemological sins of modernity give Hindu nationalists the justification for simultaneously berating and bettering "Western science" and "Western secularism."

### Science and "Vedic Science"

Before we plunge into the details, it will be useful to define what I understand by modern science and what is meant by "Vedic science" by its votaries.

What makes a theory, a method, or a concept scientific? My own view on this matter comes pretty close to the definition provided in the *amicus curiae* brief of 72 Nobel Laureates and notable scientific organizations submitted to the United States Supreme Court in 1986 in the case challenging the state of Louisiana's decision to allow equal time for creation science in biology curricula in public schools. One of the purposes of this brief was to help the court draw boundaries between science and religion. This is how the *amicus* brief defined science:

> Science is devoted to formulating and testing *naturalistic explanations for natural phenomena*. It is a process for systematically collecting and recording data about the physical world, then categorizing and studying the collected data in an effort to *infer the principles of nature that best explain the observed phenomena*. . . . Science is not equipped to evaluate supernatural explanations for our observations; without passing judgment on the truth or falsity of supernatural explanations, science leaves them to the domain of religion. (quoted here from Shermer 1997, 167, emphases added)

This definition is not very different from Michael Ruse's testimony before the Arkansas Supreme Court in 1981 in another creation science challenge to evolution. According to Ruse (1996, 300), "the single most important element of our modern understanding of science is that science is limited to naturalistic processes that do not rely upon, or permit, the intervention of supernatural forces. Scientists seek to understand the empirical world by reference to natural law and naturalistic processes."

In both cases, methodological naturalism, or the exclusion of the supernatural as an explanation of natural phenomena, is where today's scientists draw a line between science and other ways of knowing nature. This demarcation obviously only holds true *after* modern science has already disenchanted nature. When scientists in the past invoked God's intent or powers as an explanation of nature, they were not smuggling in their personal faith, but invoking what they took to be an evidentially grounded background belief of their scientific paradigm; after all, prior to Darwin, the order of nature was generally taken as a proof of God's design. So while exclusion of the supernatural from the natural is necessary for demarcating science in modern times, it is neither necessary nor sufficient for understanding premodern or non-Western scientific traditions. But that does not mean that the latter can continue to bring in their own conception of God and divine action as a causal force in nature *today*. One can make a concession for the Vedic people who imagined the Brahman as an active force in nature. But in the light of what we know about nature today, such an assumption would be entirely unnecessary and unwarranted.

What else is required for an intellectual enterprise to be distinctively scientific? It is here that the second half of the above definition comes in. To repeat: science seeks to *infer the principles of nature that best explain the observed phenomena.* This attempt to explain the observed world is of course not foreign to religion, myths, and plain old common sense. But science provides a distinctive mode of explanation which is capable of public, inter-subjective testing, and attempts to hold beliefs tentatively in proportion to the empirical evidence available. Scientific explanations succeed because scientific method is devised, at least as an ideal, to control the intervening factors in a way that allows the causal structures of nature to push back against the conjectures of the scientific community, leading to a gradual approximation of our knowledge claims with the underlying structures in nature.[3] This attempt to infer the explanatory principles of natural phenomena through empirical investigation is fundamental to all science, even though the modern breakthrough came with Galileo and Newton who combined sense observation with mathematization and measurement.

Vedic science literature eschews definitions in favor of grand and sweeping statements. Overall, the tendency is to classify *all* systematic attempts at acquiring knowledge, whatever the method or the subject matter, as "science." Two descriptions of the enterprise come close to a definition. Writing in the introductory volume of Maharishi Mahesh Yogi's *Journal of Modern Science and Vedic Science*, Kenneth Chandler writes:

Maharishi's Vedic Science and modern science are complementary methods of gaining knowledge of the same reality—the unified field . . . that silent level of awareness where all laws of nature are found together in a state of wholeness. . . . Vedic science is a reliable method of gaining knowledge, as a science in the most complete sense of the word. It relies upon experience as the sole basis of knowledge, *not experience gained through the senses only*, but experience gained when the mind, becoming completely quiet, is identified with the unified field." (1989, v, emphasis added)

Vedic science, according to this definition, includes a dimension of reality that cannot be experienced by human senses in the ordinary, everyday state of consciousness, but can be experienced in an altered state of consciousness. To put it in Kantian terms, Vedic science does not distinguish between the noumenal and the phenomenal worlds. By claiming to know the noumenal world, Vedic science displays a hyper-scientistic hubris which disregards all limits on what human beings—given the kind of creatures we are—can know.

The second description gives a better idea of what this super-sensory experience is an experience of. The object of experience of Vedic scientists is Brahman, the Absolute Consciousness of the entire universe, which, according to the Vedāntic teachings, is the material and the efficient cause of all nature. Swami Vivekananda, the nineteenth-early twentieth-century monk who was one of the founding fathers of Vedic science, and who we will encounter again later, explained the nature of this quest: "The Hindu does not want to live upon words and theories. If there are existences beyond the ordinary sensuous existence, he wants to come face to face with them. If there is a soul in him that is not matter, if there is an all-merciful universal soul, he will go to him direct. He must see Him, and that alone can destroy all doubt" (Vivekananda 1968, 10).

Thus, a "direct experience" of existences beyond the ordinary sensuous existence is the primary motivation of Vedic modes of thought. Knowing this super-sensory existence, it is assumed, will explain all worldly phenomena. This is essentially a nature-mystical approach to knowledge in which the "ultimate reality" hidden behind appearances is "seen" in the mind's eye without artificially dividing it up in separate, distinct units. The concern of mystical knowledge is with the "being-ness" of the phenomena. Activities like differentiation, definitions, and experimentation are considered as ephemeral "lower" truths, not worth wasting effort upon (Jones 1986, 41–47).

Despite the stark differences between the two, proponents of Vedic science claim that traditional, mystical Vedic knowledge systems are of the same nature as modern science; they ask the same type of questions and provide equally rational answers. The line between nature-mysticism and natural science tends to get more easily blurred in Hinduism because of its monist ontology. Because the World Soul, or Brahman, exists in the substance of all matter, animate or inanimate, knowing it in one's own soul through mindful meditation, one can know its laws as they manifest themselves in nature. That is the promise of Vedic sciences.

### Vedic Science in Action: "The Pizza Effect"

We have already encountered two instances of Hindutva-style bricolage in our examination of dharma and the bomb in the previous chapter. We saw how the modern idea of secularism is being reinterpreted to be functionally the "same as" (i.e., serve the same function of providing pluralism) the Hindu ideal of "tolerance" or hierarchical inclusion. We saw how the term "secularism" is emptied of its history of disenchantment and secularization and retrofitted into a conceptual vocabulary which, in fact, denies secularization.

We also saw how India's nuclear bomb was enveloped in the discourse of dharma. The nuclear tests were celebrated as a religious event and even the physics behind the bomb was fitted into the discourse of *vigyan*, or Hindu science.

In this section, I want to revisit how dharma took over the bomb, or how Vedic metaphysics claimed nuclear physics. The rationale behind this episode is symptomatic of a much larger problem. Treating modern science as just "another name" for Vedic science and vice versa has become the state's justification for introducing Hindu precepts and superstitions—Vedic astrology, priest-craft, and faith healing, for example—as part of science education in colleges and universities. Like "creation scientists" in the United States who have been trying to smuggle the Bible into public schools, Vedic science proponents are borrowing the prestige of science to smuggle in their own peculiar interpretation of Hindu scriptures into schools and other institutions in the public sphere. The situation in India is far more frightening because this Hinduization of education is taking place in the context of extreme Hindu chauvinism directed at Muslim and Christian minorities.

To recapitulate from the last chapter, the test-explosion of nuclear devices in 1998 was experienced as a religious event by a large proportion of Indian people. Nuclear weapons were justified and packaged in dharmic terms by Hindutva ideologues allied with the ruling Bharatiya Janata Party. They claimed that the bomb was foretold by Lord Krishna in the *Bhagavad Gita* when he declared himself to be "the radiance of a thousand suns, the splendor of the Mighty One. . . . I am become Death." They celebrated the bomb by invoking gods and goddesses symbolizing *shakti* (power) and *vigyan* (science). Even the idols of Ganesh turned up with atomic halos around their elephant-heads, and guns in their hands! Invocation of gods in the context of nuclear weapons has become a constant feature of public discourse. During the 2002 stand-off between India and Pakistan, India's most popular newsmagazine, *India Today*, prefaced its tasteless warmongering with references to the *Mahabharata* and the "thousand suns." The net result of these references is to turn these ugly developments into something like the *Mahabharata*, in which God Krishna sided with the virtuous.

The invocation of the goddesses of *shakti* and *vigyan* was not fortuitous. It was claimed at that time that the bombs were a symbol of India's advanced science and technology, the roots of which could be traced back to ancient Vedic texts. The idea of constructing a temple to the goddess of learning at the site of the explosion was meant to propagate the age-old popular myth that the Vedas presage all important discoveries of science, especially quantum and nuclear physics. A popular version of this myth was reported by Jonathan Parry in his ethnography of the holy city of Benaras: "In Benaras, I have often been told—and I have heard variants of the same story elsewhere—that Max Müller stole chunks of the Sama-Veda from India, and it was by studying these that German scientists were able to develop the atom bomb. The ancient rishis (sages) not only knew about nuclear fission, but they also had supersonic airplanes and guided missiles" (Parry 1985, 206).

Finding physics in the Vedas is a good illustration of Hinduism's peculiar dynamic: its tendency to claim for itself those elements of alien traditions that it needs for its own aggrandization. Agehananda Bharati, aka Leopold Fisher, a Viennese who spent many years as a Hindu ascetic in India,[4] coined a new term to describe this peculiar cultural dynamic. He called it the "pizza effect" (Bharati 1970, 273). The pizza was originally the staple of Italian and Sicilian peasants. It became a part of haute cuisine in Italy only after the Americans popularized it around the world. Like the humble pizza, Bharati argues, any traditional Hindu idea or practice, however obscure and irrational it might have been through its history, gets the honorific of "science" if it bears any resemblance at all, however remote, to an idea that is valued (even for the wrong reasons) in the West. Thus, obscure references in the Vedas get reinterpreted as referring to nuclear physics. By staking a phony priority, modern science gets domesticated; it was always contained in India's "wisdom" anyway. Whatever good they might do for national pride, such claims cannot cover up the fact that Indian people remain mired in a view of the world that is deeply irrational and objectively false.

The pizza effect is at work in the current government's policy toward research and education in science, as the following two examples show.

In April 2001, the Indian Space Research Organization made history by successfully putting a satellite into the geo-stationary orbit, 36,000 km above the earth. In July 2001, the University Grants Commission, the central body overseeing the funding of higher education, announced its plans to offer courses in Vedic astrology as science courses in India's universities and colleges. Astrology has been declared to be a legitimate subject of scientific inquiry, worthy of new research and training, complete with funds for libraries, laboratories, and faculty. The same space power that takes justified pride in its ability to touch the stars, will soon start educating its youth in how to read our fortunes and misfortunes in the stars and how to propitiate the heavens through appropriate rituals. For all we know, the satellites launched by India's own launch vehicles might some day carry internet signals that will make horoscopes easier to match! The pizza effect was in action; the education ministry defended its decision to offer Vedic astrology in colleges and universities on the grounds that it was gaining new adherents in the West, and India needed to match the demand with supply. Astrology is not the only "science." There are other new courses, including, training in *karma-kanda* (priest-craft), Vedic mathematics, "mind sciences," including meditation, telepathy, rebirth, and mind control that are being planned.[5]

In May 2002, in the middle of the dangerous military build-up along the border with Pakistan, with careless talk of nuclear weapons in the air, the Indian government started funding scientists in premier defense research institutes to develop techniques of biological and chemical warfare based upon the *Arthashastra*, a 2,300-year-old Sanskrit treatise on statecraft and warfare. The book is supposed to include recipes for "a single meal that will keep a soldier fighting for a month, methods for inducing madness in the enemy as well as advice on chemical and biological warfare" (Rahman 2002). Space scientists and biologists are trying to replicate the ancient formulas for feeding the soldiers a ration of a special herbs, milk, and clarified butter that will keep them going for a month without food. Other projects include "shoes made of camel skin smeared with a serum from owls and vultures that can help soldiers walk hundreds of miles without feeling tired. . . . A powder made from fireflies and the eyes of wild boars that can endow night vision . . . a lethal smoke by burning snakes, insects and plant seeds . . ." Scientists next plan to turn their attention to other ancient manuscripts which "claim to provide secrets of manufacturing planes which cannot be destroyed by any external force and remain invisible to the enemy planes" (ibid.).

The sacralization of war has meant a simultaneous scientization of sacred Hindu texts. Technological modernization is being encompassed into the traditional, religiously sanctioned understanding of the natural world. New technology, old idea.

These two examples make up only the tip of the iceberg. Nuclear physics is not the only science for which Vedas are being given the priority. Here is a partial, and somewhat dated list of Hindutva claims for priority in scientific discoveries taken from D. B. Thengadi's much-cited 1983 lecture, "Modernization without Westernization":

1. The well-known theorem of Pythagoras who was described by King Clement of Alexandria as "pupil of a Brahmin."
2. The atomic theory of the West which was anticipated thousands of years ago by pramanuvad of Kanaad.

3. Dialecticism of Hegel and Marx, which was first envisaged and systematized by Kapil Muni.
4. The fact that it is the earth that moves around the sun ... which was proved by Copernicus, and more than a thousand years before Copernicus by Arya Bhatta.
5. Materialism of Democratus (sic), of which the first ever sutra was written by Brushaspati centuries back ...
6. Scientific concepts of space and time explained by Einstein and enunciated first by Vedānta philosophers.
7. The scientific definition of matter given for the first time to modern science by Heisenberg and to Hindus by Patanjali.
8. The relativity of space and time, the unity of the universe, a space-time continuum, established in ancient times by Vedāntic thinkers and proved in this century by Einstein.
9. The process of scientific philosophical thinking initiated by Parmesthi Prajapitha of "nassdeeya suktha" and climaxed by Einstein. (Thengadi 1983, 5)

In addition, just about every "miracle" like the Ganesha idols "drinking" milk, every age-old ritual (fire sacrifices, or *yagna*), and Sanskrit chant (e.g., the Gayatri mantra) and every superstition like fire ceremonies for rains and *vastu shastra* (architecture that tries to align built structures with cosmic energy, the Indian equivalent of the Chinese Feng Shui) has been declared to be "scientific."[6] Just about every verse of the Vedas, which has anything even remotely to do with elements of nature, whether literally or metaphorically, has been declared to contain "scientific" facts, including the speed of light, the distance of the sun from the earth, and other such cosmological constants—corresponding to the last decimal point!—to the values obtained by modern physics. It has been claimed, for example, that the *Rig Veda* had discovered the Newtonian laws of gravity as well as Einstein's theory of relativity *and* calculated the velocity of light, discovered cosmic rays, so on and so forth. Nearly all important discoveries of biological sciences are right there in the sacred books as well, from the discovery of photosynthesis, the knowledge of molecular receptors of Ayurvedic medicines, microscopy, even test-tube babies, etc. If the apologists are to be believed, the Vedas were actually engineering manuals, describing precisely those technological advances that have already taken place in the West, from airplanes to submarines, running on everything form solar power to nuclear energy. Everything of value that Western science and technology has produced, even if the "value" lies in warfare and destruction, is claimed to be presaged by Hindu holy men of a bygone era.

The examples in the above paragraph are taken from the recent output of only two think-tanks—the Dharam Hinduja Foundation with centers in India as well as in Columbia and Cambridge universities, and Prajna Bharati, a "national forum for thinkers with a nationalist orientation."[7] There are many academics, with degrees in sciences, many of them working in the United States and Canada—notably, Subhash Kak, David Frawley, N. S. Rajaram (now in India), Ram Mohan Roy, the "Vedic creationists" associated with the Hare Krishna movement and the Ramakrishna Mission, the "unified field" "physicists" associated with Maharishi Mahesh Yogi, and the monks of Ramakrishna Mission's Vedānta centers around the world—who are publishing books claiming that Vedic literature is actually a record of scientific discoveries. Needless to say, all the discoveries are invariably *exactly* those that modern science made later on! Compared to the texts from which the above examples are taken, the work of these later scholars has at least a hint of an argument for the purported "scientificity" of the Vedic scriptures. They are able to connect their arguments to postmodern and New Age critics of science and modernity in the West, on the one hand, and with the previous generation of Hindu nationalists who practically invented the pizza effect.

Apart from these (and countless other) private initiatives, the current Hindu nationalist government is generous with grants for "research" and propagation of such esoteric ideas as *vastu shastra*, Vedic mathematics, transcendental meditation, and even faith healing and a host of "urine therapy" research programs ranging from cow to human urines, to say nothing of Ayurveda and yoga, which are relatively better established (even though their findings have not been tested in double-blind experiments). To this list, the Indian government has now added defense research. But "research" in these areas is really a sideshow; it has not come at the cost of research and development, especially in the areas of information technologies, biotechnology, and space and defense research.

The real purpose of Vedic science is the establishment of Hindu supremacy. The main targets are the schools, both public and parochial, where a massive Hinduization of history and science curricula is going on. The fantastic claims of Hindu science enthusiasts are dangerous because under the current regime, they have a very good chance of finding their way into school textbooks. The Hindu nationalist groups together run some 20,000 low-cost private schools, teaching 2.4 million children, with nearly a thousand new schools coming up every month (*The New York Times*, May 13, 2002). These outfits also run special residential schools in tribal areas and urban slums where they openly indoctrinate disadvantaged children into hardcore nationalist ideology. These schools are the Hindu equivalent of *madarsa*s in Pakistan. Science teaching in these schools is already heavily Hinduized. According to Tanika Sarkar (1996, 243) who has studied urban schools run by the Rashtriya Swayamsewak Sangh, "scientific education, whether on physics or mathematics, is always concluded with Hindu textual approximation mentioned as the real source of that knowledge. There is a confident disregard of authenticated detail, and of boundaries between myth and reality that postmodernists would appreciate."

The Rashtriya Swayamsewak Sangh's agenda of Hinduization of education is now close to becoming the official policy of the Indian government. The new National Curriculum Framework for School Education announced by the current government in November 2000 promises to inculcate patriotism and national pride by "indigenizing education." A major component of indigenization will be highlighting "India's contribution to world wisdom" which will include all the usual items, from Ayurveda to yoga. In addition, the new curricula will require that religious/spiritual teachings be "judiciously integrated" with all subjects so as to raise the "spiritual quotient" of the students.[8] After two years of court challenges, in September 2002, India's Supreme Court allowed the government to go ahead with the new framework.

The real threat of Vedic science is not to research and development in science, but to the educational system that is gearing up to produce a Hindu supremacist mind-set.

### An Objection Confronted and Answered

Notwithstanding the dangers of nationalism, one could still legitimately ask: What is wrong with the Hindu nationalist desire to claim the glories of science for Hinduism? Doesn't this show that they accept the value of science and want to cultivate it in India? Aren't they paying modern science a backhanded compliment by wanting to claim it for Hinduism, even as they distort the history of both modern and ancient Indian science?

This is indeed the burden of the postcolonial criticism of Hindutva. Postmodernist intellectuals have taken Hindu nationalists to task for accepting the modern "logocentric" and materialist Enlightenment ideals introduced by their colonial masters and distorting the "innocent" faith-based traditions of the nonmodern masses to fit it into a Western mold. The whole burden of the much celebrated works of Ashis Nandy and Partha Chatterjee is to criticize Indian nationalists of *all* political hues alike—from the socialist Nehru to the Hindu chauvinist Bankim Chandra—for daring to reinterpret the inherited Hindu worldview from a Western scientific perspective. They see the very idea of assessing Indian ways of knowing from a modern rationalist perspective—either to criticize it, as with the secularists, or to celebrate it, as with Hindu chauvinists—as a sign of mental colonialism and self-orientalization.

I disagree with the postcolonial position. It fails to examine how Hindu nationalists are merely using the rhetoric of science to boost the prestige of traditions. What is worse, the postcolonial critics themselves have been the loudest champions of culturally distinct sciences. The problem is not that the nationalists accept the importance of modern science—it would be foolish and blind not to do so. The real problem is not even that they try to find analogues for scientific thought in India's own history—it is necessary to anchor new ideas in elements of inherited cultural traditions in order to maintain some sense of continuity with the past. *The real problem is that Hindu nationalists are claiming to find scientific tradition in the wrong elements of Hindu tradition.*

There *are* rationalist, skeptical, and naturalistic strands in the Indian tradition, but they are *not* in the poetic myths of the Vedas, or in the mysticism of Vedānta. On the contrary, natural sciences in India have had to struggle *against* the Vedāntic traditions of Brahmin priests, who looked askance at such worldly knowledge, and looked down upon the practitioners of natural sciences as impure and lowly. The notion that Vedic literature was the source of Indian science flies in the face of everything we know about the intellectual history of India.

Take, for example, the two best-known sciences of pre-Islamic India: medicine and mathematics. The original doctrines of Hindu medicine—Ayurveda—have nothing in common with the spirit-centered, yoga-based "quantum healing" peddled by Deepak Chopra and Maharishi Mahesh Yogi and their innumerable clones. The original *Charaka Samhita* and *Susruta Samhita* do pay lip service to the orthodoxy of atman and karma, which had already solidified around the time when these two major texts of Indian medicine and surgery, respectively, were composed (between 600 BCE and 100 CE). But when it came to the actual details of human physiology and explanation of the causes and cures of disease, Ayurvedic texts completely set aside any notion of disembodied atman and karma working on matter. As the painstaking and detailed study of Indian medicine by Debiprasad Chattopadhyaya (the well-known historian of Indian materialism, or *Lokāyata*, who followed the tradition of Joseph Needham) has shown, the physicians who compiled the Ayurveda were committed to a purely naturalistic understanding of all life. They considered everything in nature, including human beings and their intelligence and consciousness, as constituted of five elements, or *bhuta*: earth, water, air, fire, and ether (*akasa*), each with its own innate quality, or *svabhava*, but capable of change of form under the effect of digestive fire. The curative value of foods and herbs was determined by the preponderance of the *bhuta* corresponding to the diseased organ. The importance of this tradition lies not in its content (which obviously is terribly outdated), but in its rejection of all supernatural causes and cures of disease. Moreover, these ancient medicine men (they *were* mostly men) were militantly empiricist in their outlook, practicing precisely those sensory ways of acquiring knowledge—including dissection of dead bodies, considered most impure by the orthodox—that are constantly derided in the Vedāntic traditions as "false knowledge" or *avidya*. Those who claim, as Hindu apologists do, that Ayurveda "developed from scriptural or Vedic tradition and that the physicians . . . worked out their science in the cool of orthodox piety . . . [are creating] a fairy tale, not history" (Chattopadhyaya 1978, 7).

Likewise, there are good historical reasons to believe that the much-celebrated "Vedic mathematics"—especially the geometric theorems contained in the *Sulva*

*Sutras*—received no substantive *intellectual* input from Vedic ontology and epistemology. The magico-religious ethos of the Vedas is irrelevant to the actual reasoning displayed in the theorems of geometry in these Sutras. The term *sulva* refers to rules relating to sacrificial rites, as well as the rope or cord used for measuring the sacrificial altars. The main text of these sutras consists of rules and instructions governing the measurement and construction of altars required for carrying out the fire rituals or *yagna*s, especially those spectacular, large-scale *yagna*s sponsored by kings, nobles, and other notables among the twice-born castes. Special, multi-layered altars with complex shapes depending upon the purpose of the *yagna*, all requiring unbroken and unblemished bricks, were prescribed. For example, a falcon-shaped altar was required for those desiring to go to heaven, a chariot-wheel shaped altar for those wishing to vanquish their enemies, an altar shaped as a trough for those wishing for food, etc. (see Chattopadhyaya 1986, 152). The theorems of the *Sulva Sutras* are meant to figure out how to fit a prescribed number and shape of bricks in an altar of a specified shape and size. Some of these are fairly complex and sophisticated, presaging the Pythagorean theorem ("the cord which is stretched across the diagonal of a square produces an area double the size of the original square") and solving complex construction problems like how to turn a square into a circle (for a complete list of propositions, see ibid., 159–160). The exact date is not known, but these texts are generally dated around the time of the Buddha, somewhere between 800 and 600 BCE.

Because these mathematical texts are a part of the ritualistic literature, it is assumed that priests were mathematicians as well. But Chattopadhyaya makes a persuasive case that the priest could only have contributed the specifications of what was needed to be done (the ritually acceptable size and the shape), but *not* the technical-mathematical knowledge of how to actually build the altars. For one, the *Sulva Sutras* completely lay aside the magico-religious terminology of the bricks and deal with them as material objects. They are concerned with such nitty-gritty details as shrinkage on drying and the equipment needed for measurement—details that could only be known to those who actually dirtied their hands. "Whatever might have been the motivation of the priests dictating certain orders to be carried out for their sacrificial rituals," Chattopadhyaya writes (ibid., 163), " that had no bearing at all on the making of science in ancient India as found in the Sulva texts." The entire magical-religious ethos of the priests, based upon an associative logic, is not just irrelevant but actually opposite to the ethos of accuracy and measurements that the mathematics of the *Sulva Sutras* display. There are good sociological reasons why this should have been so. By the time these texts were written (around the time of the Buddha), caste divisions had already solidified in most of India. Priests by this time had become the "administrators of superstition," cut off from the "organization of production" (ibid., 146).

Based upon the fact that the nomads who wrote the Vedas were not a brick-making people and the fact that fire-baked bricks flourished in the pre-Vedic, Indus valley civilization, as shown by the archeological remains of Harappan cities, Chattopadhyaya has conjectured that the *Sulva Sutras* could well contain the systematization of the geometry implicit in the brick-making techniques of the pre-Vedic, Harappan people. This conjecture has found some support in the more recent work of Frits Staal. (See the next chapter for details.)

The separation of intellectual workers (the priestly castes) from the manual workers (the Shudras and the untouchables) has also meant a separation of theory from practice; the methods and findings obtained by actually working on nature could not be applied to an abstract and systematized understanding of the workings of nature. This theologically sanctified and ritually enforced separation exacted a heavy price on the growth of science in India. Because of the superiority accorded to the ultimate knowledge that can be acquired by pure reason, the knowledge of the working people, the impure castes, was not just deprived of any theoretical illumination from the learned classes, it was simultaneously derided as inferior knowledge.

The point I am trying to make here is that both the neo-Hindu apologists and postcolonial critics are wrong. They are both overlooking the contradiction that lies at the heart of Hinduism. The contradiction between the philosophical idealism of Vedānta and the philosophical materialism of the heterodox, non- Brahminical castes who practically made up the entire class of workers—peasants and craftsmen—who actually came in contact with the elements of nature. Scientific developments in India took place not because of, but in spite of the philosophical idealism of the Vedāntic orthodoxy. The neo-Hindu attempt to claim the glory of science for Vedānta amounts to a denial of even the existence of non/anti-Vedāntic systems of thought. At the same time, contrary to the postcolonial critics, to judge the development, or lack, of science in India by the standards of "Western" science is not Eurocentric. Insofar as Indian scientists in antiquity were trying to understand the workings of nature, they did try to develop a naturalistic metaphysics. They, too, tried to depend only upon ordinary human sensory powers and ordinary everyday logic to grasp the workings of nature. Underneath the culturally distinct idiom of their science, one can clearly identify a way of conceptualizing nature which *any* scientist, past or present, from India, or Africa, or Europe, can understand. The scientific enterprise, as we understand it today, is *alien only to the idealistic and gnostic tendencies of Vedāntic Hinduism.* The physicians of Ayurveda and the bricklayers and masons who made the fire altars using the geometrical theorems of the *Sulva Sutras* would find themselves perfectly at home with modern science.

To understand what is at stake, why it is important to demystify rather than to celebrate the Vedic worldview and why it is vital to recover the real scientific traditions in India, we need to take a short detour into the content and social function of the Vedic understanding of nature.

### Two Cheers for Disenchantment

There is a common thread running through all the various claims of Vedic sciences. They all treat the supernatural—the spark of consciousness, Brahman—as a constituent of material nature and interpret mystical knowledge as providing genuine laws of nature. Just as India declared itself secular without secularization, Hinduism is now being turned into a "science" without allowing for a diminution of the divine presence in nature.

The erosion of the supernatural is a vital component of the secularization of consciousness. Secularization, to paraphrase a central insight of Charles Taylor's monumental *Sources of the Self* (1989), produces individuals with new moral intuitions about what it means to be human. These moral intuitions radically alter the social goods that modern individuals and societies come to value. While all societies, in their own ways, value human dignity, and flourishing, modern individuals come to see human dignity as a matter of the universal right for each to develop his/her own unique potential, the fulfillment of this conception of human beings puts a high premium on reducing all avoidable suffering and on affirming the ordinary life of here-and-now.

In Taylor's account, the naturalization or disenchantment of nature is a sine qua non for the creation of modern identity. For men and women to find their own meaning and develop their own unique potential, they first have to break away from (or in Taylor's words, "become disengaged" from), the God-ordained cosmic order. For human beings to be able to give their own meanings to their lives, and to trust their own abilities to understand the world, the natural order has to first become devoid of preordained meanings, final causes, and moral purposes. The cosmopolis has to break: the affairs of men and women have to be set free from the cosmic order. It is only after morality ceases to have cosmic meanings, and conversely, natural phenomena cease to have moral significance, that human beings can exercise their autonomy without the fear of cosmic consequences and divine (which is also societal) punishment.

If this disenchantment of the world has meant a sense of homelessness and even nihilism, it also has a positive promise for millions of women and men around the world, especially those whose humanity has not been fully recognized by traditional religions. Superstition is as much (if not a greater) a source of tyranny as any despotic power. As Lucretius wrote more than two thousand years ago in his *De Rerum Natura*: "Prophets frighten believers with tales of eternal punishment. *Therefore*, men must learn the laws that govern the sky and the earth." This is nowhere truer than in India, where systematic propagation of superstition was, for centuries, the fundamental basis for priestly power which worked to keep the majority ignorant and submissive. (This propagation of superstition is not a thing of the past. It has become more technologically sophisticated in some circles, but continues unabated in most of the country.)

When "men learn the laws that govern the sky and the earth," they do not learn new facts, but acquire a new grammar and new metaphors that structure their thoughts. The world gets disenchanted not because people become better educated, learn science, and become less credulous. There are simply too many examples of modern people believing the most dreadful nonsense to continue with the positivist faith that better supported facts will always be more acceptable. Besides, a culture's ontology—its understanding of what exists—is not merely an intellectual position comparable to a scientific theory. It is *also* an existential response, which includes some propositional beliefs about nature, which help make sense of existential anxieties and provide resources for ethical values.[9] Even though religions ground their claims in the natural order of things, their *primary* goal is not to explain how the world works, but to enable human beings to be at home in it.

Disenchantment, therefore, is not a matter of scientific truths driving out superstitions. Yet, developments in natural science and their circulation in a society through education, demonstration, and a purposive popularization of naturalistic and verifiable explanations are not irrelevant to the cultivation of moral intuitions that underpin the modern outlook about self, nature, and society. What the growth of scientific reason and its cultural acceptance does is to alter the background assumptions, or the social imaginary, regarding matter, causality, and knowledge, which in turn, have a tremendous impact on how we think of ourselves, our neighbors, our relationship with each other and with nature, and who we consider a legitimate source of knowledge and authority.[10]

The mechanical philosophy of Descartes, Boyle, and Newton had the long-term—and mostly intended—effect of divesting nature of all agency, moral qualities, and final ends. In the name of the greater glory of an all-powerful, radically sovereign God, whose will was the only active principle in the world, the mechanical philosophers rejected the Aristotelian belief that inherent tendencies in natural objects determine their ends. The presence of God was to be discerned in His laws, which the natural world obeyed, but God himself was not immanent in the substance of the objects He created.[11]

The change in the way of conceptualizing nature and God that the Scientific Revolution initiated was gradually integrated into the popular culture of the West through the efforts of mainline churches, scientists, and laypersons themselves in predominantly Protestant England and in the United States, and through the anti-clerical writings of the philosophers in France. Under the combined impact of the Scientific Revolution and the Enlightenment, religious institutions themselves undertook the task of reinterpreting the Bible to bring it in accord with a scientific conception of the world. The supernatural was interpreted as mythic, or re-interpreted to accord with scientific findings. God was not driven out; He remained the creator and the sustainer of the physical world, but his logos could not contravene the understanding of the physical world obtained through the work of science. As Arthur Peacocke, a liberal theologian, puts it, the epistemological revolution in theology has created a generation of "wistful agnostics who respect Christian ethics and the person of Jesus but also believe that the ontological baggage of traditional religion can be discarded as not referring to any realities" (2000, 120).

Growth of scientific knowledge and strides in technological progress are necessary, but not sufficient for altering the governing metaphors and the grammar of a society. The assumption of the modernization theory, in both the liberal and vulgar-Marxist versions, that disenchantment will follow, more or less automatically, in the wake of the development of science and industry, is simply not borne out by careful examination of the evidence of past and present societies. Just two examples will suffice to show the poverty of a materialist theory of secularization. The first comes from Marshall Sahlin's (1999) recent essay that dwells upon the so-called "oil-age Eskimos," who have revived the social customs of their traditional subsistence economy using the higher productivity made possible by the modern technology of fishing, communication, and such. The other example comes from Keith Thomas (1971) who concludes from his historical study of the role of religion in the decline of magic in sixteenth-seventeenth-century England that the decline in magical thinking came before there were technological solutions to life's problems and misfortunes in place. He argues in a Weberian vein that rather than technological growth eroding magical thinking, it was the "abandonment of magic that made the upsurge of technology possible" (ibid., 657). Thus, one can safely conclude that the optimist vision of the classical modernization theories of inevitable and universal disenchantment is simplistic and false.

What made it possible for Western Europe to abandon magical thinking more readily than other cultures is an enormous subject, and this is not the right place to delve deeper into it. The point I want to make is that disenchantment requires an active and intentional translation by scientists, humanists, philosophers, and teachers, of scientific discoveries into a language of everyday life for ordinary people. Without this active work, science will remain merely an enclave, separated from the rest of the society. In such a situation, the dangers of reactionary modernism will always remain with us, where ancient superstitions and prejudices will hitch the instrumentalities of science to their own survival.

This active work of translation of the facts of science into a worldview is the work of the Enlightenment. As Jose Casanova (1994) has argued in his important work, *Public Religions in the Modern World*, secularization or disenchantment will follow different dynamics in different societies, depending upon the strength the four chief carriers of secularization have, namely, the Protestant Reformation (or its cultural homologues), the formation of the modern nation state, capitalism, and the cultural impact of science. The last factor, that is, the spread of the scientific worldview, or the Enlightenment critique of religion, Jose argues, "be[comes] an independent carrier of the processes of secularization wherever the established churches become obstacles to the modern process of functional differentiation. By contrast, wherever religion itself accepted, perhaps even furthered, the functional differentiation of the secular sphere from religion, the radical Enlightenment and its critique of religion become superfluous" (ibid., 30). Thus, in Protestant countries, ideas of Newtonian Enlightenment were accepted by mainline churches, the educated public, and even in the royal court and were not considered particularly radical. The same ideas became seditious and sacrilegious in France and other parts of Europe under absolutist states.

It is my contention that in a country like India, where neither the religious establishment, nor the presumably secular state, nor the bourgeoisie has taken the challenge of secularization seriously, it falls upon the public intellectuals, the clerks—the scientists, the men and women of letters, journalists, teachers—to act as carriers of disenchantment. Nationalism, coupled more recently with anti-Enlightenment views made fashionable by postmodernism, has detracted secular intellectuals from a serious engagement with Hinduism. But this is getting ahead of the story. We must first look at the hurdles Hinduism itself poses to the disenchantment of the cosmos.

### The Spirit in Nature, or the Mythos as the Logos

While the Scientific Revolution contributed to disenchantment in the West, neo-Hinduism and its philosophical inheritor, Hindutva, have followed the exact opposite route. Rather than divest the natural world of miracles, spiritual powers, and moral qualities, neo-Hinduism has treated the most mystical and otherworldly teachings of Hinduism—Vedānta and Advaita Vedānta—as the crowning glory of modern science of nature. Rather than re-interpret the Vedāntic postulate of the immanence of the divine spirit in nature as a mythic, or psycho-spiritual hunger for connectedness with the rest of the creation, neo-Hindu intellectuals reinterpret the available scientific evidence to "prove" that the divine power, or the Brahman, acts as an actual force of nature, operating in a dimension which is extra-sensory and supra-rational but nevertheless "real." Rather than a "mere" analysis of causes and effects, Vedic sciences promise a "direct realization" of the unity of the knower and the known, matter and energy (the latter, for some unexplained reason, is treated as a synonym of "consciousness").

To understand why this trend is extremely problematic for the growth of a genuinely secular and liberal culture in India, it is important to understand what the Vedāntic teachings actually are and the role they have played in perpetuating superstition and social hierarchy in India. Presenting a succinct summary of the main ideas of Vedāntic Hinduism is not a simple task. For one, the corpus of writings is vast, multi-stranded and often self-contradictory. There are four Vedas, with the earliest, the *Rig Veda* dating back to about 1500 BCE, followed by nearly 200 Upanishads (800–400 BCE), of which 13 are considered to be of philosophical importance. The philosophy of Vedānta, literally the end or the essence of the teachings of the Vedas, is contained in the *Vedānta Sutra*, a part of the Upanishadic literature. Vedānta itself is the subject of many later commentaries, the most important being that of Advaita Vedānta, developed by the ninth-century (CE) philosopher, Shankara. It is Vedānta that has been propagated as the true teaching of Hinduism by neo-Hindu intellectuals. The second difficulty in presenting a succinct overview of Vedāntic philosophy is that it conflates the concepts that modern minds have become accustomed to separating. It does not separate God from nature, nor the knowledge of God from the knowledge of nature. Notwithstanding these difficulties, I have attempted a brief outline of the Vedāntic view of nature and knowledge.[12]

The Upanishads are the product of the Axial Age in India's antiquity (800–200 BCE), the period when the ritualism of the Vedas came to be questioned. No longer satisfied with divinizing the forces of nature and trying to propitiate them through rituals and sacrifices, inquiring minds in ancient India began to ask questions about the reality hidden behind appearances, the ultimate source and the ultimate goal of all that exists. There was no doubt a scientific spirit at work among the wise men who composed the Upanishads. They were trying to find connections—indeed, the word *upanishad* means, among other things, "connection" or "equivalence" (Olivelle 1996)—between this all-transcendent source of all things, the cosmic realities and rituals. Rather than unquestioningly accept the magico-religious rituals of the Vedic priests, the Upanishadic thinkers tried to find rational explanations by connecting the prescribed rituals to the natural world and human existence.

The Upanishads contain two sets of teachings: the materialist and the idealist. Contrary to the image of Hinduism as a spirit-based idealistic religion, materialists and skeptics who denied survival after death and the existence of other worlds have been a part of the Vedic religion, from the earliest *Rig Veda* to the later Upanishads. Even the most revered Hindu texts, the *Ramayana* and *Mahabharata*, contain elements of materialistic teachings. Like the pre-Socratic Ionians who replaced the Homeric and Egyptian gods with elements of nature, Indian materialist schools—called Lokāyata or Chārvāka—declared that not Brahman or consciousness, but the four material elements of water, earth, air, and fire were the basis of the whole universe, including intelligent beings like humans. These materialists declared, to quote from *Svasanved Upanishad*, "There is no incarnation, no God, no heaven, no hell. All traditional religious literature is the work of conceited fools. Nature, the originator, and Time, the destroyer, are the rulers of things and take no account of the virtue or vice in awarding happiness or misery to men. People deluded by flowery speeches cling to god's temples and priests when, in reality, there is no difference between Vishnu and a dog" (quoted here from Damodaran 1967, 86).

But materialism was clearly the road not taken: passages like this one are not considered the main message of the Upanishads. Far from it. Materialists were ridiculed and caricatured as simple-minded, working-class hedonists, out to destroy the very basis of civilized life. None of the original materialist texts has survived. The only references exist in critical commentaries by Vedāntists determined to discredit the materialists. Yet, the fact that so many orthodox Brahmins spent so much energy in criticizing and ridiculing the materialists shows that these ideas were not entirely marginal. Materialism was considered threatening enough to require constant vigilance.

The mainstream of Hinduism chose the road of idealism—and that has left a distinctive stamp on Indian intellectual history, including its sciences. The central question the Upanishadic thinkers set out to answer was: "what is the stuff nature and the entire phenomenal world is made up of—" The answer they came up with was: "This whole world is Brahman. Brahman is the hidden mover within all that moves, breaths and winks" (*Mundaka Upanishad*). Brahman, *Taittriya Upanishad* teaches, is "that from which [all] beings are born, that by which they live, that into which when departing they enter. That is Brahman" (quoted here from ibid., 181).

What is the nature of this Brahman, the primary source of all existence? It is a non-material life-force, pure spirit, the Absolute Consciousness of the universe that permeates all space and manifests itself in all things. It is as indistinct from nature as salt is from sea water, or as the spider's web is from the spider, and as vital as the seed is for the plant. Brahman, or spirit, is the material and the efficient cause of nature. Unlike the monotheistic God who is separate from its creation, Brahman is "not merely a creator, externally related to the world, but also constitutes its very substance" (Hiriyanna 1996, 16). The order of nature is a chain of being, connecting the humblest organism to the rest of nature through the presence of this divine force within them. Advaita Vedānta, the eighth-century interpretation of Vedānta philosophy adds a radically idealist monism to Vedānta: Brahman is not just the ultimate reality, but the *only* reality. The pluralities of objects we see in the world around us are illusions. On this account, real knowledge lies in the realization that what in our ordinary lives we call "reality" is an illusion.

There is no denying the spiritual solace Vedāntic teachings have provided to so many, for so long. The God of the Vedas is not a distant law-giver, but an all-pervading presence that human beings can experience within their own souls. But as a philosophy of natural science and as the philosophical base of society and ethics, the identity of matter and spirit is extremely problematic. *As a philosophy of natural science, Vedānta denies the autonomy of nature from the divine consciousness, and as social ethics, it denies the autonomy of individuals from the totality of nature and society.* Everything and everyone is a hierarchic, morally coded, and fleeting manifestation of the one permanent, non-material essence.

While the illusionism of the Advaita school (that holds that the spirit is the *only* reality) is attenuated even by those who pay lip service to it, there is no dispute over the fact that *all* Vedāntic schools treat the supersensory, ineffable, supramaterial Brahman as the animating force actually present in all matter. The entire cosmos, including inanimate objects of nature and animate beings, from the most humble viruses and plants to human beings, are all permeated with this life-force.

All are permeated with the same life-force, *but not in equal measure.* They all partake of the Brahman in proportion with their karma (deeds in previous lives), forming a chain of being that "extends from (the highest individual god) Brahma to the tufts of grass," as the standard explanation goes. All human beings do not stand at one level, or form a single link in the chain. Rather, they are seen as different hereditary and intermarrying groups (varnas and *jati*s, commonly lumped together as castes) occupying different positions, depending upon the amount of purity, goodness, or luminosity (*sattva*), or conversely, pollution, evil, and darkness (*tamas*) encoded in the very matter (*prakriti*) that makes up their nature. Human beings and indeed, all animate and inanimate objects in the entire cosmos, come with innate moral qualities, acquired through the workings of their karma from past lives on the *prakriti*, or nature: the balance of moral and immoral actions accumulated over this and previous births influences the qualities of nature, and these qualities in turn determine moral predispositions.

This law of karma, furthermore, has the status of a scientific theory in classical Hinduism where it serves to predict, postdict, and otherwise explain the common facts of human conditions such as social station, predispositions, memories, and good and bad fortune (Potter 1980). Karma acts simultaneously as a moral law and as a law of nature, because your moral or immoral karma have a direct impact on the *guna*s, or qualities, your body and your soul will partake of. Karma is the vehicle through which actions, and even thoughts, in the societal/moral/spiritual realm are transferred on to the material/physical realm in the entire cosmos. *Thus moral or immoral actions of humans, in this and in any number of past lives, carry ontological consequences*: diseases, misfortunes, or good fortune (e.g., your birth as a Brahman or as a man) are just deserts of karma which acts upon matter. Human morality, in other words, sustains the natural order and the natural order is the archetype of the moral order. As Wendy O'Flaherty (1976, 95) has put it succinctly, "the moral code in India is nature." Virtue, or dharma, consists of living in accordance with your own innate nature, "however worthless and destitute of qualities" it may be, as the *Bhagavad Gita* teaches. The only way to eventually move up the chain of beings is to accept with detachment the rung you currently occupy and fulfill the duties required of you at that rung. Thankfully, the law of karma is only imperfectly followed in practice simply because it amounts to blaming the victim for his/her lowly station or misfortune: it explains why misfortune befalls, but brings very little psychological comfort to the victim. Yet, for all its psychological harshness, the doctrine forms a part of the common sense of most Indians (including even non-Hindus) and provides "a shared frame of reference that is potentially applicable in any situation calling for interpretation of destiny" (Babb 1983, 171).

So much for the ontology of Vedānta. Not surprisingly for a philosophy that treats spirit as the primary cause and constituent of nature, Vedāntic epistemology privileges intuitive experience of the totality of things, rather than an empirical analysis of the causes of phenomenon. The *real* reality is Brahman which is "beyond all sensations of sound, touch, color, taste and smell; it is beginingless and timeless . . . it is the subtlest of the subtle and the greatest of the great . . ." Such a reality, "cannot be known by learning, scholarship or sharp intellect. . . . Any man can know with his senses, but it is only when he turns away from the senses that he perceives it (Brahman)" (Dasgupta 1969, 31).

Using the criteria described by Bertrand Russell (1935), traditional Hindu epistemology is better described as mysticism rather than science because it *(a)* values intuition over discursive analytical knowledge; *(b)* is monist; *(c)* denies the reality of time; and *(d)* believes that evil and good are mere appearances. The knowledge of the mystics is valuable to the mystics and to the rest of us—as poetry and as a source of metaphysical pathos for those who value it. But to confuse it with propositional knowledge about the world is to devalue mysticism and science at the same time.

It is this fundamentally spiritual/non-material conception of nature and a mystical way of knowing that are defended by the "intellectual Kshatriyas" as being consonant with the findings and methods of modern science. One cannot call this enterprise a "*re*-enchantment" of nature, for Brahman, or God/gods have never vacated nature in India in the first place. Nothing comparable to the Western experience of disenchantment of nature and a reinterpretation of the sacred has taken place. But it is clear that the claims of "Vedic sciences" are riding the wave of the postmodernist disenchantment with disenchantment in the West. It is the disaffection of the West's internal critics with the mechanistic conception of nature, ruled by a distant and patriarchal law-giver god, that has provided an opening for Vedānta-centered neo-Hinduism to present itself as a paradigm of "alternative science" which can, presumably, correct all the failings of modern science. (Another example of the pizza effect, one wonders?)

However appealing it may be for the postmodern West, glossing a Vedāntic understanding of nature as scientific is a huge impediment to cultural reformation at home. It only adds the prestige of science to a worldview which has, in fact, *retarded* the growth of science and social equality in India. Because the orthodox Hindu conception of God and the good is so intertwined with what it takes to be the order of nature, conservative Hindu nationalists have no choice but to defend the Vedāntic understanding of nature from scientific criticism. And what better way to immunize this worldview than to claim it to be in accord with, and even superior to, the modern scientific understanding of the laws of nature? That seems to be the unstated compulsion behind the unique obsession with science that Hindu theologians, priests, and assorted gurus have.

It is clear that those who are agitating for the scientificity of Vedāntic ontology and epistemology are also at the forefront of defending the righteousness of the traditional Hindu social order which gives priority to the social whole over the individual. Even as they set up re-conversion camps that seek to "purify" the untouchable and tribal converts to Christianity and Islam, Hindutva intellectuals remain committed to the fundamental soundness of the fourfold caste order. The philosophy of "integral humanism," which treats castes as different limbs of a body, all performing their respective functions in "harmony" and for the good of the whole body, is the official ideology of all parties committed to Hindutva, from the "moderate" Bharatiya Janata Party to its parent, the Rashtriya Swayamsewak Sangh. A defense of the holistic, chain-of-being ontology as "scientific" is thus being propagated together with the virtues of an illiberal social ethic, in which the individual counts for nothing, while duties to the collective are paramount. Holist thought in India has always been in an unholy alliance with social conservatism. Glossing this holism as "scientific" is crucial for the conservation of conservative traditional values in a technologically modernizing society.

### Historical Roots of Vedic Science:
### British Orientalism and Indian Nationalism

How Vedānta, the heart of Brahminical orthodoxy, came to be celebrated as the hope of salvation for India—indeed, for the whole world—is an aberration of the complex love-hate relationship between the British (and to a lesser extent, German) Orientalists, the orthodox Brahmins, and the rising Indian middle classes in nineteenth- century India. Very simply, British Orientalists enabled the largely upper-caste bourgeoisie to uphold the mystical and socially conservative elements of Vedic traditions as constituting Hinduism's "golden age." Revitalization of *these* traditions became the dominant goal of nationalist modernizers and traditionalists alike. The contemporary project of Hinduism-as-science belongs to this tradition.

After Edward Said's influential *Orientalism* (1978), the term "Orientalist" is often used as a pejorative epithet, denoting colonial manipulation of the East by the West. Western Orientalists are supposed to have interpreted the East through

Western conceptual categories and declared it to be wholly different from—and entirely inferior to—the West. This is the gist of major postcolonial theorists. (See Inden 1986, as a paradigmatic example of this genre.) What such a one-sided analysis overlooks (or explains away as manipulation by another name) is the "affirmative Orientalism"[13] of the British and German Sanskritists who believed that India possessed an ancient and uniquely spiritual civilization. *Even as they tried to understand the East through their own conceptual categories, and declared it to be Eurpoe's opposite or the "other," Western Orientalists did not necessarily find the East to be inferior.* They did not treat India's spirituality as archaic, but as a necessary complement and a corrective to post-Enlightenment Europe's materialism and secularism. (Postcolonial theorists do acknowledge the positive value Orientalists put on things Indian. But they see it as distorting the perception of Indians who came to see their tradition through the colonial lens.)

The British Orientalists under the employ of the East India Company, including the well-known Sanskritists William Jones (1746–1749), Charles Wilkins (1749–1836), and Henry Colebrooke (1765–1837), were men of the Enlightenment and shared its cosmopolitan and tolerant outlook. Their interest in ancient Hindu Sanskrit texts was in part sparked by a conviction that Hinduism could prove the existence of a "natural religion" shared by all cultures, which could challenge the universalist and absolutist claims of Christianity. Apart from translating major Hindu sacred texts (notably, the *Bhagavad Gita, Manusmriti*, and selected Upanishads) and scientific treatises (notably, the mathematical works of Brahmagupta and Bhaskaracharya) from Sanskrit into English, British Orientalists made two other major contributions which have had a lasting influence on the self-understanding of Indians. The first was the emphasis on the Vedāntic conception of God, Brahman, who is immanent in nature as the most important discovery of Hindus. William Jones and Colebrooke both singled out the Vedāntic monism as the crowning achievement of the Hindu golden age. They claimed that India's contemporary weakness stemmed from forgetting this tradition and that India could recover her glory by revitalizing it. The second major contribution was William Jones's discovery of the similarities between Sanskrit, Greek, and Latin, which led to the disastrous search for the homeland of the original "Aryan race"—a search that is still going on among the Hindu nationalists in India today who claim that the Indian subcontinent was the original homeland of the "Aryans."[14]

If the British Orientalists were attracted to Hinduism out of their Enlightenment outlook, the German Orientalists were driven by a romantic reaction against the Enlightenment. As is well-known, the idea of India assumed mythical proportions within the German Romantic Movement. India represented the opposite of the quantifying, mechanical, and merely rational way of viewing the universe that the romantics hated about the Age of Enlightenment. The Indo-philia of German intellectuals, including Herder, Schelling, Novallis, the Schlegel brothers, and Schopenhauer is well documented. Max Müller (1823–1900), the well-known German translator of the *Rig Veda*, popularized the idea of India as the original home and the childhood of the West (see Halbfass 1988). This tradition of romantic Orientalism later contributed to the anti-Judaic elements of the national socialist religion which sought a less alienating, non-transcendent "religion of nature" in Vedāntic pantheism, occult, and theosophy (see Pois 1985).

This affirmative Orientalism set the tone for the revivalist elements of the so-called Bengal Renaissance, the beginning of the spirit of reform which spread all over India, lasting about one hundred years ending with the First World War. The encounter with the British was not the only cause of the awakening; there were many heterodox sects and religious movements within Hinduism and Islam that denounced polytheism, idolatry, and caste distinctions (see Panikkar 1995). But while the indigenous reform movements could be assimilated into the Hindu mainstream, the encounter with the British was an encounter with radically different forms of knowledge, technology, and social organization. The encounter literally shook the Indian elite out of their slumber for it gave them a whole different vantage point to judge their own present and the past from. Thus, the nineteenth century, stretching into the early parts of the twentieth century, finds India rife with Hindu, Islamic, neo-Buddhist (among the untouchables and the Shudra castes), and rationalist-humanist (a minority) reform movements.

The intellectual awakening to the West, under the circumstances of colonial domination created a dilemma for the Indian nationalists. How could their rational appreciation of modern science be reconciled with their need to defend their own heritage against their colonial masters? How could they distance themselves from the tradition which they had begun to see through the conceptual categories of modern science, and simultaneously, defend (if not glorify) it against the colonial critics?

The mainstream of Indian intellectuals, apart from a minority of socialist-humanists, resolved this dilemma by subsuming science into the Vedic tradition of Hinduism. They declared the latter as containing all the essential ingredients of modern science. This was the solution favored by such renowned neo-Hindu intellectuals as Ram Mohan Roy (1771–1833), Swami Vivekananda (1863–1902) and Sri Aurobindo (1872–1950). This culminated in neo-Hinduism, which basically restated Vedic Hinduism in a "scientific" idiom. They declared the Vedic tradition to contain all the elements of the "Greek miracle" that had presumably given birth to science and technology. Science was subsumed in the Vedic tradition.

But the Indian Renaissance turned out to be a veritable Hindu revival; not a birth of something new, but an aggressive restatement of the old in a "scientific" idiom. Clearly, nationalism was the over-riding impulse of neo-Hindu intellectuals. Under the rising tide of nationalism by the end of the nineteenth century, even those who accepted the Western ideals of equality between castes and genders, began to "return to native intellectual resources . . . for finding traditional shastric, scriptural justifications" for their agendas (Heimsath 1964, 24). These nationalists concluded that "India's essential spirit had been smothered by the centuries' accumulation of the refuse of degraded customs and untruths, and Indians must restore *it*, not build afresh on alien foundations. Not an evolutionary progress toward a new society based upon 'social efficiency,' but regeneration of purified Hindu society based on a spiritual revival should occur" (ibid., 25).

Going back to the original sources in order to clean up the present is a common feature of reform movements. The problem lies with what elements of the multifaceted and highly diverse Hindu tradition were conjured up, and whether the chosen traditions ever had the potential to encourage a secular and rational view of nature and society. The Renaissance and Enlightenment in Europe led to the rediscovery of the ancient Greek and Hellenic materialists and humanists—Lucretius, Seneca, and Cicero over Plato and neo-Platonist tendencies in Christianity. The philosophers of the Enlightenment were "modern pagans," who in Peter Gay's memorable words (1966, 8), "used their classical learning to free themselves from their Christian heritage, and then, having done with the ancients, turned their face toward a modern world view."

The exact opposite happened with neo-Hinduism. As the reformist movements turned more nationalistic, they embraced the Orientalist views of the Vedic golden age as a model of reform. As a result, the Indian Renaissance rediscovered the most world-denying and reason-defying philosophy—the Indian equivalent of Plato—over the world-affirming and empirical philosophies and vernacular traditions, which were *also* a part of India's ancient heritage. Indeed, Vedāntic philosophies were a closely guarded secret knowledge of the minority of most elite Brahmins throughout India's premodern history. The mystical flights of Vedānta were (and still are) perhaps much more foreign and counter-intuitive to the masses of Indian people than the "Western" science of Galileo and Newton, which slowly began to trickle down to them through the activity of the missionaries and through the concerted efforts of lower-caste and untouchable intellectuals.

As Margaret Jacob (1988) has argued persuasively in her study of Newtonian science in seventeenth-century England, it needed the active support of the emerging bourgeoisie, in alliance with the Anglican Church to make the mechanical philosophy of the Newtonians a part of the cultural values of their time. In India, the emerging bourgeoisie was both the main beneficiary of British education, but also, over time as they gained more self-confidence, the main adversary of the British in the name of national independence. By the later part of the nineteenth century, intellectuals from this class increasingly began to turn to ancient Hindu religious texts—which many of them read for the first time in English translations produced by the British Orientalists!—to find justification for even those Western ideas that they accepted as reasonable and desirable (for example, the end of the practice of sati or widow immolation, raising the age of consent for marriage, and the end to at least the more egregious aspects of untouchablity, though not an end to caste system). The need to assert their pride and dignity in the face of the colonialists led Indian nationalists to reinterpret modern ideas in the language of the ancients: they could not be seen as imitating their colonial masters, even when they in fact were. In a stark contrast to the liberalizing bourgeoisie in Britain, the Indian bourgeoisie, even as they consented to liberal reforms, ended up defending hoary Hindu traditions, especially those admired by the Western Orientalists. Being mostly from twice-born castes, they were comfortable with these idea and at the same time, socially cut off from the rationalist currents in the lower-caste and untouchable movements that were critical of the Hindu golden age. As David Kopf concludes in his well-known study of the Bengal Renaissance, the nationalist reformers "used the ideas of the West as means for modernizing their own traditions . . . for *pouring the new wine of modern functions into the old bottles of Indian culture*" (Kopf 1969, 205, emphasis added).

A good example of the limits of upper-caste bourgeois nationalism is the "evolution" of Ram Mohan Roy, the "father" of Indian modernism. Like many of his generation, he started out as a rationalist, subjecting the contemporary social-religious dogmas to the twin test of reason and social comfort. His first extant work, *Tuhfat-ul Muwahhiddin* published in 1803, challenged popular miracles and superstitions and cut through centuries of dross to purify the core of a universalist, natural religion consisting only of three core beliefs: a single God, soul, and after-life. But as he got more drawn into Orientalist influences, he replaced reason with Vedāntic golden age as the criteria of how to judge the present. The contemporary irrationalities could then be decried as the corruption of the originally pure wisdom, but the original doctrines themselves were protected from rational critique and reinterpretation. Not surprisingly, Roy himself, and Brahmo Samaj, the reformist movement he started, continued to openly practice caste rules and the dietary and sartorial habits of upper castes.[15]

Roy's pattern was followed time and again. As the social historian of Indian nationalism, Tanika Sarkar puts it, "the compulsions of Hindu nationalism overrode and displaced earlier liberal concerns . . . [the reformers'] self-critical, radical sensibilities were transformed into an authoritarian and intolerant voice" (2001, 156). The two most famous heroes of the Indian Renaissance, the novelist and man of letters, Bankim Chandra Chattopadhyaya and the nationalist monk, Swami Vivekananda display this capitulation to nationalism. As young men, both were deeply influenced by the ideas of John Stuart Mill and Auguste Comte. Both gradually turned away from self-critique to cultural nationalism and declared the acceptance of alien traditions under colonialism to be the true cause of India's plight. Vivekananda, more than Bankim Chandra, declared Vedāanta to be the source of science and reason which could teach a thing or two to modern Western scientists. (We will encounter Vivekananda's arguments in defense of Vedic science in the next chapter.)

Instead of an uncompromising self-critique of the inherited worldview in light of reason and improved knowledge of the world, intellectuals beat a hasty retreat into the comfort of tradition. Nationalist pride and (in most cases) unconscious or conscious caste-belongingness, overcame the demands of reason. Consequently, a myth was created that the original doctrines themselves had nothing to do with innumerable popular superstitions and the social customs of caste, patriarchy, and deeply ingrained authoritarian anti-individualism. This is a deeply conservative myth for not only does it protect the core of the tradition from demystification, but also presents the revival of the tradition as the solution to present-day ills.

This myth has got a new lease of life in the current theories of Vedic science, which we examine in the following chapter.

## Four

# Vedic Science, Part Two

## Philosophical Justification of Vedic Science

*Ekam sad vipra bahudha vadnati.*
*(Truth is one, wise men call it by different names.)*
—The Rig Veda

*Conclusions of modern science are the very*
*conclusions the Vedanta reached ages ago . . . only in*
*modern science they are written in the language*
*of matter.*
—*Swami Vivekananda, Mission of the Vedanta,* Collected Works

*India is the guru of nations, the physician of Europe's*
*maladies.*
—*Sri Aurobindo,* India's Rebirth

The postmodern condition, as Jean-François Lyotard (1984) famously defined it, is nothing if not a suspicion of all metanarratives. All overarching universal theories about history and truth-claims about nature are to be resisted, for they do "violence to the heterogeneity of language-games" by imposing their own parochial categories of thought on other cultures. "All we can do," Lyotard advises his postmodern followers, "is to gaze in wonderment at the diversity of discursive practices around the world, just as we do at the diversity of plant and animal species" (ibid., 26).

We can all "simply gaze in wonderment" because all sciences are systems of myths, stories, interpretations and none is any more true to the *really* real—which is not accessible to us anyway (for all "reality" comes to us only through our inherited conceptual categories). The new postmodernist philosophies of science do not deny that modern science discovers some truths about nature (e.g., Newton's laws) that are universally valid. Yet, even these universal truths are seen as the product of metaphysical assumptions and cultural prejudices of the Western culture. Interests and prejudices of other cultures, it is assumed, are equally capable of producing equally universalizable truths about nature, which, all things considered, correspond equally well, or even better, with the full nature of reality. Given this new understanding of knowledge, postmodern intellectuals can only serve, to use Zygmunt Bauman's (1987) metaphor, as "interpreters" of local knowledges, and not as "legislators" judging their validity in the light of modern science.

Like defenders of other faiths,[1] Hindu intellectuals are making an opportunistic use of the postmodernist repudiation of the universality and objectivity of science. They decry the repudiation of *all* metanarratives, for that will obviously deny the metanarratives of traditional religions. But they welcome the postmodern skepticism toward modern science. The denial that the scientific method exemplified by modern science is the sole measure of truth, makes room for asserting truths that are beyond ordinary reason and beyond normal sensory experience. The claim made by many postmodern theologians and feminists (for example, Griffin 1988; Merchant 1980; Harding 1998) that the naturalistic assumptions of modern science reflect only the historically contingent experience of disenchantment of Protestant Christian societies, opens the door for claims of mystical experiences of the unity of nature and the divine for producing alternative but "equally valid" descriptions of nature. If modern science is seen as a product of the dualist metaphysics of Christianity, then other societies are free to develop their own sciences based upon their own religious experiences and their own metaphysical assumptions. The search for truths about the natural world that starts out from an enchanted, animated, divinized view of nature is as legitimately scientific as the search for truths in modern science.

This is the edifice on which the defense of Vedic science rests. There are three clusters of arguments. The first defends the holism of Vedic thought-style as the mark of a true and complete science, as compared to the dualism and reductionism of Western science. The second is marked by a neo-Spenglerian insistence upon the cultural distinctiveness of all sciences. On this account, India must cultivate a science that is in accord with the nation's cultural "soul" or "essence." The final move consists of establishing a relationship of homology, or likeness, between scientific empiricism and the Vedāntic view of experience and reason, leading to a declaration of equality between the two. The combined thrust of these arguments is to present the Vedic tradition as the original inspiration and the final culmination of modern sciences.

In this chapter we will examine these arguments in detail. But before we proceed, it will be useful to define two terms that we will encounter again and again in this chapter. Vedic science stakes its claim for superiority on the fact that it provides a "holistic science" against the "reductionist science" of the West. What is meant by these two kinds of sciences?

## *"Reductionist" and "Holist" Science*

The crucial metaphysical assumption behind reductionism is that parts of a system determine the character and behavior of the whole. Parts of a whole are distinct entities, each with its own distinctive make-up. Each element of a system can enter into a relationship with other element(s) without being constituted by that relationship. This fundamental assumption justifies three related theses described succinctly by Nancey Murphy (1997, 12–18) and paraphrased here. *Ontological reductionism* holds that higher level entities are nothing but a complex organization of simpler entities. Nothing extra is needed to explain higher level functions like consciousness or life, over and above the relationship among the building blocks that constitute the brain or the body, respectively. *Causal reductionism* holds that the ultimate causes of the physical universe are bottom-up; laws operating at the level of the parts are more fundamental than the laws operating at higher levels. In the famous example of the piece of chalk, used first by T. H. Huxley and more recently by Steven Weinberg (1992), observable qualities like color can be explained by the properties of the components down to the levels of the quarks, electrons, and photons that make up the chalk. *Methodological reductionism* holds that the proper approach to scientific investigation is an analysis of entities into their parts and that the laws governing the highest-level entities should be reducible to the laws of the lower levels. Reductionism has been the overarching philosophy of modern science for most of its history. It is largely responsible for its tremendous success in figuring out those fundamental principles that can explain larger phenomena.

Holist science finds favor among the critics of modern science. The crucial metaphysical assumption embodied in all varieties of holism is the exact opposite of reductionism. Holism states that the whole determines the character and behavior of the parts that constitute the entity or the system. The fundamental intuition here is that the whole is more than just the parts and that something vital is left unexplained when an entity is taken apart into its components.

Holism is practically a trademark of Hinduism and Vedic science. When proponents of Vedic science claim their science to be holistic, they mean that it is concerned with the study of the system or entity as a whole in which "each entity is inseparably connected with the whole universe" (Jitatmananda 1993, viii). In this view, as described succinctly by Pratima Bowes (1977, 2), "nothing is totally distinct and separate from other things . . . relationships between things belong to the inner dynamics of their nature. The laws that define relationships between things define the nature of things themselves." Causality and explanation, in other words, flow form the top to the bottom, from the system to the elements.

As it is deployed in the Vedic science discourse, holist science makes additional assumptions, very similar to the ones described by Anne Harrington in her 1999 book, *Reenchanted Science*. One, holistic science must answer the teleological question, "what for" and not merely "how." Two, the apparently discrete processes have to be understood in their contribution to the function and purpose of the whole. The third assumption is the denial of a separation between mind and body, or matter and spirit. Finally, holism is meant as a metaphor for an organismic or communitarian society, as compared to an atomistic and individualistic society. The general assumption is that a reductionist way of thinking legitimizes an atomistic society that puts rights of individuals first, while a holistic way of thinking will put the claims of the whole community before the rights of individuals. In the Vedic science discourse, a preference for holistic science is accompanied by a distinct preference for a holistic, hierarchical but harmonious, organism-like community.

### Philosophical Anti-Semitism:
### The Hindutva Critique of "Dualist" Science

Neo-Hindu and Hindutva literature offers an outstanding example of doublespeak when it comes to the subject of the origin and nature of "Western" science. Being the nationalists that they are, Hindu ideologues spare no effort in decrying the "mental colonization," the godlessness, the social alienation, and the ecological devastation allegedly created by modern science and technology. But being the Hindu supremacists that they are, they also want to claim the glory of science for Hinduism. The West is condemned as a menace to non-Western traditions, and at the same time, accepted and even praised, but only as a junior, immature partner of the "Indic civilization." The treatment of "Western science" thus runs the gamut between pronouncements declaring that "today's all round imbalance is the direct result of Western thinking," and others proclaiming Hindu India to be a "sister civilization" or even the "mother" of the West in view of Hinduism's purported "emphasis on reason, rule of law and spirit of inquiry." The first statement is from K. S. Sudarshan (1998), a one-time high official of the Rashtriya Swayamsewak Sangh and the second from Girilal Jain (1994) the former editor of India's leading newspaper, *The Times of India*, and a well-known voice in the Hindutva camp.

This balancing act between the condemnation and appropriation of Western science is accomplished through a passionately argued philosophical anti-Semitism; all that is objectionable about modern science is attributed to the Semitic part of its parentage, and all that is truly universal and enduring is claimed to come from Hindu influences. Any attempt to contradict the Vedāntic teachings on how this world is constituted is decried as a symptom of "Semitic dogmatism," one of the many pernicious Western imports. Thus, while Hindutva intellectuals readily grant universal validity to scientific knowledge, they see the "truly" universal parts of it stemming from the wisdom of ancient Hindu sages. Their affirmation of science's universalism is but only a form of the self-aggrandization of Hinduism.

There are two distinct civilizations, so goes the refrain, the Semitic or Abrahamic and the Eastern. The Judeo-Christian branch of the Semitic family is given some credit for having evolved, even though it eventually declined because of its crass commercialism, materialism, and secular humanism. Their Islamic cousins, on the other hand, are declared to be "incapable of change," a people of "unmitigated stagnation," reveling only in bloodshed and mayhem (Gurumurthy 1993, 163). Open hatred of the Muslims and the undisguised abuse of Islam make for chilling reading. Muslims in India are portrayed as the enemy within, and Islam is made into a religion of intolerance, despotism, and ignorance. While "the Hindu" only builds "cultural empires," "the Muslim," who is compared to a hunting animal, creates "military empires" (Kak 1994a, 25). All the contributions of Islam to Indian science, literature, arts are simply wiped out. It is as if the entire history of India, from the establishment of the Delhi Sultanate in the early eleventh century to the end of the Mughal Empire in the eighteenth century, never happened.

But there is a distinctive philosophical content to this Semitic-Eastern divide. It is the nature of all Semites, be they Judaic-Christian or Islamic, to separate, atomize (thereby rendering everything uniform and comparable) and annihilate: "the essential thrust of the Semitic civilization is to enforce uniformity, and failing that, to annihilate," while "our [Hindu] society had a more evolved mind . . . a greater spiritual power . . . (Gurumurthy 1993, 161). It is the nature of Hindu spiritual culture to connect, harmonize, and to let a thousand flowers bloom: "the Hindu mind sustains a connection with the cosmic mind and the blueprint of creation and evolution of the physical world, as well as our connection to worlds more subtle and spiritual. . . . The Hindu mind does not seek to impose itself upon [other] people through force or persuasion . . . [but accepts] whatever name or form we might want to approach the higher consciousness" (Frawley 2001, 12).

But this polar opposition does not exhaust the relationship. The Eastern—the "Hindu mind," to be more precise—is to absorb all that is worthwhile in the Semitic, but serve as the latter's guru in order to lead it out of its errors. In the domain of natural sciences, "the Hindu" is to teach "the Semite" how to re-establish the broken connections between natural forces and "the cosmic mind," or between science and spirituality. Let us see how this claim is sustained.

Nearly all the social ills of modernity—from environmental degradation, imperialism, patriarchy and even caste oppression, and of course, the rise of religious intolerance in India—are laid at the doorstep of Semitic monotheism brought into India first by the Muslims and then by the British. The basic claim is that the one supreme God of semitic faiths stands apart from nature, allows nature to be treated as dead matter, separates human consciousness from the rest of the nature, and gives the believers faith in an absolute truth which allows no contradictions. Epithets like "despot," and even "jealous gangster" with qualities of "brutality" abound in Hindutva descriptions of all monotheistic conceptions of God, while these religions themselves are accused of "heresy-hunting" and "intolerance." The content of some of the specific complaints, especially those regarding environmental degradation and imperialism, are very similar to those

found in postmodernist and feminist literature. We will examine some of these issues later in our examination of feminist and postcolonial critiques of science
(chapter 5) and ecofeminism (chapter 9).

Curiously, and paradoxically, while sharing the postmodernist critique of the
West's "hyper-rationality," Hindutva also condemns the Judeo-Christian tradition
for *not* being rational enough, for holding beliefs that are not consonant with science and for inhibiting the full development of science.

There are two basic theses. The first thesis claims that the transcendent law-
giver God of the Abrahamic traditions led to the reductionist and dualist Newtonian worldview. The mechanical philosophy of Newtonians reduced nature to a
machine without a soul, in which all elements of nature obeyed the same laws of
cause-and-effect without any meaning or purpose of their own. Human beings became spectators to nature, without nature participating in their inner world. But,
Vedic science proponents claim, the Newtonian worldview is now outdated, for it
has been overtaken by quantum physics which presumably restores meaning
and consciousness to a nature which human beings can participate in, reducing
the dualism between the thinking human subjects and the dead objects of nature.
Those who refuse to let go of the mechanical understanding of nature are wed-
ded to a Semitic worldview, either by birth or through mental colonialism. Indeed, there are writers in this genre who believe that the reason Einstein did not
accept quantum physics was because he was blinded by his Jewishness.

The second thesis asserts that the Judeo-Christian and Islamic God is dead,
killed by the development of science. The only conception of God that can withstand a rational scrutiny of science is the God of Vedānta. Because the God of
Vedānta is immanent in nature, it is but another force of nature. Advances in science, especially quantum mechanics and mind-body medicine, Vedic science proponents claim, support the existence of the spiritual element in nature.

The seeds of these ideas were sown by Swami Vivekananda. In a much-cited
lecture he delivered in England, titled "Reason and Religion," Vivekananda declared that "if there is any theory of religion that can stand the test of modern scientific reasoning, it is the Advaita." He arrived at this conclusion negatively by
establishing that Judeo-Christian faiths, which believe in a transcendent God, fail
the test of scientific reason. Vivekananda argued—correctly—that scientific explanation of natural phenomenon cannot invoke any supernatural beings, forces,
and entities: "science wants its explanation form inside, from the very nature of
things" (Vivekananda 1965, CW I, 371). Because they hold their God to be outside and entirely separate from the universe, Judeo-Christian religions have no
choice, he argued, but to invoke something extra- and supernatural to explain the
creation and the working of natural phenomena. They all therefore, clash with
science. But the Hindus see God as an impersonal force which acts not by laying
out laws form the outside, but by expressing itself "in and through every particle

of matter and force." A God immanent in nature allows Hinduism to explain natural phenomena without bringing in any power outside nature, making it more "scientific." This, among other criteria,[2] is Vivekananda's logic for presenting Hinduism as the universal religion of the future which can combine scientific evidence with faith in God.

Interestingly, Vivekananda's view that an immanent God is better suited for the scientific age has found a sympathetic echo among some Christian process theologians, including such notable figures as Ian Barbour (1997) and Arthur Peacocke (2001). Rather than conceive God as an omnipotent, supernatural being who created the world ex nihilo, process theologians see God as working through the laws of nature discovered by modern science. God, as Arthur Peacocke puts it, "continuously gives existence to all the natural events, entities, structures and processes . . . revealed by the sciences" (ibid., 138). This panentheistic naturalism seems to me a having-it-both-ways compromise, which tries to accept the understanding of nature as uncovered by natural sciences, without accepting the overall worldview of materialism that denies the presence of God, meaning, and values in the structure of nature. I am personally not convinced that making God responsible for whatever science finds out about the natural world does any real work, either for science or for faith (except that it reaffirms the faith of the already faithful). God seems to be an unnecessary overlay on natural phenomenon if His/Her presence in nature has not produced any alternative explanations of nature which are as well-confirmed and convincing as scientific explanations that we already have.

Be that as it may, these immanentist trends in Christian theology should not give Vedāntists much comfort. For one, they retain an ontological distinction between God and the processes in nature: God creates these processes, but these processes are not God. As Philip Clayton (1997, 90–91) makes it clear, for Christian panentheism, "the world is within God, and at the same time, different from God, there is *both* an identification and inclusion *and* distinction of God and the world. . . . We are not God because we are different *in our fundamental nature* from God." Moreover, the sociological context is totally different. The panentheistic trends in Christian theology remain more of an academic pursuit, with not much presence in the actual community of worshippers. In India, on the other hand, a conception of God, as not just present in matter but as constituting its very substance, is the majority view. Under the influence of Vivekananda and other nationalistic intellectuals, this view is now being invested with false pretensions of scientificity.

While Vivekananda was inordinately keen to confront the Biblical accounts of creation and other miracles with scientific reason, he (and his contemporary followers) simply refused to accept that the Hindu metaphysics positing a nonmaterial life force in nature is *also* without any scientific evidence whatsoever.

The presence of the extra-sensory, noumenal Brahman that pervades and constructs "every particle of matter" is totally unnecessary to explain natural phenomena, all of which can be more than adequately explained by purely material elements and forces. There is no evidence whatsoever of human actions in non-material spheres of life—in social relations, in ethics, in spiritual matters—influencing matter. There is no evidence of cycles of life and death—except in the most materialistic sense of circulation of chemical elements. There is no atman, no karma, and no reincarnation. Trying to defend them as scientifically confirmed *actual forces of nature*, shaping the physical qualities of living beings and active in the physical evolution of the universe (see below for vitalism and Vedic creationism) amounts to intellectual dishonesty of a very high order. If the Biblical account of the universe cannot stand the scrutiny of reason, neither can the Vedāntic account. In fact, Hinduism has far more at stake. Because Hinduism acknowledges no distinctions between the social-moral and the natural realms—the moral quality of your karma can affect the shape, form, and qualities of the physical body that your soul will acquire—a materialist understanding of natural laws can bring down the entire house of cards. Perhaps that is why there is such a scramble among the upholders of the faith to neutralize science by all means possible.

Vivekananda's "proof" of the irrationality of a monotheistic God has grown into a veritable cottage industry devoted to showing the superior rationality of all things Hindu. Some of the contemporary "proofs" verge on the ridiculous. There is a quantum mechanical proof of the irrationality of revealed religions making the rounds. The argument is that only a completely determined Newtonian worldview allows for prophets like Jesus or Mohammad who, as messengers of God, "can claim to possess infallible knowledge of the future obtained through revelation." Quantum physics, in this view, has overthrown the Newtonian model, "taking with it the prophets [of revealed religions], and their revelations" (Rajaram 1998, 58–59). Because quantum physics is supposedly an affirmation of Hinduism, this move clears the way for crowning Hinduism as the religion of the future.

One politically pernicious effect of this style of thinking is the widespread belief that only the irrational and authoritarian Christianity (and Islam) needed an Enlightenment-style confrontation with science. Vedāntic Hinduism, being in accord with reason, has no need of a critical challenge from what we have come to learn about the world. Indeed, any rational questioning of Vedantic metaphysics is put down as a sign of "Semitic" and "Eurocentric mind-set." There are some followers of Vivekananda's brand of Hindu science (Mukhyananda 1997, 94), who go so far as to declare even the very idea of contradiction as Semitic in origin and foreign to India: "The idea of contradiction is an imported one from the West by the Western-educated, since 'modern science' arbitrarily imagines that it only has the true knowledge and its methods are the only methods to gain knowledge, smacking of Semitic dogmatism in religion . . ." In a striking resemblance to post-modernist critics, knowledge that can challenge and falsify other claims is seen only as a source of intolerance, but never as a source of clarification and growth of knowledge.

The persecution of Galileo, the burning of the heretics and witches in the West is cited regularly in order to present Hinduism as un-dogmatic and open-minded. This is a distortion of the history of science, Christianity, and Hinduism as well. It is true that there was no organized state or Church-like religious institution to enforce Hindu dogmas. But the fear of the power of priestly rituals, which could bring misfortunes upon the non-believers and non-conformists, was no less coercive. If Hinduism did not have a central Church enforcing the dogma, it is because every minute act of everyday life had become invested with ritual, conformity with which was a precondition for social acceptance in the caste one belonged to. The historical record clearly shows that around the time of the Buddha, the revealed (*shrutis*) and remembered (*smritis*) tradition had already become canonized. Strict obedience to the Vedas and tradition was enforced by instilling the fear of the ritual power of priests and the equally strong fear of social ostracism and fall from caste. *Manusmriti*, described as the "pivotal text of the dominant form of Hinduism" (Doniger and Smith 1991, xvii) declared free thinking one of the cardinal sins to be punished by ex-communication: "The Veda [shruti] should be known as revealed canon, and the teachings of religion [smriti] as the tradition. These two are indisputable in all matters, for religion arose out of the two of them. *Any twice-born man who* disregards these two roots of religion, because he *relies on the teachings of logic, should be excommunicated by virtuous people as an atheist and a reviler of the Veda*" (*Manusmriti*, chapter 2, verse 10–11, emphasis added, see Doniger and Smith 1991).

Not exactly a ringing endorsement of scientific temper! (For even more radical proscriptions against questioning the Vedas, as laid out in some of the most revered texts, see Chattopadhyaya 1976, 191–194.)

The law book of Manu and the ethos it prescribed had already become an established source of authority by the early centuries of the Common Era. Theories were rejected or accepted depending upon their agreement with tradition. The heterodox schools which did not accept the authority of the Vedas were either reduced to a caricature (especially the materialist schools of Chārvāka and Lokāyata), or absorbed into the Vedic tradition (as with the originally materialistic school of Sankhya which was assimilated into the Upanishadic teachings in the *Bhagavad Gita*, and as with the Brahminization of the teachings of Buddha). Those who sing praises of Hindu hospitality to reason and innovation turn a blind eye to the contrary historical evidence described famously by Alberuni (973–1048 CE, the renowned Islamic mathematician, astronomer, and political philosopher who has left behind a vivid record of his sojourn in India in the early years of the eleventh century. Alberuni describes how the most eminent Indian astronomers like Varahamihira (sixth century CE) and Brahmagupta (seventh or eighth century CE), knowing fully well the cause of lunar and solar eclipses, bowed to tradition and accepted the myth of a demon's head swallowing the sun or the moon.

These are well-known facts of Indian intellectual history. The myth of critical thinking in the dominant Hindu tradition can only be maintained by ignoring these facts. Unless the nationalist and postcolonial apologists want to claim (as some have) that the best-known, most cited and commented upon, most-revered texts—the *Manusmriti* itself, the central themes of which are also present in other revered texts such as the *Bhagavad Gita* and the *Ramayana*—had no effect whatsoever on practical life before the British colonialists "discovered" them, they have an obligation to examine the full record with an unblinking eye. White-washing the irrationalities or excusing them as colonial constructions will only aid and abet the forces of Hindu chauvinism.

When it comes to the cultural significance of science, the often-heard assertion is that the "Christian West needed secularization as an antidote to the Church tyranny in Europe . . . but there was never any Church tyranny in Hindu India. Thus [secularization of culture] is totally irrelevant to India with its history of tolerance and pluralism"(Rajaram 1998, 184). This in a country where, for almost all of its history, a majority of human beings were not considered pure enough, for religious and ritual reasons, to be worthy of basic human dignity!

### Vedic Science as the Expression of the Hindu Soul

Oswald Spengler (1880–1936), the author of the bestseller *The Decline of the West* (1991), was arguably one of the most influential intellectuals of the Weimar Republic. His condemnation of "Faustian science" based upon causal laws, and his ideas about the cultural relativity of all sciences contributed to the turn toward the idea of wholeness and "destiny" among physicists themselves and encouraged the pernicious theories of Aryan science.[3] Spengler saw cultures as self-unfolding essences, each expressing itself in a unique science, art, literature, etc.:

In place of that empty figment of *one* linear history . . . [I see] the drama of *a number* of mighty Cultures, each springing with primitive strength from the soil of a mother-region to which it remains firmly bound throughout its life-cycle . . . each stamping its material, its mankind, in its *own* image, its *own* passion, its *own* life, will and feeling and its *own* death. . . . There is not *one* sculpture, *one* painting, *one* mathematics, *one* physics, but many, each in its deepest essence different from the others, each limited in duration and self-contained, just as each species of plant has its peculiar blossoms...its special type of growth and decline. (ibid., 17, all emphases in the original)

The spirit of Spengler is alive and well in India today. It is also alive, albeit in a non-essentialist vocabulary of socially constructed "difference," in the burgeoning field of cultural studies of science (which we will examine in the next chapter).

The Vedic science project assumes that all cultures, like individuals, have their own natures (their *svabhava*, or *chitti*) which are innate and unchanging, but which change everything else. A culture's *chitti*, in a manner described by Spengler, leaves its stamp on everything, including all norms of reasoning, all discoveries, and all innovations. A nation has an obligation to protect and cultivate its *chitti*. It must therefore cultivate sciences which express its cultural essence. If it has to borrow from others, it must borrow only those ideas which do not go against its essential nature.

This is how a case is made for cultural essences in the Hindutva literature. Vivekananda, the pioneer of "spiritual science," put it across in a lecture he gave in Lahore in 1897: "Just as there is individuality in every man, so there is a national individuality. Just as it is the mission of every man to fulfill a certain purpose in the economy of nature, just as there is a particular line set out for him by his own past Karma, so is it with nations. Each nation has a destiny to fulfill . . . a message to deliver . . . a mission to accomplish" (1968, 39–40).

Here is another influential voice, that of Deendayal Upadhyaya, the author of *Integral Humanism* talking of Hindu India's inborn, innate nature, or *chitti*: "Every group of persons has an innate nature, or *chitti*. Similarly every society has a soul or *chitti* which is inborn and is not the result of historical circumstances. [Like the soul in human body], the nation's chitti is unaffected by history. *Chitti* is fundamental and is central to the nation from its very beginning. *Chitti* determines the direction in which the nation is to advance culturally" (Upadhyaya 1965, 8).

So, what is "Hindu India's" *chitti*? Devotion to reason, experience, and truth is innate to "the Hindu mind." Accepting the stereotype of India as a mystical or spiritual country, but *redefining spirituality itself as indistinguishable from the rational and empirical science of nature*, Hindu nationalists are keen to present "the Hindu mind" as "scientific" in a deeper, more comprehensive sense that does not split this world from the world of the spirit. *India is spiritual, but in a scientific way; spiritual experiences are experiences of the divine in nature.* To harmonize reason and spirit is Hinduism's innate nature. To lead the entire world to this new synthesis is its destiny, determined by its spiritual karma through the ages.

What makes Hindus spirituality innately scientific? The basic argument goes like this: Hindu spirituality is an empirical, experimental attempt to directly see what really lies behind appearances. Because Hindus don't separate nature from spirit, they have developed spiritual techniques to see nature directly, as it is. The science of nature is distinct and opposed to the experience of the spirit only in the West. In India, spirituality has always been and is a science. This basic claim is constantly reaffirmed by selectively interpreting quantum physics and biology to show that physical matter is sentient and that consciousness is an active force in nature.

Hindutva literature is replete with many formulations of this denial of boundaries between the science of nature and the experience of the spirit. The contemporary statements, however, are variations of the original formulation by Vivekananda who almost single-handedly began this trend of reading modern science into mysticism. Just as modern science seeks to understand the reality that lies behind appearances, Vivekananda suggested, so did the Vedic sages want to go to the ultimate source of all that exists. In his famous address to the Parliament of World Religions in Chicago in 1893, Vivekananda told the admiring crowds: "If there are existences beyond the ordinary sensuous existence, the Hindu wants to come face to face with them. If there is a soul in him which is not matter, if there is an all-merciful universal Soul, he will go to Him direct" (1965, CW I, 13).

In another lecture, this time in Lahore (Pakistan), Vivekananda claimed the innate superiority of the Hindus: the "materialist Western" culture "only" applied reason and experience to the sensory and material world, whereas Hindu sages with their "gigantic brains" persevered in their attempt to "directly see" the non-material, non-sensory force behind it all. Brahman was not a matter of faith that spoke to the heart, but it was to be directly seen through the power of the mind cultivated by yoga (Vivekananda 1968, 41).

This erosion of any dividing line between mysticism and science was picked up by the followers of Vivekananda, who have established an international network of monasteries and ashrams named after Vivekananda's guru, Rama-krishna. The Ramakrishna Missions in India and around the world propagate this scientistic idea of spirituality, and invariably end up teaching modern science as only a fulfillment of the spiritual truths of Vedānta. They present yoga as the empirical method of Hindu science which enables the practitioners to directly "see" the interconnectedness between the life force that is in them and the life force that permeates the universe. In that state, there is no subject and no object—everything becomes "omnijective," a term sometimes used in this literature (see Jitatmananda 1993) to describe the fusion of the inner spirits of the knower and that of the object under study. This holistic conception of matter and spirit, this ability to see interconnections between nature and consciousness, is claimed to be innate to the Hindu culture.

Vedic science proponents believe that this holistic thinking has been unfairly looked down upon as anthropomorphic, magical, and irrational. It is only when the fundamentally Christian monotheistic assumptions are held as universally valid that Indian ways of knowing appear irrational. Decolonization of the Hindu mind requires understanding Indian science through Hindu categories. This basically means that the disenchantment of nature brought on by the Scientific Revolution is irrelevant for Hindu science. Indian science can move straight to the quantum mechanical view of the world which, on this account, affirms the presence of spirit in matter (more on this in the following paragraphs). Powerful statements to this effect can be found in nearly all the publications of the Ramakrishna Mission, and in the writings of the new crop of Hindutva ideologues who are addressing the science question, especially Elst (2001), Frawley (2001), and Rajaram (1998).

Having established the existence of a unique and unchanging essence, Vedic science proponents draw political lessons from it. If holistic thinking defines the Hindu *chitti*, its *svabhava*, then India has an obligation to develop its sciences and design its education to conform to its innate nature. The following words of Deendayal Upadhyaya (1965, 5) are a part of the ruling Bharatiya Janata Party's manifesto: "A nation's chitti is fundamental. . . . Whatever is in accordance with chitti, is to be cultivated and added on the culture. Whatever is against the nation's chitti is to be discarded as perversion, undesirable and is to be avoided. *The nation's chitti is the touchstone on which each attitude is tested, and determined to be acceptable or otherwise*" (emphasis added).

These are not mere words. The ruling Hindu nationalists are in fact using the supposed Hindu essence as the "touchstone of all actions and attitudes" to set policy in science education and research. In the name of cultivating a science that is in tune with India's cultural ethos, the minister of human resources, Murli Manohar Joshi, has many times declared his government's intention to teach modern science that will confirm the truths of "our ancient scientific traditions." We have already examined the Hinduization of science education in schools (chapter 3).

Apart from school science, one finds cultural essentialism at work in the vitalistic theories of biology and the mystical interpretations of quantum mechanics being peddled by countless gurus as the scientific "proofs" of Vedāntic teachings. These spiritual spins on contemporary science are purposely used by religious organizations, especially the Ramakrishna Mission, as a strategy to entice the younger generation of students in scientific and technical institutes. Many "Vedic science clubs" have sprung up on the campuses of elite institutions, many with the active collusion of religious organizations, that teach a scientized version of the Vedas.

Let us take a brief look at the place of vitalism and quantum mechanics in the corpus of Vedic sciences.

## *Vitalism*

Vitalism has a special place in Vedic science. An internationally renowned physicist-turned-botanist, Jagdish Chandra Bose (1858–1937) tried to prove the existence of consciousness in plants and metals. After a promising start to his career as a physicist, Bose turned his attention to life sciences. He studied the pattern of electric current through metals and plants under different conditions—for example, tin foil being "poisoned" and cabbages being burnt—and claimed that the response was similar to that found in animal muscles. Bose claimed that these results showed the presence of sensations and irritability in plants and metals. Bose went on to interpret his findings as a confirmation of the "message proclaimed by my ancestors on the banks of the Ganges thirty centuries ago," of the presence of consciousness in all of the universe, plants and inanimate matter included (see Jitatmananda 1993). Bose's turn to Advaitic monism came through his association with Sister Nivedita, a close associate of Swami Vivekananda, who encouraged him to "reflect distinctively Indian sensitivities" in his scientific work and to use this experimental technique to prove the truth of Vedānta (Nandy 1980, 45).[4]

Bose's biological discoveries were refuted in his own lifetime, and were quickly dropped from the scientific literature as vitalism was discredited in biology. Yet, he remains a hero of the Vedic science tradition. Riding on Bose's coattails are a variety of contemporary vitalistic theories that find great favor in Vedic science circles. These theories range from the totally discredited studies of plant consciousness (the so-called "Backster effect" celebrated in popular books like *The Secret Life of Plants*), to Rupert Sheldrake's theory of morphogenetic fields with all varieties of paranormal phenomena like auras, Kirlian photography and telepathy in between. (For a celebratory exposition, see Jitatmananda 1993.) While no hard data exist, auras, Kirlian photography, and similar pseudo-scientific ideas have a substantial following in India, especially in urban areas.

None of these ideas has withstood tests with double-blind controls. Yet, they are held up as scientific proofs of the correctness of the Vedāntic vision of all pervasive consciousness, Brahman. Obviously, agreement with the cultural ethos of India is of primary importance here which over-rides their scientific merit.

## *Quantum physics*

The use—or rather, abuse—of quantum mechanics as Vedānta-in-disguise is yet another case where Hindu orthodoxy is serving as the touchstone for a cultural conservationist interpretation of one of the most fundamental developments in physics. An enormous volume of quantum mystical literature has grown around the presumed convergence between quantum physics and Eastern religions. Spurred by books by mystically inclined physicists like Fritjof Capra, Gary

Zukav, and many others, Indian Vedāntists have had a field day claiming that Heinserbeg's principle of uncertainty proves the primacy of consciousness over matter and the interconnectedness of the human mind with consciousness in the rest of the universe. Among Indian Vedāntists, three nuclei of quantum mysticism have appeared which operate globally, alone or in collaboration with each other: the Vedānta centers associated with Ramakrishna Missions that follow in Vivekananda's footsteps in "synthesizing" science and spirituality using quantum physics (for a sample see the recent books by Ranganathnanda 1991; Jitatmananda 1993; Mukhyananda 1997), the Maharishi Mahesh Yogi's school of transcendental meditation which can enable a "direct experience" of the "unified field" (Chandler 1989), and the bestselling books of Deepak Chopra, a one-time colleague of Mahesh Yogi, who promises "quantum healing" by creating "happy molecules" through "happy thoughts."

The question, what quantum mechanics "really" means, is as old as quantum mechanics itself. Physicists themselves remain divided, to say nothing of the rest of us. This is not the place, nor do I have the required qualifications, to resolve the technical issues involved.[5] But the centrality of quantum physics in Hindu science does not hinge upon technicalities either; many of the numerous swamis who discourse on quantum mysticism are barely literate in any science,[6] let alone quantum physics. It is all about interpretations, and one finds the most extreme mystical meanings, which do not have the support of the mainstream of physicists, being read into this abstruse branch of basic science.

In the context of cultural essentialism that we are dealing with, I am only raising a simple question about the criteria that are being used in claiming the convergence of Vedānta and quantum physics. Among those who call for a synthesis of quantum physics with mysticism, there are no doubt some who are genuinely trying to retain their faith while grappling with physics. But by and large, cultural nationalism and/or profit motive, plain and simple, seem to be the driving forces behind most of the hasty attempts at "synthesis" that one finds in the three schools mentioned above. This literature is replete with most radical declarations that respect neither the integrity of physics nor the authenticity of mysticism that is the heart of Vedānta: physics is turned into mysticism and Vedānta is made to sound as if it were chiefly concerned with understanding the material world, which it never was. It is claimed that the Newtonian worldview of cause-and-effect is dead; the distinction between matter, energy, and consciousness is no longer valid; the separation between object and subject is no longer valid and reality has become "omnijective" in which the subject creates the object, and therefore human consciousness creates the world; by learning to control consciousness, one can control the laws of nature; and so on. Needless to say, all that is dead belongs to the pre-quantum mechanics physics and all that lives belongs to Vedānta. The "higher" has encompassed the "lower."

If one were to patiently tease out the layers of obfuscation and ask on what basis the Vedānta is being read as containing and transcending quantum physics, one quickly runs into the familiar old "touchstone": all that is in accordance with the nation's "holistic" *chitti*, is to be cultivated and added to the culture and whatever is against it, is to be discarded as perversion. Of all the other possible interpretations of the paradoxes of quantum mechanics, only the most mystical ones are chosen because the whole purpose of the exercise is to judge science against the truths contained in Vedānta.

The lush overgrowth of quantum mysticism needs a trimming with Occam's razor. As Victor Stenger has argued in his various writings on this subject, entirely "un-conscious," materialistic and deterministic interpretations of quantum mechanics are available that are fully consistent with all observations of physics, including relativity. The extraordinary claims involving human consciousness and/or sub-quantum forces that move faster than the speed of light are not required to explain the available experimental evidence which can just as well be explained using unconscious or non-sentient alternatives. In this situation, Stenger writes: "the only rational procedure is to apply Occam's razor and reject those interpretations that are less economical than the others and to pragmatically adopt those remaining that are the most useful" (1996, 2–3).

Under this test of parsimony, clearly, Vedānta is *not* another name for physics. In fact, those who claim to value spirituality should welcome this separation of Vedānta from physics, for it can free their spiritual knowledge from the changing fortunes of scientific theories.

To conclude this section, the insistence upon a distinctive "Hindu science" is based upon a Spenglerian belief in the cultural distinctiveness of all sciences. Each culture is said to have an innate nature, a soul. All cultural products, including sciences, must be an organic expression of this soul. Holism is taken to be the soul of Vedāntic Hinduism, which, in the nationalist discourse, is equated with the soul of India itself. Hindu science proponents look at the entire corpus of natural scientific knowledge through the lens of holism and claim to find a convergence between the two, leading them to insist that Hinduism is science by another name.

### Equality of all Truths

Synthesizing science and religion is a risky business. Those who want to claim the glory of science for their religious traditions run the risk of tying the content of their faith to the ever-changing vicissitudes of science. That opens the door to empirical examination, and often falsification, of the beliefs about the natural world that religions invoke to provide ontological grounds for their teachings.

Given the potential of conflict, the standard liberal position, both among the believers and non-believers, has been one of "good fences make good neighbors."

To avoid conflict, respect the boundaries between the two; do not claim that religions have anything to tell us about nature, the province of science, or that science has anything to tell us about ultimate values, meanings, and ethics, the domain of religion. Keep the two magesteria separate, as Gould recommended in his principle of Non-Overlapping Magesteria (or NOMA) in his 1999 book, *The Rock of Ages*. (Although this is not the place to go into the details, I am not convinced of the tenability of Non-Overlapping Magesteria. I believe that a scientific understanding of the natural world does have implications for ethical values and, on the other hand, religions necessarily make propositional claims about the natural world and cannot help stepping into science's turf. Before we can talk of peace and harmony between the two, religion has to do a lot of house-cleaning, especially Eastern religions which have so far avoided an Enlightenment-style re-examination of their fundamental assumptions about nature.)

The proponents of Vedic science are juggling with the two magesteria in a very self-serving manner. They want to get all the advantages of a synthesis, but none of the conflicts. In other words, while the Vedic corpus is claimed to encompass the knowledge of nature and therefore overlap with the subject matter of natural science, it is never *contradicted* by scientific findings in any of the shared domains. As Vivekananda and his followers insist, modern science is only an echo of Vedānta, never its adversary.

How is this position maintained?

The short answer is, by declaring the equal scientificity of all ways of knowing. The method of Vedic science—associative reasoning, or reasoning by homologies—is declared to lead to an understanding of nature that adequately connects empirical experiences to the underlying structures of reality and that provides logically coherent explanations of the observed natural phenomena. The Vedic method of knowing nature's truths by turning inward, into the realm of spirit, meets the test of science in the sense that it is demonstrable and logical, "no different," therefore, from the methods of modern science. What is more, it is claimed that the conclusions about nature arrived through the spiritual disciplines of Hinduism have been proven to be true by modern science. Therefore, there is no conflict between Hindu spirituality and modern science. Science confirms Hinduism's eternal verities.

These are not mere boasts. Rather, these claims are made as justification for two major Vedic science projects, namely, Vedic physics and Vedic creationism. The first involves the reinterpretation of the *Rig Veda*, the earliest of the four Vedas, as a scientific treatise that presumably contains major discoveries in physics, cosmology, and computing, all in a coded form. The major proponents of this project are Subhash Kak, a professor of electrical and computer engineering in Louisiana State University, Raja Ram Mohan Roy, a physicist working in Canada, and David Frawley, the American-Hindu Vedāntist and astrologer. The

Vedic physics project is intimately linked with the thesis of India as the original homeland of the "Aryans," and therefore the cradle of all the civilizations of the world that the "Aryans" contributed to.

The second project is that of Vedic creationism, launched by the followers of Swami Prabhupada of the Hare Krishna movement. The massive *Forbidden Archeology* by Michael Cremo and Richard Thompson (1993), serves as a manifesto as well as a research program for Vedic creationism. Both these projects are getting a lot of favorable press in Hindutva circles, including the multiculturalists who are seeking to revise history books in American and British public schools.

In what follows, I will briefly examine the major claims of these two projects. My aim will be to unearth the cultural constructivist logic that underlies the claims of scientificity of these projects. I will argue that hiding behind the cultural relativity of reason, associative logic of sympathies, and equivalences—the logic of occult sciences and theosophy—is being presented in a scientific garb.

### Vedic physics

The goal of finding modern physics in the Vedas is to show that not just the later Upanishads, but the earliest Vedic text, the *Rig Veda*, is a "way of science" that "embraces physics, mathematics, astronomy, logic, cognition and other disciplines. . . . Vedic science is the earliest science that has come down to mankind" (Kak 1999a, 1).

All these sciences, however, exist in the Vedas in a "coded" form: the phrases which do not make any sense if understood in their commonplace meanings, *in fact* stand for scientific concepts so refined that they contain the exact values of physical constants only recently discovered by Western physicists. When the verses talk about cows, for example, they actually mean the earth (Kak n.d.). When some other verses talk about domestic animals, they are *really* referring to bosons, the sub-atomic particles that "live together," but when they talk of wild animals, they are *in fact* referring to fermions, the particles that "live alone" (Roy 1999, 115). The proponents of Vedic physics take it upon themselves to decode the Sanskrit verses to reveal the astronomy, cosmology, and other discoveries of physics hidden in them.

Finding elements of scientific inquiry in the *Rig Veda* is tied to the larger goal of proclaiming India to be the "cradle of civilization"—*all* civilizations.[7] Vedic science proponents want to claim that *Rig Vedic* India was the real birthplace of rational science, mathematics, and geometry. What is more, the "Aryan" authors of the Vedas did not come from outside, as the standard history of India's antiquity would have it, but rather the landmass which is now India (and Pakistan and Bangladesh) was the original homeland of Indo-European Aryans. Using highly dubious archeological, astronomical, and linguistic evidence—evidence which is not accepted by the vast majority of experts in these fields—Hindutva intellectuals have pushed back the date of the *Rig Veda* from around 1500 BCE to 3000 or even 4000 BCE. Using these dates, and armed with their re-interpretation of the *Rig Veda*, they claim that Indo-Aryans took their science with them as they migrated westward into the Middle East. All that the Babylonians, the Egyptians, and the Greeks knew of physics, mathematics, and geometry could only have come from the Vedic Aryans. The history of science has become one of the most politically charged subjects in India.

The pride of place in this literature goes to Subhash Kak's claims of finding "astronomical codes" in the design of the fire altars and in the literary structure of the *Rig Veda* itself. It is well-known that the shape and the design of fire altars had magico-religious significance. The *Taittiriya Samhita* of *Yajur Veda* prescribes altars in the shape of different birds, for example, to gain desired ends: an altar shaped like a falcon for those desirous of going to heaven, in the shape of an "Alaja bird" for those desiring support, etc. Kak believes that finding only magico-religious interests in these texts is a sign of the Eurocentric mind-set that equates India with superstition and magic. He claims to have proved that hidden underneath the magical formulae are astronomical findings which are based upon a record of careful observations dating back to 3000 or 4000 BCE.

Kak describes the prescribed construction of these fire altars (Kak 1994a; Feuerstein et al. 1995). From among the Sanskrit words, numbers, and numerology that crowd the pages of his books and essays, here are a couple of cases that seem to be crucial to Kak's argument. The first has to do with what he calls the "Yajnavalkya cycle." This cycle involved the building of a falcon-shaped altar in the *agnicayana* ritual. The ritual demanded that every year, a new altar be built with the same shape and same number of layers, but with a slightly larger area. This gradual increase was to go on for a period of 95 years. Kak also seems to find significance in the number of pebbles that were to surround three other kinds of less complicated altars. The third piece of "evidence" is the number of syllables and verses and their organization in the *Rig Veda*.

From this evidence Kak claims two theses: *(a)* the prescribed number of bricks for different altars encode astronomical knowledge about the length of solar and lunar years, the distance between the sun and the earth and other astronomical facts; and *(b)* the number of bricks corresponds to the number of hymns and syllables in the *Rig Veda*, or, to put it in other words, the *Rig Veda* is organized on an astronomical plan.

How does Kak establish these claims? His method is breathtakingly ad hoc and reads like numerology 101. Kak finds auspicious meanings in some numbers of bricks (which just happen to be numbers that we know today from physical constants derived from modern physical laws!) and establishes their connection with the motion of the sun, moon, and other planets. What is worse, Kak justifies this method of equivalences between altars and the skies as perfectly scientific within the holistic metaphysics of Hinduism.

Take, for example, the 95-year "Yajnavalkya cycle." He reads the textual pre-scription of making the falcon-shaped fire altar larger each year in a cycle of 95 years as evidence that the Vedic people "knew the length of the tropical year to be equal to 365.24675 days" (Feurestein et al. 1995, 203), a value surprisingly close to the modern value. What is more, Kak reads into this 95-year cycle of al-tar design a method of intercalation. He claims that the amount of increase each year represented the extra days needed to make the lunar year equal to the solar year.

There is independent evidence that intercalation—adding an extra month to the calendar—was a big problem for the Vedic people. Developing an accurate calendar was of great interest to the priests, as their rituals were supposed to be efficacious only when performed at the auspicious time. Because they divided a year into 12 lunar months of 30 days each, and also into four seasons determined by the sun, the lunar year would go out of sync with the seasons. Adjusting the lu-nar month with the seasons seems to have been a major preoccupation at that time. The consensus among historians of Indian science is that the *Rig Veda* fol-lowed a five-year cycle in which an extra month was added to the lunar year to synchronize it with the solar year (Sen 1971, 1996).

Kak believes that the five-yearly intercalation was not accurate enough and so the 95-year method makes more sense. But how does he conclude that when the relevant text talks of increasing the size of the altar each year for 95 years, it is *in fact* instructing the priests to mark the length of time and that the number of bricks corresponds with the length of time? In order to establish that, one would expect at least some independent evidence that the altar size represents the length of time. The only independent evidence that Kak gives is a verse from the *Satapatha Brahmana* to the effect that "Prajapati (the creator God) was created sevenfold in the beginning. He went on constructing his body, and stopped at one hundred and one fold one." And that "one hundred and one fold altar becomes equal to the seven fold one" (Kak 1994b, 83). Nowhere does he prove that Pra-japati was meant as a unit of time, and that the altar size corresponds with time. He simply assumes a homology between bricks, time, and God. Why does he make this assumption? Therein lies the key to associative thinking which Kak wants to claim as scientific even today. We will consider the insidious cultural relativism of Kak and his associates' logic shortly. But first, let us gather all the evidence.

Kak does not limit his evidence to altar bricks. He finds cosmological signifi-cance in the number of rosary beads and the number of hymns in the *Rig Veda*, which he says encode the distance between the sun and the earth in terms of solar diameters (Feuerstein et al. 1995, 205). From the number of stones that were to be scattered around three other commonly used altars—21 stones around the earth altar (the circular altar called *gārhapatya*), 78 around the atmosphere altar (the semicircular altar called *dhisnya*), and 261 around the sky altar (the square-shaped altar called *āhavaniya*)—Kak claims to find two numbers of great cosmological significance (the numbers 21 and 339) (ibid., 205). Another remarkable feat of ancient scientists, discovered only now by Kak, is that they knew the speed of light to be 186, 536 miles per second, a very close match with 186,000 miles per second that Roemer only found out in the seventeenth century. But it doesn't stop there. The cosmological numbers 21 and 339 just happen to be present as factors—if multiplied creatively enough with suitable numbers—in the number of syllables, verses, and hymns of the *Rig Veda* itself! The conclusion Kak derives is that the *Rig Veda* itself is homologous to the five-layered *agnicayana* altar. Like the altar, the *Rig Veda* is supposed to be organized according to an astronomical plan (Kak 1994b, chapters 5 and 6).

Kak's work has inspired another—even wilder—attempt at the deconstruction of the *Rig Veda*. Raja Ram Mohan Roy's 1999 book, *Vedic Physics*, acknowledges Kak's influence and Kak himself has written a foreword to this book. Roy grabs hold of Kak's "method" of finding equivalences—and runs with it. The result is a shameful demeaning of physics as well as the Vedas. Both are reduced to ravings of mad men.

Roy's "method" is simple: wherever the Vedic verses make no sense, one must assume that they must have intended to convey knowledge about physics! Behind the apparent meaning which may not make much sense, there is an actual meaning which contains cosmological theories, many of which were belatedly proven right by modern physics. When the Vedas *appear to* talk of animals, stones, or deities, they are *actually* talking of fundamental particles and forces of nature in a dramatic form. The conclusion: "the Rig Veda is a book of particle physics." Roy's book is a compendium of absurdities where references to animals mean bosons and fermions, animal sacrifices stand for quark containment, where annihilation of dark-skinned people means annihilation of anti-matter, food is matter-energy, and where the reference to 10 directions stands for super space, so on and so forth . . . ad nauseam.

In ordinary times, one wouldn't dignify such absurdities with a second thought. But these are no ordinary times. Kak is a stalwart of Hindutva, one of the leading "intellectual Kshatriyas." Along with Frawley and other "scientists," his work counts as "evidence" of the antiquity and originality of the Aryan people and establishes them as India's native sons. With the Bharatiya Janata Party in control of the machinery of the state, these ideas have a direct pipeline to policy-makers and textbook writers in India. But the West is hardly immune. The United States-based Kak takes pride in playing a leading role in an initiative to rewrite school textbooks in the United States: "school texts in California and other American states have been rewritten. . . . New college texts in the US speak of these new findings" (Kak 1999b). One can easily see how the multicultural academy in Northern America would gladly make room for such ideas without much fact-checking. After all, if the wildest varieties of Afro-centrism and feminist episte-mologies can thrive in the groves of the academe, why wouldn't Indo-centrism get equal time and attention?

Now to the question of how they justify the method of Vedic physics. Here Kak and his colleagues are opening the door to an insidious cultural relativism which they use not just to justify the rationality of Vedic science in antiquity, but to up-hold these irrational ideas as scientific even today. They are basically erasing all distinctions between science and associative thinking, the latter being the hall-mark of magic.

In defending their findings, they inevitably run into the question: How did the Vedic sages know all this physics? What was their method? Why don't we find any material evidence of observatories, or records of observations? Invariably, the answer one gets is that Vedic sages "intuited," "experientially realized," or "di-rectly perceived . . . in a flash" the laws of nature by altering their consciousness through yogic meditation. *By knowing themselves, they came to know the world.*

But why should knowing oneself lead to knowledge about the natural world? The answer lies in the monism of Hinduism discussed earlier, which Kak and his colleagues describe in their *Cradle of Civilization* thus: "the innermost essence of the human being is the very same essence that underlies the universe at large" (Feuerstein et al. 1995, 41). Man is a microcosm of the entire macrocosm, united by the same spiritual energy that can transform itself, through the agency of karma, from any element of nature, animate or inanimate. There is no ontological barrier between the material and the spiritual, the sentient and the insentient, the physical and the psychic, between God and man, etc. Underlying all the apparent barriers and separate entities, there is underling unity. In this organic universe, ordinary laws of contradiction do not apply, if x is x, it can also be not-x but y at the same time. Contradictions are contained in the unity that constitutes the ele-ments and therefore contradictions can coexist. Because an entity can be both material and spiritual (i.e., non-material), spiritual knowledge is a rational way of knowing the material world.

This answers the question how the Vedic scientists know the laws of physics and cosmology that are supposedly coded in Vedic literature; deep introspection allowed them to see likenesses, homologies, or equivalences (called *bandhu* in Sanskrit) between the cosmic (the *ādhidaivika*), the terrestrial (the *ādhibhuatika*) and the spiritual (the *ādhyatmika*). What other cultures, including the modern Westernized scientific culture, dismiss as magic, coincidences, or lucky guesses, our Vedic forebears saw as signs of the interwovenness of all things (ibid., 197).

It is from this perspective of holism and interconnectedness that the Vedic gods can be taken to legitimately "represent the stars and the planets, as well as the psycho-physiological centers within the body, or even the bricks in the altar. *The correct interpretation can only be obtained from the context*" (Kak 1995, 32). This last sentence, emphasized here, obviously gives a carte blanche to the most extreme interpretations, as the ones found in Roy's book and in countless lesser known works in this genre.

As the great scholar of magic and religion, Keith Thomas, described in his 1971 masterpiece, *Religion and the Decline of Magic*, seeing the existence of correspondences and equivalences between different parts of creation is the very essence of magical practices like palmistry and physiognomy, for just as man was supposed to mirror the world in miniature, the hand or the face mirrored the man (1971, 223). The doctrine of correspondences is based upon the equivalence of the microcosm and the macrocosm, which was as prevalent in pre-Reformation Europe as it is in India even today. In Europe too, the micro-macro homology was used both ways, just as it is being used in Vedic science and Vedic environmentalism today: human beings were seen as an ordered system comparable to the whole world, the world was seen as an organism endowed with vital powers, agency, and its own reason. In the West, this magical view of the world peaked around the Renaissance, and began to decline with the Protestant Reformation and the rise of the mechanical philosophy in the seventeenth century. It saw a brief revival in theosophy and holistic schools of biology in the nineteenth and early twentieth centuries, especially in Germany. It is now a province of fringe occult groups in the West.

There is plenty in Kak and his associates' writings that indicates that they will decry the above paragraph as Eurocentric. Associative thinking is occult, they insist, only within the dualist logic of Semitic religions in which the knowing conscious subject is radically different and apart from the objects in nature. But the history of Indian science has to be studied through Hindu categories; this is the battle cry of all cultural nationalists. Because in Hinduism the macrocosm is "enfolded" in the microcosm of the human mind, introspection is a perfectly rational method of seeing a connection between the two. The injunction "know thy self" takes on a whole different meaning here.

Kak and his associates want to defend associative thinking as the basis of scientific reason not just in antiquity but in the contemporary world as well. Defending the contemporary relevance of the Vedic worldview is a crucial aspect of their project, otherwise the very ground for modernization without secularization will be lost. If they grant that associative (magical) thinking was rational within the parameters of what the people in antiquity knew about their world, but that it has been surpassed by modern science, then they will have to question such things as Ayurveda, astrology, and the rationale behind many rituals. For this purpose, they have to show that associative thinking meets the standards of reason and evidence set by modern science.

Here they take recourse to the social constructivist critics of science. Modern scholarship, they claim, has shown that science is pretty much in the same boat as any other culture-bound way of knowing. Its hegemony around the world is based upon the imperialistic hegemony of a secularized culture in this godless age, when all magic and wonder has disappeared. But take away the false, power-imposed hegemony of the secularized world, and you find that science too is "rationalized mythology" like any other mythology. The old standbys—Paul Feyerabend, Thomas Kuhn, and Carolyn Merchant—are trotted out in support of the culture-boundedness of modern science (see especially Feuerstein, et al. 1995). Science is not completely relativized, but enough space is created to argue for other, equally rational ways of knowing—the associative thinking of Hinduism being as good as any. All ways of knowing are declared to be simply different names for the same enterprise of science, all equally scientific within their own metaphysical assumptions, and it is assumed that associative thinking can be substituted for experimentation and logic with no loss of truth. Cultural boundedness of all sciences works as a taken-for-granted assumption, a backdrop against which Vedic science is presented as an alternative epistemology in its own right, with as much claim for universal acceptance as modern science.

The Vedic science proponents, however, do not press relativism too hard because they need those findings of modern science that can lend support to the holistic assumptions of the Hindu cosmology. Thus, one finds a constant reiteration of the quantum mechanical ideas of David Bohm who has shown that the implicate order of the cosmos is folded into the explicate order that we see in the world as it appears around us. Quantum mechanics is supposed to lend support to the presence of the entire macrocosm in the microcosm of man's inner world. Newer discoveries from biology are brought in to show that biological rhythms (the microcosm) follow the periods of the sun or the moon (the macrocosm), thus making room for astrology and other superstitions. Modern science is brought in only to strengthen magical thinking!

Leaving aside the issue of establishing the epistemological soundness of magical thinking, there seem to be serious historical errors of facts in Kak's account. These errors become important in light of the fact that Kak's work is cited in support of the thesis that the Vedic Aryans were original to the landmass of India. As Michael Witzel, a Harvard-based Indologist who has challenged the thesis that the Aryans were indigenous to India, has pointed out, the falcon-shaped altar is *post*-Rig Vedic. To quote Witzel:

> The piling of fire altars made of thousands of bricks belong to the post-Rigvedic period and even then, occurs only in comparatively late Yajurveda material . . .

and can at best be dated to the beginning of the iron age. . . . Even the post-Rigvedic texts say only that three ritual fires represent the earth, sun and the moon, and that the offering priests walk about in space. The complicated post-Rigvedic brick pilings represent a bird that will take the sponsor of the ritual to heaven. There is no indication of any typical Brhamana-style speculation that goes beyond the identification of the sponsor of the ritual with the creator god Prajapati. Complicated astronomy is absent. . . . *If there is any older, local tradition hidden behind all of this, it may go back to local, non-Vedic sources."* (Witzel 2001, 74–75, emphasis added)

The last sentence of Witzel's statement is significant. There is sufficient evidence that supports the conjecture that the entire technology behind the *agnicayana* ritual might be borrowed from the urban, brick-making, pre-Vedic Indus Valley civilization, which in turn might have been influenced by altar design in the so-called Bactrian-Marginian Complex (BMAC) in the area east of the Caspian Sea (Staal 1999, 119–120; also Staal 2001). The Indo-Aryan speaking people who composed the Vedas were not brick-making people; being nomadic, they had no use for bricks. The Harappan civilization, on the other hand, was an urban civilization that made extensive use of fire-baked bricks. Historians have conjectured that the knowledge of brick-making was preserved among the inhabitants even after the Harappan civilization was destroyed (see Chattopadhyaya 1986 for an extensive treatment of this issue). Staal has conjectured that the *Yajur Veda* was composed by "indigenous Indians who had become bilingual by adopting the language of the incoming (Indo-Aryan speaking) tribes"(1999, 119).

Of course, the Harappan origin of the technology of brick-making does not rule out the possibility that the brick altars were still laid out according to the specifications derived from astronomical (or any other) logic of the Vedic priests who, evidence suggests, hired the local people as craftsmen and masons and used their technological know-how.[8] The origins of the so-called Aryans and their various exploits is, of course, one of the most inconclusive—and most politically charged—debates ever in archeology. That is all the more reason for exerting extreme care in assessing the available evidence. That care is completely missing from Kak who appears too eager to declare India the homeland of the Aryans.

Apart from the extreme interpretive flexibility, there is another flaw in Kak's work. The Vedic literature is entirely oral: the knowledge revealed in the Vedas was considered too sacred to be put down in writing. The entire corpus, mathematical parts included, was passed from generation to generation through memory. It was finally compiled and written down in the fourteenth century CE. Witzel correctly points out that the internal organization of Vedic literature is designed to aid memorization. Witzel asks (2001, 70) why anyone should order their text according to an astronomical pattern, when their chief concern was memorization. All these questions remain unanswered to date.

### Vedic creationism

Christian creationists are no longer alone in battling Darwin. But the new enemy of their enemy may not necessarily be their friend. Vedic creationism of the Hare Krishna movement has emerged in recent years as a challenger to Darwinian natural selection *and* Christian creationism alike.[9] So far, this United States-based Vedic anti-Darwinism has not made significant inroads in India. Darwinism is not much of an issue in India, as it has never been able to displace the traditional Hindu cosmology in the first place. Creationism in India takes the form of giving a scientific gloss to the Hindu view of transmigration, karma, and cyclical time.

Like Christian creationists, Vedic creationists oppose the methodological naturalism of modern science in general, and Darwinism in particular. Like creation scientists, they seek to use the existing scientific evidence to prove the truth of their religiously inspired picture of the natural world. But unlike Christian creationists, especially the young-earth creationists, Vedic creationists assume a time scale for evolution stretching into millions, billions, or even trillions of years, with cyclical destruction and rebirth, all in keeping with a Vedic view of the world, derived largely from the *Bhagavad Gita*.

Vedic creationism is not new. The idea has been around since 1895 when Vivekananda, following Herbert Spencer, linked the spiritual progress made by a yogi to the evolution of life forms "from a fungus . . . to a plant, then an animal, then man and ultimately God" (quoted here from Killingley 1998, 151). The linking of yoga with evolution was not merely metaphorical. Vivekananda suggested that spiritual merits obtained through yoga actually allow the yogi to enter a new body of another life form, whether through rebirth or through a change of shape in his present life. In his customary manner, Vivekananda claimed that "the idea of evolution was to be found in the Vedas long before the Christian era; but until Darwin said it was true, it was regarded as a mere Hindu superstition" (quoted from ibid., 154). With breathtaking deftness, Vivekananda succeeded in pinning one more feather in the Hindu cap, while condemning the West for its ethnocentricity at the same time. Their master's views are still echoed, almost verbatim, in countless Ramakrishna Mission publications.

Vivekananda, however, is not the inspiration behind the contemporary school of Vedic creationism. The chief inspiration is Swami Prabhupada, the founder of the Hare Krishna movement, the street name for the International Society of Krishna Consciousness (ISKCON). The authors of the controversial and massive book *Forbidden Archeology* (1993), Michael Cremo and Richard Thompson, are the scientific talent behind the California-based Bhakti Vedānta Institute, the intellectual center of ISKCON.

The agenda of *Forbidden Archeology* is to debunk the existing paleo-anthropo-logical findings which place the emergence of anatomically modern hominoids around 100,000 years ago. The testable hypothesis Vedic creationists derive from Vedic cosmology is that human beings, as a distinct species, have existed along with all other animal and plant species since the beginning of this universe, bil-lions if not trillions of years ago. According to Vedic cosmology, in the beginning was Brahman, the Absolute Spirit, which projected itself into various animal and plant forms, depending upon their level of self-awareness and capacity of spiritual advancement, ending with human beings who had the capacity to realize Brah-man. From then on, any species could directly transmigrate into any other species, depending upon their accumulated karma. For this evolutionary scheme whose telos is spiritual advancement, human beings had to exist from the very beginning for they were the culmination of spiritual advancement of lower forms that, through their qualities, could be reborn as human beings. Vedic creationism is a kind of spiritual Lamarckism which allows the spiritual efforts of creatures to improve themselves into higher life forms, where higher is defined in terms of capacity for spiritual enlightenment.[10]

Not surprisingly, Vedic creationists are keen to reinterpret the existing fossil records to make them fit with their metaphysical assumptions. The first step is to cast doubt on the validity of the existing fossil record. Here, they find the work of the sociology of science useful. "We are not sociologists," Cremo and Thompson write, "but our approach in some ways resembles that taken by practitioners of sociology of scientific knowledge (SSK), such as Steve Woolgar, Trevor Pinch, Michael Mulkay, Harry Collins, Bruno Latour and Michael Lynch . . . who would all agree with the following programmatic statement. Scientists' conclusions do not identically correspond to states and processes of an objective natural reality. Instead their conclusions reflect the real social processes . . . as much as, more than, or even rather than what goes on in nature" (1993, xxv). If what we take as scientific does not tell us much about nature but about the scientists' interests, there is no reason why the same data cannot tell another story, based upon dif-ferent interests.

Cremo and Thompson claim, in effect, that the rejection of the ancient origins of Homo sapiens by the mainstream of paleontology is a social construction. They believe that Darwinist paleontologists have selectively applied very strict standards of evidence in order to discard those fossil data that show a much ear-lier origins of modern humans. They have, perhaps unconsciously, created a "knowledge filter" through which only the data favorable to a naturalistic, Dar-winian evolution can pass through. Hence their book advertises itself as expos-ing a scientific cover up.[11]

Why would paleontologists want to cover up evidence that contradicts their pet theories? Like some among Christian creationists, Vedic creationists want to establish that evolutionary theory is no different from a religion. Scientists are led by a blind faith in the naturalistic mechanisms of evolution. If scientists act like any entrenched religious community, then religious communities have an equal right to interpret the existing data to support their own worldview. This seems to be the driving force behind *Forbidden Archeology*.

The cases of Vedic physics and Vedic creationism boil down to demanding equal opportunity for different social interests and different cultures to construct science according to their own assumptions. There is an underlying assumption that all metaphysical views are equally valid, even after close to four centuries of the tremendous advancement of learning, led by developments of science. Vedic cosmology is seen as a live option for explaining the workings of nature. It is as if the facts of biology have no implication at all for the metaphysical understanding of nature preached by Hinduism.

But, then, to protect the ontological core of Vedic Hinduism from refutation is the entire purpose of this enterprise of finding modern science in the Vedas. In that regard at least, the projects of Vedic physics and Vedic creationism have been quite successful—at least for now.

### Conclusion

There is a deep irony in declaring the rationality found in the three-millennia old Vedic corpus to be at par with how today's natural scientists go about forming and testing hypotheses. If there is one thing that is distinctive about modern science it is that it has learnt to take refutations seriously. Notwithstanding the social interests that promote conformity with the ruling paradigms, and notwithstanding the personal investment of individual scientists in their pet theories, modern science owes its phenomenal success to the institutionalization of skepticism. Paradigms *do* change; old theories and old explanations *are* thrown overboard, however reluctantly and belatedly, when confronted with better evidence, simpler theories, and more comprehensive and consilient explanations.

One would think that those who are truly interested in finding the seeds and roots of scientific inquiry in their inherited traditions would want to defend *those* elements of the tradition which allowed some room for contradictions and which took the evidence of ordinary human senses seriously. One would also think that those who are truly interested in the welfare of Indian people, and not just in flying the flag of Hindu greatness, would want to take the obvious contradictions that exist between the supernaturalistic understanding of the working of material nature that the Vedas teach and the naturalist ontology of modern science. Those who truly want to embrace science, in other words, would *want* to expose their heritage to what we have learnt about the workings of the world, and what we have learnt about how to learn about the workings of the world, through the growth of scientific knowledge since the Scientific Revolution.

What we find instead is a hybrid object—"Vedic science"—which beats into place ancient Sanskrit verses alongside and/or into mathematical formulae of modern mathematics, physics, and cosmology. Instead of throwing away at least some pieces of the old jigsaw puzzle—now that better, more clearly defined pieces, offering a bigger, more detailed picture are available—Vedic science wants to hang on to all the old pieces. The old pieces, the defenders of ancient traditions claim, are "equivalent to," "bear a likeness to," or are "another name for" the new pieces. When seen from within an organic, non-dualist frame, the old can substitute for the new. Nothing is refuted, nothing given up. There is no housecleaning. The new is fitted into the old. The old carries on, now dressed in new clothes.

At various points of our analysis, we have also got a taste of how Vedic science proponents have found help from social and cultural constructivist and postcolonial critics of modern science. Each one of the three prongs of the Vedic science project—a critique of dualist science, the idea that standards of rationality are internal to cultures, and that the rationality of modern science is as socially embedded and culturally constructed as that of any other knowledge system—is a part of the central dogma of contemporary science studies, women's studies, and postcolonial studies in institutions of higher learning in the West and in many parts of the postcolonial world as well. The idea that there is nothing special about modern science that premodern, non-Western sciences need to learn from, and that what counts as reasonable and real varies with the cultural context, has become a part of the common sense of the postmodern academia. Defenders of Vedic science count upon this widespread and diffused attitude of cultural relativism to garner sympathy for their position.

In the next part of the book, we examine these postmodernist views of science in greater detail.

*Part II*

# Postmodern Critiques of Science

*Five*

# Epistemic Charity

*Equality of All "Ethnosciences"*

> *I for one consider it unthinkable to claim that a Piaroa*
> *of the Venezuelan rain forest is irrational when he says*
> *that rain is the urine of the deity, Ofo Da`a. The*
> *Westerner asserting that rain is $H_2O$ and the Piaroa*
> *saying that it is the urine of the deity are doing so on*
> *similar grounds. . . . What is more, the one truth does*
> *not go against the truth of the other.*
> —Joanna Overing, Reason and Morality

> *Give [us] anthropologists a culture, and we will show*
> *you how utterly science and its laboratories are*
> *tangled in it.*
> —Emily Martin, "Anthropology and Cultural Study of Science"

> *All cows are grey again . . . everything is like science*
> *and science is like everything.*
> —Ernest Gellner, "The Paradox in Paradigms"

They say it is impolite to look a gift horse in the mouth. It is indeed doubly impolite if the gift was meant as a token of respect and solidarity. Yet it is precisely the task of examining a well-intended gift that I have taken upon myself in this chapter.

The gift I intend to examine is a cluster of theories that forbids evaluation of the truth or falsity of popular beliefs, especially in non-Western cultures, from the vantage point of what is scientifically known about the world. Conversely, these theories allow insiders to reject as ethnocentric and imperialistic any truth claim that does not use locally accepted metaphysical categories and rules of justification. The theories in question reject the traditional philosophical ideal of universal standards of scientific rationality, objectivity, and truth in favor of a perspectival approach that grounds these concepts, in the end, in local and particular social and cultural circumstances. All sciences are "ethnosciences," for all are equally embedded in their own cultural context. It is not as if "they" have myths and "we" have science, but our science is not devoid of myths, and their myths are not devoid of facts.

As Ernest Geller put it more colorfully, "all cows are grey . . . everything is like science and science is like everything."

This attitude of charity and open-mindedness toward all ways of knowing nature is what I call epistemic charity. I use the word "charity" advisedly, as I want to convey the connotations of condescension and leniency toward the other. Those who argue for equality of all ways of knowing, however, have noble motives. They want to celebrate the difference of non-Western, non-masculine, non-modern, non-capitalist cultures around the world. The fascination with different sciences is fueled by a belief that a recovery of these alternative ways of comprehending the world will enable modern science to recover more humane, peaceable, and ecological values presumably exemplified by women, non-Western, and non-modern cultures.

Epistemic charity comes in many different colors. The full range includes the bland and hyper-scientistic "Strong Programme" of the sociology of scientific knowledge (or SSK), the red-hued radical science programs seeking "standpoint epistemologies" of women, non-European and other subaltern social groups, the green-tinted ecologists seeking a whole different way of conceptualizing the unity of humans and nature, and lately, the bold saffron-colored program seeking to establish scientific credentials of the sacred literature of the Vedas. All these programs fall under the rubric of "social constructivism" which, when applied to natural science, claims that scientific knowledge is not autonomous or based upon universal principles of rationality, but rather, tied directly to social interests and conditions. Social constructivism in science holds the following axioms:

1. Every aspect of that complex set of enterprises we call science, its content and results, is shaped by and can be understood only in its local historical and cultural context.
2. In particular, the products of scientific inquiry, the so-called laws of nature, must always be viewed as social constructions. Their validity depends upon the consensus of "experts" in just the same way that the legitimacy of a pope depends upon a council of cardinals.
3. The standards of evaluation of evidence are relative to a culture's assumptions about nature and its relationship to the human mind. These assumptions vary with gender, race/caste, and class in all cultures. Science is therefore "politics by other means" (for an extended list, see Koertge 1998, 3–4).

In the first chapter I described the changes in the social and cultural context in the West when, starting around the late 1960s, modern science came to be seen as a part of the problem, rather than as a solution. The nuclear bombs on Hiroshima and Nagasaki, the war in Vietnam, the movements for civil rights, women's rights, and the anti-colonial struggle of Third World peoples, all contributed to a radical critique of the role science and technology had played in propping up relations of domination. I described how radical science movements gradually moved away from critiques of *uses* of science for war and profit to the critique of the *content* and the instrumental reasoning of science. I touched upon how the Western Marxist critiques of scientism of Marxism and Leninism came to be generalized to all of science, and how these critiques coalesced with the work of Thomas Kuhn. I placed the origins of social constructivism in this social-cultural context. I also showed the appeal of this view of science for Third World intellectuals who sought to recover indigenous knowledge traditions from the worldwide hegemony of modern science. This Third Worldism flowered into "postcolonial studies" which marries a postmodern suspicion of science and modernity with cultural nationalism. Postcolonial studies try to eschew the ahistorical essentialism of old-fashioned nationalists, but still remain caught in the trap of "difference."

In this chapter, I move from the *context* of the social constructivist turn in science studies to the *content* of the main constructivist programs. I will examine the core arguments of the "Strong Programme" and the cluster of programs that it has inspired, including cultural studies or the anthropology of science, feminist epistemologies, and postcolonial science studies. In all cases, I will focus on the arguments for parity between different ways of knowing, or in some cases, the putative superiority of non-Western and feminist science. Many critics of postmodernism, especially those with historicist tendencies, are satisfied with studying the social context that gave rise to the widespread suspicion of science and disillusionment with promises of modernity. I believe that understanding the social context of social constructivism is necessary but not sufficient. In order to defeat the challenge these theories pose to the project of Enlightenment in non-Western societies, it is necessary to understand—and defeat—the relativist logic of these theories.

My second objective will be to show that the denial of the objectivity and universality of science has political consequences. What looks like a tolerant, non-judgmental, "permission to be different" is in fact an act of condescension toward non-Western cultures. It denies them the capacity and the need for a reasoned modification of inherited cosmologies in the light of better evidence made available by the methods of modern science. It enjoins non-Western societies to give up their struggle for secularization and Enlightenment and to make peace with the limits their cultural heritage places on their imaginations and their freedoms. Moreover, the injunction to prefer cultural authenticity over truth, or at least, to consider authenticity as a determinant of truth, plays into the hands of religious and cultural nationalists who are sowing the seeds of reactionary modernity.

Thus, it is with the clear intention of returning the well-intended but dangerous "gift" of equality of all sciences that I explore, below, the family of social constructivist theories of science.

### The Strong Programme: I'm OK, You're OK

Before the "Strong Programme" (SP) of the sociology of scientific knowledge made its appearance in the 1970s, the field was dominated by the supposedly "weak program" of the sociology of science, largely associated with the work of Robert Merton and his students. This fairly conventional sociology of science concerned itself with questions about institutional factors (e.g., professional norms, reward structure) which promoted or inhibited the growth of objective knowledge in natural sciences. Differences in the institutionalization of the professional norms of universalism, communism, disinterestedness, and skepticism were examined to the extent they served to insulate the methodologies, theories, and interpretation of evidence from social interests and cultural biases.

The Strong Programme in the sociology of science arose explicitly as a rejection of the Mertonian sociology of science. Institutional sociology was considered "weak" because it exempted the technical content of science from sociological explanation. The manifesto of the Strong Programme, David Bloor's *Knowledge and Social Imagery*, first published in 1976, with a second edition in 1991, opens with a daring question: "Can sociology of knowledge investigate and explain the *very content and nature* of scientific knowledge?" (Bloor 1991, 1, emphasis added). Accusing all previous generations of sociologists of knowledge, from Karl Mannheim to Robert Merton, who exempted the content of science from sociological analysis and concentrated on studying the conditions of the production of science, of "lack of nerve and will," Bloor announced a new agenda, the muscular-sounding "Strong Programme."

Bloor's book was the manifesto of a group of sociologists based in Edinburgh who produced some of the founding ideas of the new post-Mertonian sociology of scientific knowledge. This group included David Bloor, Barry Barnes, David Edge, and Donald MacKenzie. Harry Collins led an affiliated program at Bath. Together these scholars developed the new sociology of science which would be, according to Bloor (1991, 7):

1. Causal, i.e. concern itself with conditions which bring about beliefs or states of knowledge.
2. Impartial with respect to truth and falsity, rationality and irrationality, success or failure (of the belief in question).
3. Symmetrical in its style of explanation. The same type of causes would explain, say, true and false beliefs.
4. Reflexive in applying the same patterns of explanations that apply to science, to the social studies of science.

These tenets, when used to study different sciences, make the Strong Programme profoundly right, and at the same time, disastrously wrong. This program is profoundly right insofar as it endorses the rational unity of humankind, but disastrously wrong when it denies the universality of modern science. It holds that all sciences are alike in that they are all equally embedded in their local cultures. The rational unity of humankind, ironically, lies in the disunity of their sciences.

Following the demand of symmetry, sociology of scientific knowledge scholars deny the modern science of nature the status of rationally justified beliefs. Modern science is assigned the same status as any collectively held belief in any tribe, in any society. If sociologists and anthropologists look for social causes for explaining why any tribe anywhere holds apparently irrational beliefs, they will feel free to find local, contingent social causes for explaining why the tribe of scientists hold the beliefs they hold. There will be no asymmetry in the treatment of science and any other collectively held beliefs.

In a twist that can only sound ironic in the context of Vedic monism claiming the status of certified "science," what makes the Strong Programme "strong" is—what else?!—its purported "monism." In a much-cited essay Bloor wrote with Barnes, his colleague at the University of Edinburgh, the Strong Programme was declared special because it "stressed the essential identity of things that others would hold separate." Whereas the rationalists clung to such dualisms as "true and false, rational and irrational beliefs," the Strong Programme would not respect these distinctions (Barnes and Bloor 1982, 25). Instead it would be "symmetrical in its style of explanation. The same type of causes would explain say, true and false beliefs." This is the famous "symmetry" principle, the third axiom of the Strong Programme.

A concrete example may help understand what symmetry is all about. A well-known social anthropologist, Frederique Marglin, has argued that the widespread Hindu belief that smallpox is caused by the anger of Goddess Shitala is perfectly rational within the context of their "non-logocentric" metaphysical beliefs. These beliefs do not draw clear boundaries between human beings, animate, and inert elements of nature, but rather consider all of them as different forms of the spirit or consciousness. In this view, because the spirit is embedded in matter, and matter in spirit, a prayer is no different from rational efficacious action in the material realm. Marglin has gone on to declare that the introduction of the modern cowpox-based vaccine by the British, and later the Indian state, was an affront to the local practice of variolation which was embedded in a ritual prayer to the goddess of smallpox. Even though the traditional practice of disease control was more likely to cause the disease rather than cure it, Marglin upholds it as rational in the context of non-dualist assumptions of the Indic civilization (Marglin and Marglin 1990; Marglin and Marglin 1996).

If confronted with this case—and the Strong Programme gives many examples of similar cases—the proponents of symmetry will not claim that the goddess theory of smallpox is as true as the modern virus theory of smallpox. What they do assert is that those who believe in the goddess theory do so for no less rational reasons than those who hold the virus theory. And those who believe in the virus theory do it for no less social reasons. It is not as if the goddess theory is just blind faith, the lore of ancestors passed on, without any contact with nature or without any attempt to explain and control the disease. And contrarily, it is not as if the virus theory is just pure observation of nature in itself, without any grounding in the lore of ancestors, or connections with the rest of the society. People in both knowledge systems are equally close to nature *and* equally embedded in society.

In what sense is this charitable view of equality profoundly right? It is right because it resolutely turns its back on the notions of non-Western or non-modern peoples as "pre-logical" or mystical or in any way less rational in their interaction with nature. All people, everywhere, interact with nature with an interest in explaining and controlling nature. All people, everywhere, bring their culturally transmitted, socially embedded conceptual schemes as they interact with nature. Explaining why they believe what they believe as facts of nature (e.g., the smallpox goddess visits those who have angered her, versus the smallpox virus spreads by contact), will require an equal attention to both nature and society in both cases, regardless of the truth and falsity of the belief.

This is indeed a liberating insight, for it brings folk beliefs to the same level as scientifically attested beliefs. One cannot any longer say that the goddess worship has no cognitive component and is only part of the mystical lifestyle of India's villagers. The two have broadly similar capacities, broadly similar interests and purposes in acquiring knowledge. The two are not apples and oranges, but only different varieties of apples and can be compared against each other.

But the sociologist giveth and the sociologist taketh away. From the perfectly sensible reading of symmetry described above—that is, that people in both knowledge systems interact with nature, while also participating in their societies' rituals and meaning systems—the Strong Programme concludes that people in all societies comprehend nature through *equally* socially embedded and culturally approved categories. The same external nature that all societies share, takes on infinitely different representations, *and there is no way of telling which is fact and which fiction.* What turns a representation of nature into a "fact" or a "law" of nature, in all cases, is nothing more and nothing less than following the community's standards for gathering and evaluating the evidence of our senses and the commands of reason. As long as scientists in different cultures and subcultures are applying, "carefully and honestly," the community-sanctioned rules of classification and evaluation, they are being rational and the question of "incorrectness" or irrationality cannot even be raised (Barnes 1991, 327). The point of all this symmetry is to insist that modern science is in the same boat as any other knowledge system; it is no more differentiated from the cultural common sense of Western societies than sciences of other societies are from theirs.

This position is forcefully argued by Barnes and Bloor by drawing parallels between the Strong Programme-brand of relativism and the situation in two primitive tribes, T1 and T2. They believe that when confronted with an alien culture, members of T1 and T2 will "typically" and "naturally" prefer their own familiar and accepted beliefs and their local culture will furnish them the norms and standards to defend such preferences, should the need for defense arise. In this situation, the words "true" and "false" are simply idioms for expressing cultural preferences of the community; different cultures will declare different statements about the state of affairs to be true or false, and they will all have acted rationally. Barnes and Bloor urge a similar methodological relativism on the sociologist of science who is trying to assess the beliefs of the tribe of modern scientists:

> A relativist [ought to] accept that [all] preferences and evaluations are as con-
> text-bound as those of the tribes T1 and T2. He accepts that none of the justifi-
> cation of . . . preferences can be formulated in absolute or context-independent
> terms. In the last analysis, he acknowledges that all justifications stop at some
> principle or alleged matter of fact that has only local credibility. For the rela-
> tivist there is no sense attached to the idea that some standards and beliefs are
> really rational as distinct from merely locally accepted as such . . . the words
> "true" and "false" provide the idiom in which these [context-bound] evaluations
> are expressed . . . (Barnes and Bloor 1982, 27)

This view of rationality as obeying community standards automatically grants rationality to all communities. As long as community members hold beliefs that are in accord with the community's view of what is real and reasonable, they are being rational by definition. But this is a very weak or minimal view of rationality ("minirat," as Newton-Smith calls it), because in Newton-Smith's (1981, 241) words, "it does not include a normative assessment of the goal or an evaluation of the truth, falsity, reasonableness or unreasonableness of the beliefs."[1] All that minirat demands is that the community in question prefers and acts upon those beliefs which appear most likely to achieve their goal, regardless of the objective truth or falsity of the beliefs in question. (Thus, in the small pox example above, prayer is rational action to ward off disease.) For the Strong Programme sociologists, to worry about objective truth or falsity in assessing the standards of rationality of a community is to be asymmetrical.

Given the history of colonialism and racism, there are plenty of liberal White guilty consciences on the one side, and plenty of anti-imperialist thin-skins on the other side to take comfort in this epistemological egalitarianism. The even-handed ascription of rationality serves as a poultice for both the victors and the vanquished. It feels good to level the purported superiority of modern science and to revel in the amazing variety of beliefs around the world. It feels radical, also, to turn the tables on the West by challenging "its" science with that of the victims of the West's colonialism.

But this political correctness is an unintended effect of the Strong Programme. The primary goal of the Strong Programme was not political at all. Instead, it set out to oppose philosophical rationalism, empiricism, and realism with a new account of experience and reason in which social biases and cultural meanings were treated as essential ingredients rather than contaminants to be removed. Indeed, its lack of political fervor has come under fire from more politically motivated critics who find the Strong Programme to be too scientistic and not sufficiently critical of "modern science as a social problem" (Restivo 1988), and too concerned with micro-processes in the lab rather than the larger issues of gender, race, and class (Rose 1996). But ideas often overreach their creators' intentions. By treating standards of rationality as social conventions, the Strong Programme opened the way for social conventions of all cultures and subcultures to vie for the status of science.

The problem with this reasoning is that while all people crave recognition of their heritage and culture, they *also* crave truth, or at least they crave a sense of progressing toward a better understanding of the world. While a sociologist or an external observer of another culture can bracket the validity of the accepted/institutionalized beliefs, those who live in that (or any) culture do not, and can not. When people in any culture ask questions, or when they manipulate their environment, they do not seek an affirmation of institutionalized beliefs. Rather, they seek to know the real state of affairs. Even such cruel practices as torture in medieval Europe, and such irrational practices as magic and divination were, and still are, used to uncover the truth about the real state of affairs (Goldman 1999, 32). Given the accumulated anthropological data, it is hard to disagree with Alvin Goldman's conclusion that "truth is a vital concern of humankind across history and culture, not an idiosyncratic concern of modern white Europeans. Despite the heterogeneity of truth-pursuing practices. . . . A single concept of truth seems to be cross-culturally present" (ibid., 33).

Because all cultures, everywhere, seek a growth in their stock of true beliefs, and not just accepted beliefs, it follows that they also have an interest in learning and institutionalizing those rules and methods of learning which increase the chances of acquiring true(r) beliefs. This means that it is in their interest to take reasonable steps to acquire the relevant evidence and to evaluate the evidence satisfactorily.

In the pre-postmodern philosophy of science, modern science was preferred over other sciences precisely for this reason. For a variety of contingent historical reasons, Western European societies had succeeded in institutionalizing certain methods of theory selection and theory-testing which, over the long term, could pick out theories of higher reliability. That is, acting on beliefs acquired through the winnowing processes of modern science better enabled us to predict, control, and manipulate nature. Although there are a variety of answers to why the scientific process makes beliefs more reliable, it was generally accepted that processes of science (controlled tests, double-blind tests, for example, aided with precise and repeatable observations) enabled the independent causal structures of nature to push back against our own hypotheses, aiding self-correction over the long term. This more discriminating monitoring of empirical evidence, under a more controlled manipulation of causal structures of reality expressly designed to test the influence of biases and expectations on experimental results, created a context in which reasons for holding beliefs could be reasons for everyone, regardless of his/her own inclination, social position, ideology, and so on. The hallmark of science, in other words, was considered a higher degree of monitoring and self-correction of our initial, culturally embedded, and interest-inflected hypotheses. If rationality is the application of reasons to tasks, then science was legitimately considered the epitome of rationality, as it had learned to apply reason to the application of reason to tasks, that is, it had learned to pick out those rules and procedures which disciplined what kind of inferences and causal connections could be legitimately made between different components of nature.

With the Strong Programme, science is deprived of this comparative cognitive superiority which is ordinarily supposed to give it its universal appeal. In the rush for even-handedness, the Strong Programme declared modern science to be no closer to nature—that is, no more truthful an account of nature—than any other science. Science is as much "a move within a game" (Barnes and Bloor 1982, 30), a game whose rules are ultimately as arbitrary and as relative to the ruling paradigm, as other modes of belief formation are to their own cultures. Just as a culture socializes and constructs a worldview for its inhabitants, so does a paradigm socialize and acculturate scientists, with the result that in modern science, as in premodern knowledge systems, "justification always stops at some principle or alleged matter of fact that only has local credibility" (ibid., 27).

In the end, the Strong Programme accomplishes this leveling by self-consciously downgrading the input from the theory-independent real world and by equally self-consciously upgrading the creative force of culture. External reality, Barnes and Bloor have argued in their numerous publications expounding and defending their program, is present in all sciences of all cultures, but only as a silent partner who is indifferent to how it is described. Reality, Barnes writes (1991, 331), "will tolerate alternative descriptions without protest. We may say what we will of it, and it will not disagree." (When a child dies because prayers to the goddesses of smallpox or AIDS or a myriad such diseases do not work, is this not reality speaking, loudly and clearly?) Bloor is even more explicit and announces that reality simply drops out of the consideration of how different cultures come to accept some beliefs as facts of nature; because those who get it right and those who don't both live in the same natural world and interact with the same reality, "there is a sense in which the electron itself [or the extra-theoretical nature, more generally] drops out of the story because it is a common factor behind two different responses and it is the causes of difference that interest us" (Bloor 1999, 93). The possibility that even though different cultures interact with the same brute reality, yet the details of the processes, social institutions, technologies of interaction may be very different, is set aside by the Strong Programme. To believe that some interactions could lead to a less biased and a more truthful picture of reality than other kinds of interactions is seen as offending the principle of symmetry.

Reality is allowed to drop out of the picture because, following a radically relativist interpretation of Kuhn, the Strong Programme sociologists believe that it is the paradigm or the culture that channels all the sensory and experiential inputs from reality into concepts. No input from brute nature is capable of self-correcting the existing fund of knowledge because all inputs are channeled along different pathways by existing social convictions. Thus, all new experiences can be described as consistent with any extant body of ancestral knowledge. Ultimately, it is the existing local conventions where the definition of reality and justification of beliefs stop.

The proponents of the Strong Programme did not set out to debunk science. In fact, they saw their demand for symmetry as imitating scientific neutrality and objectivity. But the logic of their arguments cannot but lead to a skepticism about *all* sciences: If scientific results are ultimately justified by social conventions, there is no factual or rational reason to prefer one over the other. This is obviously a disabling stance even for those politically engaged movements that want to create a better science. Yet some in science studies find the assignment of equal localness to all sciences, and equal rationality to all local ways of knowing, as politically empowering. The argument (Hess 1997, 161) is that seeing all ways of knowing as scientific in their own social context will enable those with lower credibility to level the playing field. Depriving modern science of special privilege, in other words, can be used rhetorically to put forth claims that are prima facie irrational.

But this argument assumes that those who lack credibility *also* lack political power. That is not always the case. As our study of Vedic science has shown, those who want to claim the status of science for beliefs lacking in credibility may actually belong to dominating social groups, and may well use the label of science to further a dangerous political agenda. We have examined in the last chapter how Hindutva ideologues justify the scientificity of the "methodology" of occult arts using the argument of equal localness of all knowledges.

The connection between the thesis of epistemic parity and Hindutva and indeed, Hinduism itself, runs deeper still. The inclusivism and eclecticism of Hinduism is well known; we saw one example of it in the so-called "pizza effect" (chapter 3). When confronted with heterodox ideas, which may actually challenge its own world–conception, Hinduism responds by refusing to recognize their difference. Instead, it claims that the other teaching is more or less equivalent to its own teachings: the other is claimed to be an aspect of, approach to, or aberration from the truth contained in its own doctrine. That is, the Hindu tradition evades the threat posed by foreign ideas by simply claiming them for itself, including them in its own hierarchy of truths, as always-already there as flawed expressions of the ultimate truth of Vedānta. This penchant for "claiming for one's own religion what really belongs to an alien tradition" has been described well by Paul Hacker (1995) and elaborated by Wilhelm Halbfass (1988), both highly respected Indologists.

This inclusivism is mistaken by Hindus generally, and touted by Hindu chauvinists, as an example of Hindu "tolerance" and "open-mindedness." It is true that Hinduism does not proscribe differences of doctrine, and does not clamp down upon heterodoxies. But that in itself does not amount to "tolerance," for tolerance requires that differences are accepted in their difference and still accepted as legitimate points of view, and even used to correct one's own errors. Hindu inclusiveness only tolerates what it can reduce to its own likeness and put in pre-existing slots legitimated by its own overall conception of the world. In the guise of tolerating the heterodoxy, Hindu inclusiveness quietly absorbs the dissent into itself, in the process claiming the virtue of the new ideas for itself, while overlooking and denying the contradictions and challenges it poses. Such inclusiveness is tolerant only in name. It has, in fact, served as a means of self-universalization of Hindu doctrines, as they claim to find everything from modern science to modern political virtues like tolerance, secularism, and democracy in the ancient teachings of the Vedas. In this surreal world, anything can be connected to anything, anything can become anything, and everything can be declared "Vedic." This does make it easier for Hinduism, as compared to more doctrinal religions, to accept foreign ideas into the religious tradition. But the underside is that foreign ideas are accepted only in name, while they are understood as extensions of the tradition. In the name of "Hindu modernity," hoary old traditions get institutionalized.

The social constructivist view of knowledge is a philosophy tailor-made for this kind of "tolerance" for it allows the unlike to be treated as like, while disregarding the contradictions between them. If all truth-claims, in the final instance, rest on social conventions, then one can substitute for the other and neither has the right to challenge the other. If what we take to be objective truth affirmed by modern science is a product of Judeo-Christian metaphysics and Western interests, then Vedic science is also a legitimate candidate for producing its own objective truths about nature. Because Vedic metaphysics does not separate matter, man, or society from God, the objective truths acquired through the Vedic method in fact encompass the lower truths of "mere matter" of modern science. The process of inclusion is complete, the pizza has been equated with humble peasant bread.

These kinds of arguments have long been the staple of Hindu apologetics. With the popularity of social constructivism among secular intellectuals, these arguments have found new respectability. What is worse, with the popularity of social constructivism among secular intellectuals, there are no opposing voices left in the public sphere that can insist upon respecting the distinctions between myth and science. If anything, it is the secular intellectuals who have led the charge in India for finding salvation in local knowledges, regardless of their actual validity.

### Cyborgs in the Citadel: Science as Culture

Once upon a time, so goes the origin story of the cultural studies of science, in the bad old days of modernism, there was something called "science" and something else called "society"—the two were supposed to interact, as two neighbors do, across a fence. In the bad old days, there was something called "nature" and something else called "culture"—and again, the two were supposed to interact, as two neighbors do, across a fence. Indeed, in those bad old days, there were "human beings" and there were "machines"—and the two were also supposed to interact, as two neighbors do, across a fence. The modernists often quarreled about the nature of these fences. The old-fashioned empiricists and realists insisted that a strong leak-proof fence existed between science and society, nature and culture, men and machines; good fences, they said, made good neighbors. The social constructivists took the contrary position and denied that the fence was leak-proof enough to stop the one from shaping and making the other. But both sides respected the existence of the fences, and all agreed that what is inside and what is outside the fence must be understood as ontologically distinct, as two neighbors with their own needs and goals.

But the parable for cultural studies continues, we live in postmodern times now. We—the successors of Kuhn, the Strong Programme, Foucault, feminism, postcolonialism—don't respect fences. Whereas our modernist forefathers saw such binaries as "science" and "society," "nature" and "culture," "humans" and "machines" as distinct entities, we see these pairs as made up of each other. Whereas our modernist predecessors respected fences as natural distinctions, we see that all fences are social conventions, put there to keep out whatever disturbs the existing balance of power and/or the existing pattern of meanings. We question all dualisms between context and content. We don't see the social context as an external influence on science but as a constituent part of all aspects of science. All aspects of science—what scientists take as "nature," what projects they think worthy of engagement, what experimental results they pay attention to, what they accept as justified to believe—all become manifest only through the significance their society's culture endows them with. Science *is* culture. (See programmatic statements by Rouse 1996, chapter 9; Hess 1997; and Martin 1998.) The theoretical inspiration for this culturalist turn in science studies comes largely from the work of Donna Haraway, an American feminist and Bruno Latour, a French social theorist.

Science, in the culturalist account, is not a citadel separate from society, glorying in its well-guarded moat that keeps out all "polluting" cultural values. Instead, like the rest of the human condition, science is a cyborg—part nature, part culture; part reason, part myth.[2] Culture is not a pollutant to be isolated, turned into a testable hypothesis and removed if false. But it is the condition for doing science. Reality and reason only become manifest against the background of cultural meanings, and find expression only through the metaphors and imagery of culture. Scientists are, after all, not that different from those outside the citadel of their laboratories. "What if," Emily Martin a well-known cyborg-anthropologist, asks rhetorically, "important forceful processes flow into science as well as out of it? What if nature is not simply what natural scientists tell us it is? What if instead people who call themselves scientists are continuously interacting with, and being profoundly affected by, people who do not call themselves scientists?" (1998, 28). The boundary between scientists and others is also conventional. Science has to be studied not as a culture of no culture, but as a culture of the culture of the people in the streets.

Irony of ironies, this Every man/Every woman anthropology stands shoulder-to-shoulder with the Hindu chauvinists we met in the last four chapters. Like the proponents of Vedic science, cyborg anthropologists are bricoleurs, or pastiche-makers in their epistemology, putting subjective meanings and cultural preservation ahead of objective validity of beliefs. And like the proponents of Vedic science, they are substance-monist in their ontology, replacing the Absolute Spirit, atman, with the equally ineffable Collective Spirit of culture. Nature is enchanted again, with the meanings and agency imputed to it by culture. (It is not clear who is borrowing from whom, although I am inclined to give priority to the nationalists of all stripes. After all, there is much in the agenda of showing the "entanglement" of science and its laboratories in culture [see Emily Martin's dare at the beginning of this chapter] that Oswald Spengler shared.)

Cyborg anthropologists see science as the work of bricoleurs and totem-makers. Bricoleurs are jacks-of-all-trades, who take whatever material is easily available and fit it together into a structure. Claude Levi-Strauss brought this concept into anthropology with his *The Savage Mind*. He held a bricoleur to be a representative figure of the non-modern myth maker, while an engineer was representative of modern thinking. While bricoleurs work with whatever patterns that flow from their work, engineers work from first principles. Thus a bricolage can have discordant and contradictory elements, while modern thought, starting from first principles, aspires for greater coherence and an elimination of contradictions.

Such distinctions between non-modern tribes and modern scientists are looked down upon in the cultural studies literature as a mark of Eurocentrism. The consensus is that "we have never been modern," as Bruno Latour put it pithily in the title of his 1993 book. David Hess (1995), a prominent cultural anthropologist of science, contends that modern science is no less a work of bricolage which "co-produces" nature and social orders. Just like bricoleurs and totem-makers in premodern societies, modern scientists see equivalences and homologies between their social identities and what they find in nature. They make nature meaningful by mapping their social identities, experiences, and metaphors on the world of nature. Gradually, social distinctions originally read into nature by mostly elite and mostly male scientists, come to be defined as being a part of the natural order and therefore pre-ordained and natural. Social distinctions read into nature by modern scientists become "technototems" of modern societies. Social interests produce the technototems, but technototems in turn, (re)produce the existing social divisions. Thus society and nature are "co-produced;" they are chimeras, cyborgs, each constituted by the active intervention of the other.

Hess's view of scientific facts as technototems dovetails with Bruno Latour's theory of "actor networks" and "radical symmetry." Latour has emerged as a major influence on the cultural studies of science. Professional anthropologists accept Latour's view of co-construction of content and context through actor networks, but try to embed it in a more class, gender, and race-inflected cultural history. While cultural studies of science try to explain the work of cultural imagery in establishing homologies between nature and culture, Latour's actor-networks explain how such homologies construct inscriptions and statements in the laboratories, and how these inscriptions get translated into "facts" of nature. This translation apparently takes place through formation of a heterogeneous network in which "inscriptions, technical devices, and human actors (including researchers, technicians, industrialists, firms, charitable organizations, politicians) are brought together and interact with each other" (Callon 1995, 52).[3] The gist of this model is that what we accept as a "fact" (e.g., the structure of DNA is a double helix) is not a representation of nature, but only the last link in a chain that, from translation to translation, refers to other inscriptions, embodied skills, and technical devices. Those actor networks which are able to mobilize more and more of actants—inscriptions about DNA molecules, scientists, corporations, politicians—get to have their version of the state of things labeled as "factual" or "truthful." Truth and facts are the result of social negotiations.

What distinguishes Latour and cultural studies from the sociology of scientific knowledge is the demand for radical symmetry. Whereas the sociology of scientific knowledge sought social explanations for the construction of "facts" about nature, Latour insists on a more symmetrical study of how natural "facts," in turn, explain social structures. In this reading, social power and cultural metaphors construct the heterogenous actor networks which lead to scientific "facts," while scientific facts, in turn, uphold social power and cultural meanings. Science and culture co-construct each other. Thus, Darwin mapped capitalist class struggle into nature in his theory of natural selection, while social Darwinists in turn argued that capitalism is in accord with nature's way of ensuring the "survival of the fittest." White male scientists, comfortable with being at the top of the social hierarchy, were comfortable assigning all power to one "master-molecule," namely, DNA, while Black and women scientists mapped their discomfort with power onto the more dispersed control of cellular processes in the cytoplasm (see Hess 1995, chapter 2 for more details).

Modern science, in other words, like any other ethnoscience, is a "seamless fabric" weaving together culture and nature, social order, and representations of natural order. While we *think* that we modern people have learned to separate the two, we *in fact* have never been modern. We *think* we are discovering the laws of nature which transcend all cultures, while we *actually* fabricate these laws in the laboratory (Latour 1993, 31). Cultural studies hope to achieve the same kind of analysis of modern science as the anthropologists' reports of the cosmology of "savages" in which there are no fixed, pre-existing boundaries between "how people regard the heavens and their ancestors, the way they build houses and the way they grow yams . . . and the way they construct their government" (ibid., 7). As in the associative, holistic style of thought defended as the rational basis for a distinctive Hindu science (see the section on Vedic physics, in chapter 4), cultural studies see modern science as an essentially holistic, hybrid product in which all distinctions are artificial.

With this view of modern science, the entire project of science studies shifts. Explaining the content of science (to paraphrase Geertz 1973, 5), ceases to be a problem of "[social] science in search of a law, but an interpretation in search of meaning." Rather than treat social variables such as class, race, gender, nationality as independent variables that influence and shape the content of science, cultural studies of science reject the idea that there is a fixed content to be explained. Content is what the differently marked social actors, whether scientists or not, interpret the various practices of science to mean, where they draw the "boundary between what is inside science and what is outside . . . with no authority beyond what gets said by whom, when" (Rouse 1996, 255). Or as David

Hess explains, whereas sociology of science only opens the "black box of science," cultural studies open "red boxes, pink boxes, purple boxes and a rainbow of other boxes. The fundamental SSK insight that the technical is the social/political . . . is retained, but recast in a different light. Divisions among facts, methods, theories, machines and so on are seen as culturally meaningful and as interpretable in terms of locally constitutive social divisions. In short they are 'technototems'" (1997, 160).

In less chromatic words, people who inhabit different cultural worlds will find different questions more or less worthy of engagement, different methods more or less worthy of use, different evidence more or less worthy of attention, different facts more or less justified. They will map their existential meanings onto nature and the practices of science differently. Different cultures, different sciences—this is the obvious conclusion. The obvious question, are all sciences equally *justified*, is considered less important than the fact that all are equally *meaningful* to the cultures and subcultures concerned. We are back, through a circuitous route, to Feyerabend's demand that a "free society" should allow "all traditions equal rights and equal access to the centers of power . . . not because of the importance (the cash value) it has for outsiders, but because they give meaning to the lives of those who participate in them" (1978, 9). Feyerabend was courageous enough to accept the obvious conclusion that "if the taxpayers of California want their state universities to teach Voodoo, folk medicine, astrology, rain dances, then this is what the universities will have to teach" (ibid., 87).

While old-fashioned rationalists who see science as a transcultural enterprise lament this relativism, cultural studies scholars celebrate meaningfulness as a resource for the reconstruction of science for two reasons. One, if modern science is a technototem of masculine, capitalist Western culture, then alternative technototems can be created. Thus, Hess believes that it is the more empathetic, less dualistic Japanese culture that does not separate animals from humans that is responsible for advances in primatology which are similar to the feminist reconstruction of the subject. Likewise, Indians who worship monkeys as representatives of Hanuman, the monkey god, tend to study their interaction with humans rather than with each other (Hess 1995, 51). Two, culture constructivism allows the preservation and continuation of cultural meanings to take precedence over the truth or falsity of the metaphysical beliefs that underlie different cultural practices. Only those aspects of modern science that do not challenge the existing cultural meanings are accepted and fitted into the existing bricolage. This is not very different from how Vedic science gives quantum mechanics a Hindu inflection, absorbing it into the Vedic lore, and de-emphasizing or overlooking all contradictions. Cultural studies provide a sophisticated, learned justification for the priority of cultural continuity that the Hindu nationalists simply take for granted as their right.

But Hindu science advocates find still more support in the ontological monism of cyborg anthropologists who see the world of nature and the world of people as equally sentient and active. In this brave new world, bacteria "act" as much in "their own interest," as bacteriologists act in theirs. All of reality—subjects and objects—are alive and have agency and interests. To separate active thinking, subjects (human beings) from objects of scientific inquiry (whether they be machines, animals, or rocks) is seen as a part of a dangerous Western, modernist self-deception which has only served to objectify and take over the objects of scientific inquiry, namely, the rest of nature and those human beings who could be equated with nature (e.g., women, "primitive" people, etc.). We are back to the "omnijective reality" of Hindu sciences we encountered earlier, with the collective consciousness of a society replacing the Absolute Consciousness, or Being, of the Hindu worldview.

This new holism that comes dangerously close to the enchanted, spirit-saturated world of Hindu nationalists follows from the work of the two thinkers whose work has been most influential on cultural studies of science, namely, Bruno Latour and Donna Haraway. This is not to suggest that Hindu nationalists are reading Latour, or Haraway, or even their popularizers like Hess, Martin, Rouse, and others. They are not. On the contrary, it is the theoreticians who are re-discovering a non-dualist view of nature very similar to the one that Hindu nationalists find in Hinduism. When Latour announces that "we have never been modern," he is affirming the Hindu nationalist boast that "Hinduism has always been modern!" Hinduism, in fact, comes out ahead, for it openly mixes up the "being" of nature and humans, while the moderns only do it surreptitiously, all the while pretending to keep the two separate!

Let us briefly examine the arguments given in support of this new primitiveness that Latour calls "amodernism" and that Haraway likens to being a cyborg.

First thing to note is that both Latour and Haraway deny that they are advocating a premodernist holism which justifies an organist society. They are aware of the dangers of an undifferentiated whole and want to distance themselves from such thinking. Haraway's punch-line that "she would rather be a cyborg, than a goddess" is perhaps the most cited line from her over-cited "Cyborg Manifesto" (Haraway 1991, 181), and is used endlessly by her followers to "prove" that she is no romantic primitivist. Likewise, Latour is adamant that he is only "amodern" and not premodern in that he rejects the absolute interconnectedness of nature and society (the cosmopolis) that the primitive people insist upon. Notwithstanding their disclaimers, both *require* ontological parity between natural objects and people in order to explain how the two get co-constructed. Without an ontological parity, it would not be possible to explain how elements of nature and elements of human cultures can hybridize so intimately, as both Latour and Haraway think they do, in *all* sciences alike.

Latour and Haraway agree on two starting assumptions. They believe that the self-perception of contemporary scientists and philosophers of science that modern science succeeds in keeping subjects and objects, or cultural meanings and natural objects separate, is deeply deceptive and dangerous. It is deceptive because while modern scientists *think* they are discovering the laws of nature in their laboratories, they are *actually* constructing these laws out of their inherited cultural meanings. The assumption that you can, and should, separate cultural influences from the investigation of nature is dangerous because, as Haraway argues in her famous essay "Situated Knowledges," such thinking is a part of the Western tradition of logocentrism which, by denying all agency to the object of knowledge, "objectifies" it and turns it into lifeless resource for exploitation (1991, 197–198).

In Latour's and Haraway's reading of the history of science, the modern attempt to separate subjects and objects, culture and nature, are aberrations from what has been the norm for most of human history, and which is still the norm in non-Western ways of knowing. With our contemporary condition of postmodernity, Western science is finally in a position to retrieve many aspects of that norm, without fully returning to it. Latour's "amodern constitution" and Haraway's "Cyborg Manifesto" are meant as guides to the perplexed and the angry, looking for a way to end the alienation between the facts of nature and the world of emotions, values, and lived life.

What is the amodern outlook? It is, above all, a world without ontological fences. It is a world in which natural objects (say, electrons) and knowing subjects (say, Einstein) do not stand apart, each in a separate space. Rather, Einstein gets to know the electron because he recognizes his own subjective meanings in it. But the recognition goes both ways: Einstein has the subjective meanings he has at the time he is thinking of and exploring electrons because these meanings have been put there by what is known about the electron. Thus what exists are not just two kind of objects in the world, dead objects of nature on the one hand, and thinking social subjects, on the other. This is the modernist myth of a binary, dualist world. Instead, what exists at any given time, in any given society is the "Middle Kingdom" stretched over the Yawning Gap between the two poles of Pure Subject and Pure Objects. This Middle Kingdom is populated by "quasi-objects" and "quasi-subjects" that are hybrids co-produced by each other by making a dense network that extends the laboratory to the society in both directions. Quasi-objects become objects because they carry a symbolic load put there by knowing subjects. Knowing subjects are quasi-subjects, because they only develop their symbolic meanings by what the quasi-objects have put there (Latour 1993, 49–51).

Donna Haraway's famous metaphor of the cyborg, a part-human, part-machine creature—Cyborg: [n., cyb(ernetic) org(anism)], the *Webster New World Dictionary*—connotes a world very similar to Latour's Middle Kingdom inhabited by quasi-objects/subjects. "The cyborg," writes Haraway, "is our ontol-ogy . . . it is a condensed image of both imagination and material reality, the two joined centers structuring any possibility of historical transformation." Like La-tour again, Haraway invites us to experience the joy of honestly admitting our kinship with things. She presents the figure of cyborg as an "argument for *pleas-ure* in the confusion of boundaries and for *responsibility* of their construction" (1991, 150, emphases in the original). The modern worldview erected fences be-tween the realm of knowing, culture-creating human subjects on the one hand, and lifeless nature, devoid of all agency and subjectivity on the other. All that could be exploited—women, Third World people, natural resources—were de-clared to be part of nature, mere raw materials for the production of culture by the privileged nations, classes, and gender. Haraway believes that with the rise of new genetic and information technologies, the breach between humans and non-humans, between dead matter and thinking matter, is breached. Likewise, the globalization of technology has also globalized an oppositional consciousness of women and Third World peoples; no longer are they content to be assigned to the nature pole. The result is an anti-essentialist view of the world in which "nature and culture are reworked [so that], the one can no longer be a resource for ap-propriation of incorporation by the other" (1991, 151). Haraway challenges the boundaries between nature and culture in order to show that it is conventions, and political power, and not inherent essences that decide what is nature and what is culture.

The question arises, what is the nature of Latour's quasi-objects/subjects and Haraway's cyborgs? What stuff are they made of? How do the objects and sub-jects hybridize—or more classically, how does the mind know the mindless na-ture—if the two are so radically different and apart? Both Latour and Haraway answer this question by moving one step closer to an ontological monism which characterizes all premodern worldviews and which has found its most systematic expression in Vedic Hinduism. As we observed in the previous section, the Strong Programme sociologists are *epistemological monists*, that is, they demand that true and false beliefs be explained symmetrically. Actor-network theorists and their cultural studies followers are *ontological monists*, that is, they demand that nature and society be treated symmetrically. What this translates to in real terms is that people and things are semiotically equalized. The ability to act and to make moral choices, an attribute that modern humanism reserved for human beings, is redistributed symmetrically and equally over all the "actants" in the ac-tor networks. Not only people and institutions, but natural actants like electrons and bacteria have to "give their consent" in order for a statement to become a fact. In this strange world, we are asked to suspend our belief in the usual dis-tinctions between say, Pasteur and his microbes, and treat them both at par as ac-tants who chose to align themselves in a network resulting in new facts (i.e., pasteurization). We are back to an enchanted nature in which natural objects have purposes, interests, and volition. Latour seems to suggest that such attribution of agency to objects of nature opens the way for a truly symmetrical treatment of modern science and all other ethnosciences, for it enables the anthropologist to see that "premoderns are like us" in that "we" mix up humans and non-humans as much as "they" do (1993, 103). This is epistemic charity taken to new heights altogether.

This kind of ontological symmetry is purchased at a very high price indeed. The price, of course, is negating the hard-won ability of the moderns to disenchant and naturalize nature, to free themselves from the fear of cosmological disorders brought upon by their breach of moral law. Latour is fully aware of the dangers of premodern monism (see his "nonmodern Constitution," in 1993, 138–142). But he seems to believe, naively, that he can insist upon mixing up humans and non-humans and yet somehow manage to retain sufficient freedom for humans to act without the fear of disturbing the entire cosmological order. He wants the comforts of the premodern sense of being one with nature, without any of the accompanying unfreedoms that have invariably thrived in such societies in which social relations have been naturalized. I can not but agree with David Bloor's assessment of this strange new primitivism of Latour: "it is obscurantism raised to the level of a general methodological principle" (1999, 97).

Donna Haraway's cyborgs, too, end up dancing at the edge of a re-enchanted nature. Following the lead of feminist critics of science, Haraway believes that exploitation of the weak by the strong is built into the binary logic of Western philosophical tradition "that turns everything into a resource for appropriation, in which an object of knowledge is in itself only a matter for the seminal power . . . of the knower. . . . It—the world—must be objectified as thing, and not as an agent [in order to be treated], as the raw material of culture" (1991, 197–198).

Haraway suggests a "deceptively simple maneuver" to end this exploitation of natural objects and those objectified: "Situated knowledges require that the object of knowledge be pictured as an actor and as an agent, not a screen, or a ground or a resource . . . for the authorship of 'objective' knowledge" (ibid., 198). She proposes that natural sciences should adopt social and human science as a model and allow the agency of the "objects" studied in nature to transform the production of knowledge. Natural science should not be a driven by a "logic of 'discovery,' but by a power-charged social relation of 'conversation'" (ibid., 198). Haraway exhorts scientists to pay heed to the myths of the native Americans and see that "we are not in charge of the world. We just live here and try to strike up non-innocent conversations by means of our prosthetic devices, including our visualization technologies" (ibid., 199).

To recognize the limits of human knowledge in face of the complexity of nature is of course an important reminder to scientists who may let their success go to their heads. But Haraway is arguing that a respect for nature demands that we embrace a kind of non-anthropocentric anti-humanism and treat objects of nature fully at par with human needs and interests. This perspective leads us back into the enchanted worldview that Hindu sciences celebrate.

There are of course differences. What unites the subject and the object in Hinduism is the Absolute Consciousness, "Being," which exists as the inner essence of all things, from rocks and lowly worms, to humans. The quasi-objects and the cyborgs do not partake of such supernatural consciousness, even though Latour and Haraway come close to flirting with elements of process theology and the ontological phenomenology of Mahayana Buddhism.[4] The nature-culture, object-subject hybrids in cultural studies have agency attributed to them by the meaning-making human beings. Yet, objects of nature in both cases are to be understood as "omnijective" agents in their own right which exert their own will on the actor-networks of scientists, institutions, politicians, as much as these networks impose their own meanings and interest on them. This is precisely the rationale for Vedic science that gives consciousness priority over matter and allows correspondences between nature and human affairs to pass as scientific.

### Feminist Epistemology, or the Importance of Being an Underdog

Helen Longino, a well-known feminist epistemologist, described the motivation of feminist epistemology in these words: "One of the tenets of feminist research is the valorization of subjective experience" (Longino 1990, 190).[5]

Helen Longino is correct. "Valorization of subjective experience" of women is indeed the raison d'être of feminist epistemology. Feminist scholars have claimed that since the beginning of the Scientific Revolution, the subjective experiences of women—empathetic understanding, relational thinking, and practical knowledge—have been written out of the enterprise of natural science.[6] Feminist epistemology seeks to show—correctly—that the exclusion of women has impoverished modern science, and that opening the practice of science to women of all classes, colors, and nationalities is good not just for society, but good for science as well. Additionally, but more controversially, feminist epistemology seeks to show that opening the gates of the laboratories to women is necessary but not sufficient. Science has to be made not just physically but conceptually hospitable to women by considering the "sex of the knower [as] epistemologically significant" (Code 1981, 1991). This requires that women's subjective experiences—what I call "interactionist values"—be allowed to provide a new conceptual understanding of what constitutes nature and what makes for objective knowledge of nature. A feminist conception of nature and objective knowledge will allow women's subjective experiences of womanhood under patriarchy to serve as cognitive values in choosing the scientific problem, evaluating the experimental evidence, and justifying the accepted inferences. This is the promise of feminist epistemology. (See Code 1998; Longino 1999; and Harding 2000 for recent reviews of the status of feminist epistemology.)

I will argue that the cluster of interactionist values that feminist epistemologies celebrate as feminist epistemic values has profoundly conservative implications for non-Western cultures. As we shall see below, the interactionist values, whether derived from the standpoint of the underdogs (notably by Sandra Harding), or as a matter of conscious political choice (notably by Helen Longino), invariably value the community over the individual, the whole against the part, and unity of the subject with the object over their mutual differentiation and separation. The celebration of these values opens the door to a conservative Third Worldism that can aid and comfort *any* traditionalist regime. Given that an ontological and social holism is at the heart of traditional Hinduism, feminist epistemology, especially in its ecofeminist version, has found an exceptionally snug fit with Hindu nationalist views of ecology and women. Invocation of non-Western values of non-dualism, nature-culture holism and ecological consciousness may provide a vantage point from which to criticize modern "Western" science, but it has nothing whatsoever to do with "emancipation" of women, especially in non-Western contexts.

A big part of the problem is the fundamental asymmetry between facts and values that lies at the heart of feminist epistemology. *While all statements of facts about nature are seen as value-laden, social and cultural values themselves are conceptualized as cultural givens, and beyond the pale of rational criticism and reasoned change.* This asymmetry explains the radical challenge feminist and other identity-based epistemologies pose to the spirit of the Enlightenment. The project of the Enlightenment is premised on the assumption that new facts about the world and new methods of justification of beliefs will help revise inherited values and make them more conducive to human autonomy and equality. The fundamental task of philosophy was seen as the evaluation of values in the light of modern science. Feminist epistemology renounces this task.

The asymmetry between facts and values is not a regrettable oversight. Rather, it follows from the central dogma of the post-Kuhnian Strong Programme of sociology of scientific knowledge that we examined in the first section of this chapter, namely, that true beliefs in science are as much caused by the interpretation of data through our socially embedded and race-, class-, and gender-differentiated metaphors and values, as are false beliefs. Thus, Sandra Harding has insisted, our most well-confirmed sciences—our "best beliefs"—are no more exempt from race, class, and gender relations than the social beliefs and behaviors of "health profiteers, the Ku Klux Klan or rapists" (Harding 1991, 12). Because scientists bring the dominant values of a society into the laboratory, scientific facts encode these values, despite the provisions for transformative criticism that have become institutionalized through the evolution of modern science (Longino

1990; Harding 1991).[7] Because accepted "facts of nature" are constructed out of dominant social values smuggled into the justification process of science, they can only affirm and legitimize, but can never challenge the dominant social values, either inside the lab or in the society at large. Not better-warranted facts, but only more politically progressive values imported from outside the mainstream can challenge the mainstream values. In plain language, "science is politics by other means" (Harding 1991, 10), or "the technical is political" (Hess 1997, 160) and requires a political intervention for advancement.

Having accepted the social constructivist dogma, feminist critics face a dilemma. If all knowledge systems are cultural constructs, so are women's sciences, and non-Western sciences. How, then, can one argue for a *feminist* epistemology, preferably with a postcolonial flavor?

The solution to this dilemma is the well-known feminist standpoint epistemology, a feminist version of the Biblical promise that the meek shall inherit the earth. "Women of color," having suffered devaluation and oppression under the joint assault of androcentrism and Eurocentrism, now joined together in an "oppositional consciousness" to Enlightenment humanism, will develop a "successor science" that will make it possible for women to become knowers without compromising their womanliness (see especially Harding 1986, 1991; and Haraway 1991).[8]

These underdog successor sciences will be marked by all those characteristics that the conventional sciences presumably lack—receptivity to complexity and an attitude of holism, non-differentiation, or non-dualism that does not separate parts from the complex whole, nor separate mind from the matter, or the subjective (emotions, values) from the objective description of nature. Despite challenges from postmodernist feminists who question *all* claims of "better" accounts of reality (Hekman 1997), interactionism continues to be recognized as a hallmark of feminist knowledge which gives it authority as a better and/or less distorted *science* rather than just another perspective.[9]

Interactionism as a feminist virtue claims that sciences done from a feminist standpoint—that is, a standpoint that treats women's social experiences under patriarchy as relevant to warranting scientific facts—will find dynamic, interactive, non-dualist relationships between elements of nature, rather than the reductionist, and hierarchical, "master-molecule" models of control presented as "facts" by science-as-we-know it. The meek shall put together what the powerful have torn apart.

But why should the underdogs see the whole in its totality and dynamism? The arguments for interactionism are familiar. Because material life structures consciousness, women's subordinate role in the sexual division of labor—as mothers, daughters, and wives—has epistemological consequences. Women's "relationally defined existence" results in a "world-view to which dichotomies are foreign" (Harstock 1999, 120). Consequently, women as subjects do not separate themselves form the objects under their care. Their relationships with the world are necessarily less differentiated, more relational, more acutely sensuous than the merely instrumental interchange men have with nature. Rather than demand that women fit into the masculine ideal of objectivity, women should legitimately use their relational approach to the world as a cognitive resource.

Interactionism opens the door to Third Worldism. The standard argument is that women from traditional, non-Western societies, are more "epistemologically privileged," for their bonds with family and nature haven't yet been severed by modernity. Third World women are the original postmodernists, the first cyborgs; unlike Western/Westernized scientists who abstract and objectify nature, they experience forces of nature as a continuum with their own everyday experiences in producing and sustaining life. Abstraction is male, oppressive and *Western*, while interaction is female, liberatory, and *Eastern*.[10]

Even those feminist epistemologists like Helen Longino who are weary of privileging any special female experience, biological, social, or both, end up endorsing the interactionist agenda on purely political grounds. If all science is inescapably shaped by background assumptions of a culture, Longino has argued, then feminists should feel perfectly justified in bringing in values that are "consistent with the values and commitments [they] express in the rest of [their] lives" (1990, 191), without having to argue that these values reflect a special female experience. In her more recent work, Longino (1995, 1996, 1997) has refined this choosing-to-do-science-as-a-feminist argument by drawing up a list of six "feminist virtues" (accuracy, novelty, ontological heterogeneity, mutuality of interaction, applicability to human needs, and diffusion of power). These virtues are meant to operationalize the feminist interactionism as found in the work of Harding, Keller, Haraway, postcolonial ecofeminists, and others. Longino urges feminist scientists to use these values, instead of the standard Kuhnian values for theory choice,[11] to interpret the evidence from scientific experiments. Because all epistemic values carry political valence, and Kuhn's values apparently have conservative political implications (Longino 1996, 54, 55), feminists must use feminist cognitive values so that they can "reveal gender" in nature, science, and culture.

Longino argues, for example, that feminists need not aim for the time-honored value of simplicity which seeks to explain a maximum of observations with a minimum of entities and laws. Treating simplicity as a truth-enhancing value reflects reductionist, conservative, and generally masculine political attitudes, Longino claims, because it reduces the entities and phenomena being explained to mere epiphenomena of more fundamental ("privileged") laws. Because *women as social beings* have found their individual subjectivities subsumed under masculine traits parading as universals, *women as feminist scientists* should refuse to seek simple theories in the domain of nature as well. They must instead actively look for complex interaction between distinct particulars which cannot, by definition, be explained by any other more fundamental entity.

Even more startling is Longino's advice to feminist scientists to actively seek out novel theories and models that "depart from accepted ones," purposefully "disregarding consistency with other theories" (1997, 124, also 1995, 1996). Her argument is that it is legitimate to jettison the traditional value of consistency of new findings with what we already know because "mainstream traditional frameworks have been used in accounts that neglect female contributions . . . or treat as natural alleged male superiority" (1997, 122). In other words, because existing science is a patriarchal construct, feminists need not worry if their scientific claims contradict the existing body of knowledge in any field. As long as feminist scientists can adequately explain the experimental data using feminist background assumptions, they are no less justified in holding those theories, rather than the ones justified by the traditional cognitive values that Kuhn and others favor.

This invitation to disregard external consistency with the existing body of knowledge lethally weakens Longino's earlier emphasis on shared public standards as a constraint on subjectivism in science. If feminists, and *mutatis mutandis*, any other community of scientists, are free to construct scientific theories by postulating "different entities, processes, different principles of explanation, alternative metaphors" (1997, 122) which are chosen *because* they contradict the existing stock of knowledge and the existing background assumptions, how can there be *any* shared, publicly recognized standards at all that can be used to evaluate the claims of these sub-communities? If feminists disregard the coherence and reasoned acceptance of their models and theories by the rest of the scientific community, are they not opting out of the process of "transformative criticism" that was meant to keep subjective biases under check? With the gradual loosening of the constraints of accepted standards, feminist epistemology has in fact become an argument for alternatives *to* science, and not for alternative science in a feminist vein.[12]

Longino's invitation to local communities to adopt cognitive standards that "express their aspirations" (1996, 55) runs the risk of opening the door to the worst excesses of Third Worldism and pseudo-science. Consider the following:

- David Hess (1997, 49–51) has interpreted Longino's call for novelty to argue that spiritism in Brazil need not concern itself with how it contradicts existing science. Longino's "feminist virtues" enable Hess to wrap up magical thinking in a scientific cover.
- Postcolonial science critics, Ashis Nandy and Shiv Visvanathan (1990), have deployed the distinctiveness of non-dualist values to defend the occult medical practices of theosophists like Helena Blavatsky and Annie Besant as examples

of "ethnoscience at its most autonomous." These supposedly "feminist" occultists were staunch supporters of the most conservative factions of early twentieth-century Hindu nationalists and also served as intellectual guides to the Nazis.[13]

- Indeed, Hindu nationalists themselves justify teaching Vedic astrology as a science in Indian universities and colleges by arguing that from within the Hindu non-dualism of nature and spirit, astral influences are part of nature and therefore a legitimate subject of scientific study.

- Similar examples can be encountered many times in Harding's latest work on multicultural science (Harding 1998) and in the writings of ecofeminists who have raised the back-breaking labor of peasant women into a resource of interactionist values for a new feminist science (Shiva 1988).

If different communities are free to use contextual values that express their social and cultural aspirations for justifying knowledge claims, regardless of coherence with what is already known about the world through modern science, then all these folk ways are legitimate sciences. How will feminists challenge the proponents of Vedic astrology? How will they respond to the goddess theory of disease?

The problem with the above scenario goes beyond epistemic relativism. The problem lies in immunizing non-dualist, holist values from a critique by glorifying them as a resource for better science. Feminist valorization of non-differentiated connected knowing suffers from a most grievous misunderstanding. It is simply not the case that holist ways of knowing are always and everywhere progressive or emancipatory. On the contrary, modern liberties—feminism included—became possible in the West only with a *separation* of the natural and the moral orders.

This misunderstanding has had the most disastrous significance for new social movements especially in India where *holism lies at the very heart of caste and gender hierarchy*. The essence of holism that both feminists and Hindu ideologues agree upon is a lack of separation between the subject (mind, consciousness) and the object (nature). Hindu holism introduces a supernatural element: nature and human societies are embodiments (or illusions, in the Advaita tradition) of the Absolute Spirit, Brahman. This imposes the claims of the natural *and* the sacred orders on human subjectivity, ethics, and morality; transgressions against social codes simultaneously become transgressions against the natural and sacred orders. This element of spiritual monism is missing from the mainstream feminist epistemologists, even though there are some spiritualist elements within feminism. But the Indian example should nevertheless serve as a cautionary tale for those who are keen on merging the subjective identities of women with nature or with any larger whole. This kind of merging has not been kind to women. Feminists should have nothing to do with it.

*Postcolonial Science Studies: Ending "Epistemic Violence"*

In her essay, "Can the Subaltern Speak?" which became an instant classic when it appeared in 1988, Gayatri Chakravorty Spivak identified a new species of violence which works without guns and without armies, but is, presumably, every bit as deadly. Invoking Foucault and Derrida, she called this genre of violence "epistemic violence"—that is, the violence of knowledge, or more properly, the violence of "discourse," which includes the complete apparatus of knowledge-production. She used the example of the British colonial administration's attempt to ban sati (widow immolation) in nineteenth-century India to illustrate how epistemic violence works. Spivak claims that in outlawing the practice of sati—that is, by classifying sati as a crime—the British committed an act of epistemic violence against the natives who tolerated and even "adulated" the practice as a sacred and heroic ritual. The self-immolation of widows, Spivak argues, "should have been" read in accord with local interpretations of Hindu scared books and Hindu warrior traditions which allow it be understood as "[an] act of martyrdom, with the defunct husband standing in for the transcendental One; or with war, with the husband standing in for sovereign or state, for whose sake an intoxicating ideology of self-sacrifice can be mobilized. In actual fact, it was categorized [by the British] with murder, infanticide and the lethal exposure of the very old" (Spivak 1988a, 302). In simpler words, Spivak accuses the British of violence against traditions because they re-classified what tradition condoned as a noble act, as an act of violence. The British, on this account, stand indicted of epistemic violence, because they tried to prevent actual violence against flesh-and-blood women.

This redefinition of a ritual into a crime amounts to violence against the "brown women who the white men claimed to save form brown men" (to paraphrase Spivak), for it denied them even that thin veneer of courage and self-determination that the sacred traditions of Hindus conferred upon them, in however self-serving a manner. Women, however, were not the only or even the main targets. Rather, epistemic violence which works "not by military might or industrial strength, but *by thought itself*" as Partha Chatterjee (1986, 11), another key figure in postcolonial studies, put it, is supposed to lie at the very heart of *all* colonialism, past, present, or future.

Although Spivak's is one of the most influential voices connecting concerns about the "colonies" with fashionable post-structuralist theory, her idea of epistemic violence was not new. Starting around the early 1980s, a new crop of "amodern" or "non-modern" intellectuals, upholding Mohandas Gandhi as an icon of alternative modernity, had already begun to emerge in India. In his well-known 1983 book, *The Intimate Enemy*, Ashis Nandy, the most prominent of these intellectuals, ascribed the political and economic power of the British imperialists to their "destruction of the unique gestalt of India" (1983, 73) which he explicitly identified with a non-modern, part-classical, part-folk Hinduism. Colonialism, according to Nandy, "won its victories [by] creating secular hierarchies incompatible with traditional order." Thus colonialism was an act of violence against the non-secular, non-differentiated, non-dualist view that had given meaning to ordinary people in India for centuries.

Following Nandy, another well-known book, again by an Indian intellectual, created a stir among postmodernist intellectuals around the world. This was Partha Chatterjee's 1986 book, *Nationalist Thought and the Colonial World: A Derivative Discourse?* In this book, Chatterjee argued that nationalist intellectuals, whether left-wing socialists like Nehru or right-wing religious nationalists like Bankim Chandra, were prisoners ("derivative") of a colonial mind-set, because both sides accepted the intellectual premises of the superiority of modernity and the necessity of modernization. The Fabian-Socialist Nehru is as much a colonized mind as the Hindu nationalist Bankim because both accept the Western Enlightenment belief in progress through application of reason, with the difference being that the former looked to the future for the evolution of reason in Indian society, while the latter looked to the heritage of the Hindu nation. On Chatterjee's account, both positions are equally problematic because both impose a Western teleology of progress and a Western conception of scientific reason on Indian culture. *The problem with the Hindu right, in other words, is not its backward looking, reactionary modernism, but its aspiration for modernity itself.* The only way to break out of the colonial mind-set is, according to Chatterjee, to break out the Western "thematic"—that is, "Western justificationary structures, epistemological and ethical rules"—itself (Chatterjee 1986, 38). True anti-colonialism lies in refusing the Western problematic (goals, possibilities) and Western thematic (ways of thinking). On this account, India's non-modern conceptual categories, epistemological and ethical rules should be used to study India and to allow India to chart its own future.

In yet another influential formulation by Gyan Prakash, a subaltern historian, to truly overcome the Orientalist biases, post-Orientalist historiography of India must take a postmodern turn and "repudiate the post-Enlightenment ideology of Reason and Progress" (1990, 404). Echoing Ashis Nandy and Partha Chatterjee, Gyan Prakash essentially comes down to the idea that India can only be understood through its own indigenous conceptual categories. In a classic example of self-Orientalization, Prakash believes that a genuinely Indian history can be written only as "mythographies" which understand India from the Indian point of view. Again, colonialism is seen as acting through colonial knowledge.

Spivak, Nandy, Chatterjee, Gyan Prakash, and other postcolonial theorists were following in the footsteps of Edward Said's pathbreaking 1978 book, *Orientalism*. Following the ideas of Michel Foucault, Said had argued that colonialism should not be seen merely as a project of territorial conquest and economic exploitation. Instead, colonialism should be seen as a project that constructs a new subjectivity—a new sense of what is real, normal, and good—among the colonized people that makes them available for control by the colonial powers. In the guise of producing the "objective knowledge" of the people and places they colonized, the Western powers created an Orient that could be "judged, as in a law court . . . studied, as in a curriculum, . . . disciplined, as in a school or a prison . . . illustrative, as in a zoology manual." The Orient that the West constructed was "irrational, depraved, childlike, different [making the European appear in contrast as], rational, virtuous, mature, normal" (1978, 40). The West was able to control the Orient by "worlding" the subject peoples by substituting their version of reality with its own mode of understanding and structuring the world. It is through its claim to superior, objective, and universal knowledge that the West exerted its power.

This, then, is the pedigree of postcolonial studies, the latest entry to the jungle of postmarked areas of scholarship. The "postcolonialists" see themselves as creating a clearing where the power-knowledge of the West can be deconstructed and the colonized allowed—again—to see reality through "their own" conceptual frameworks. "Postcolonialism," in the words of Dipesh Chakrabarty (2000) is a project of "provincializing Europe," showing that what the West claims as universally valid categories are actually provincial ideas of Europe which have acquired the status of universal truths because of Europe's economic and military power. Universal truths, like language, Dipesh Chakrabarty argues, are only "[provincial] dialects backed by an army" (ibid., 43). If truth does not represent a reality outside the discourse, alterative, non-Western truths, if backed by the required trappings of power, can become alternative universals.

How is this task of provincializing and decolonization accomplished? The short answer is by deconstructing the universality of modern science. In this deconstruction, postcolonial theory joins hands (wittingly) with the social constructivist and feminist critiques of science on the one hand, and (unwittingly) with the right-wing defenders of Hindu science, on the other.

*At its most fundamental level, the postcolonial project is an epistemological project.* The postcolonial critique of the West is premised upon a repudiation of the objectivity, progressivity and universalism of modern science. In general, postcolonial theorists take for granted what the assorted social constructivists, cultural studies and feminist critics (see preceding parts of the chapter) labor so hard to argue, namely, scientific truths are social constructs; and that norms of justification depend upon the conceptual categories of a culture. Postcolonial theorists take it for granted that (to paraphrase Edward Said 1978, 272 and passim) there cannot be true presentations of *anything*, because *all* representations are embedded alike in language, culture, institutions, and the political ambience of the representer. Truth, postcolonial theorists agree, is a dialect backed by an army; a "fact" becomes a fact because it is "co-constructed with social power."

Lately, the alliance between postcolonialism and science studies, especially cultural studies and feminism, has become even more explicit. In her latest book, *Is Science Multicultural?* Sandra Harding (1998) has argued that because modern science is both Eurocentric *and* androcentric, it is in the common interest of non-Western peoples and feminists everywhere to join forces to confront it. Because the colonized peoples and women alike have been driven to the margins of the modern world, they together offer a "standpoint epistemology" of the oppressed which can expose the blind spots of modern science which are not visible to the beneficiaries of modernity.

Harding is only belatedly echoing the postcolonial position of Ashis Nandy (1983) who has long held Gandhi as an exemplar of how to build an alliance between the "other" West and the non-modern India. Like the generations of anti-Enlightenment romantics before him who found in India an escape from the instrumental, rational tendencies of the Western world, Nandy has argued that non-modern India holds in "trusteeship" the softer, feminine, but repressed side of the West, namely, a relational view of the world, a non-instrumental search for knowledge that advances spiritual ends (see Nandy 1987, chapter 2 and passim). This provides grounds for a civilizational alliance between the marginalized selves of the West and the non-Western worlds. Like Harding and other feminists, Nandy believes that this alliance of the oppressed will rediscover the less dualistic, more interconnecting readings of nature which can then serve as alternative technototems of a more interdependent society.

The question still remains, in practical terms, how will postcolonial science differ from science-as-we-know-it? What difference will starting thought from the standpoint of postcolonial cultures' science make on the ground?

Exhortations apart, real examples of ethnosciences in postcolonial science studies tend to be disappointing and limited mostly to defending the rationality of the local knowledge of peasants, shamans, midwives, herbalists, astrologers, and other ritualists within their own conceptual universes. The literature is vast and growing (see Hess 1995; Nader 1996; and Goonatilake 1998, for relatively recent overviews). The objective of these new comparative studies in the postcolonial mode is to argue that "Western 'rationality' and 'scientificity' [cannot any longer] be used as the bench-mark by which other sciences can be evaluated. The ways of understanding the natural world that have been produced by different cultures and at different times should be compared as knowledge systems on an equal footing" (Watson-Verran and Turnbull 1995, 115). The theorists of alternative sciences quietly sidestep the issues of whether the local knowledges can stand up to rigorous empirical testing with adequate controls. As examined in earlier sections, the standards of what constitutes evidence and what is a robust enough test are themselves treated as a part of the local context in all cases. How this position can escape a serious epistemological relativism is left largely unanswered, or answered with an attack on the political motives of those who raise the relativism question.

Reading this literature it becomes clear that the underlying motivation is to give the subaltern or marginalized social groups everywhere the right to challenge what the experts, bureaucrats, and those in power tell them. Scientific experts are seen not as serving truth but as glossing dominant interests in Western capitalist societies as facts of nature. Thus, postcolonial theorists insist upon, in the name of genuine openness and radical democracy, the right of the pre-scientific traditions to question the scientifically established knowledge. Dipesh Chakrabarty (1995) reads it as a sign of a colonial heritage of "hyper-rational" Indian intellectuals who expect that only the peasant who understands the world in terms of ghosts and spirits has anything to learn from modern science and never the other way around. Nandy speaks for many in postcolonial science studies when he argues that the non-Western cultures hold in trusteeship an alternative conception of knowledge which does not tear apart cognition from affect, facts from values, and does not reduce all knowledge for control and manipulation (see especially, Nandy 1987, chapter 2). Clearly, the exhortation for cultural resistance against what modern science describes as normal and sane is motivated by commendable political motives of enabling the weak to become the creators of their own symbolic worlds.

The problem, however, is that *what appears as marginal from the point of view of the modern West, is not marginal at all in non-Western societies* which haven't yet experienced a significant secularization of their cultures. Local knowledges that Western critics assume to be standpoints of the "oppressed" are in fact, deeply embedded in the dominant religious/cultural idiom of non-Western societies. Using local knowledges to challenge Western science may, however dubiously, illuminate the blindspots of modern science in the West. But in non-Western societies themselves, such affirmation of the scientificity of local knowledges ends up affirming the power of the dominant cultural-religious institutions. Even worse, while the left-wing critics of science invoke local knowledges "strategically" in order to fight what they see as the bigger evil, that is, the West, the right wing can use the same logic and invoke the same local knowledges much more "authentically" and "organically", for it can mobilize all the traditional religious piety and cultural symbolism that go with local knowledges.

This is precisely what is going on in India today. It is not a mere coincidence that the Hindu right-wing supporters of Vedic science declare themselves to be a part of postcolonial studies. It is not a coincidence that the Hindu nationalists claim important postcolonial scholars, especially, Ashis Nandy, Vandana Shiva,

Claude Alvarez, and Ronald Inden as their own (see Elst 2001; Jain 1994). The right-wing Hindu intellectuals have good reason to join the ranks of postcolonial scholars for they, too, see themselves as fighting for the "decolonization of the Hindu mind." It is the de-Westernization of the national imaginary that they profess to fight for when they decry secularism and individual rights-based feminism and liberalism as signs of colonized minds. It is the cause of a "deeper" decolonization of the mind that they profess to fight for, when they rewrite science books to include astrology and *vastu* as sciences, and when they find the Vedas to be the real source of all sciences.

There are, of course, clear differences between right-wing and left-wing postcolonialism, with two worth noticing. One, the whole point of social constructivism, feminism, and postcolonialism has been to deny that there is any such thing as "the Hindu mind" or "the scientific method" that can escape the contingency of history and politics. While there are many examples of cruder varieties of gender and Third Worldist essentialism in the writings of some postcolonial theorists (especially Vandana Shiva and some other difference feminists), most postcolonial and feminist studies scholars have made strenuous efforts to distance themselves from essentialism. Scholars like Haraway, and more recently Spivak, have taken care to define subalternity or marginality not in racial, gender, or national identities, but in terms of "oppositional consciousness" (Haraway 1991; Harding 1998), or in terms of the ability to speak (Spivak 2000). Spivak defines the subaltern as "everything that has limited or no access to the cultural imperialism is subaltern—a space of difference" or as all those "cut off from upward—and in a sense 'outward' mobility" (Spivak 2000, 325). Other postcolonialists share this non-essentialist view of subalternity as well. Gyan Prakash, for instance, wants to define subalternity as a "variety of shifting positions," that develop as an "effect of power relations expressed through a variety of means, linguistics, economic and cultural" (1990, 400) rather than in static, essentialist terms like the "proletariat," "caste system," "Third World," or "Eastern," etc. Thus defined, subalternity can cut across class and national lines, leading to a new "subaltern international" of sorts which defines itself in opposition to modernization and its constituent theme of reason and progress.

Yet, for all these disavowals of essentialism, even the best of postcolonial scholarship ends up taking shelter in what Spivak has called a "strategic essentialism." Strategic essentialism is essentially strategic. It amounts to saying that while we *know* that there are no real essences, no pure subjectivities, we can nevertheless "strategically use a positive essentialism in a scrupulously visible political interest" (Spivak 1988b, 13). The "visible political interest" is to give voice to the subaltern groups who have not been allowed to represent themselves. In this larger political interest, nearly all postcolonial studies fall back, strategically or otherwise, on a view of the subaltern knowledge practices as essentially cyborgian, undifferentiated, non-dualistic, and monistic (see the ideas of Latour, Haraway, and feminist epistemologists discussed above). In postcolonial studies proper, subaltern rationality is accepted as a domain of "innocence" (Ashis Nandy), "community" (Partha Chatterjee), and an unproblematic unity of effect and analysis, values and facts (Ashis Nandy, Frederique Marglin, and Dipesh Chakrabarty)—all the traits that are supposedly lacking in the Western, Enlightenment rationality. Take the last example: Dipesh Chakrabarty berates Indian critics of Hinduism as "hyper-rational" creatures of Western Enlightenment who cannot empathize with the religious imagination of their countrymen. He complains that the secular critics simply fail to understand the Hindu rationality which does not split analysis form emotions, or nature from the supernatural. In this reading of difference, to separate these domains, to value the separation of facts from values, to aim for objectivity in knowledge itself is a mark of the colonial mind-set. If we grant the very foundations of objectivity to the West, are we not back to the old stereotypes of irrational, emotional natives?

The Hindu right wing is, of course, unabashedly essentialist. What is only "strategically" essentialist for the left-wing postcolonial theorists becomes a part and parcel of the eternal "Hindu mind" in the right-wing postcolonialism. The additional windfall for the right-wing postcolonialists is that all the traits of non-differentiated, interactionist knowledge that the left-wing science critics defend fit very well with the holistic, undifferentiated anti-dualism of the Vedic worldview. The left-wing postcolonial scholars in other words, have been doing the spadework for the right-wing postcolonial scholars. The highly visible scholars of international repute in progressive academic circles have been defending the ideas which turn out to be the staple of a conservative, neo-Hindu understanding of Hinduism and Hindu sciences.

There is a second issue on which the left-wing postcolonials can legitimately deny any overlap with the right-wing Hindu nationalists. The left-wing postcolonials have no sympathy at all with the right-wing attempt to justify Hindu beliefs and practices as "scientific," as the right-wing defenders of Vedic science do. In fact, Ashis Nandy and Partha Chatterjee have been explicit in their denunciation of Hindu nationalists as mentally colonized by the West for justifying Hinduism against a Western episteme. Nandy and other left-wing postcolonialists want to reverse the direction of justification: they want the modern West to examine its conceptual categories in response to the subaltern categories. They see Hindu nationalists as distorting India's non-modern gestalt by defending those elements of the Vedic tradition which can outdo the West in their "scientificity." For the left-wing postcolonialists, the right-wing discourse of Hindu science is modernist, and therefore a "derivative" discourse of the colonial masters.

But as I have been arguing, the Hindu right wing is modernist in a reactionary, anti-Enlightenment way. Hindutva is gobbling up modern science by declaring the Vedic knowledge systems to be at par with modern science in rationality and credibility. Proponents of Vedic science claim the Vedas to have presaged all the advances in modern science without admitting, that in fact, modern sciences challenge the metaphysical foundation of the Vedic view of the world.

The varieties of social constructivist critiques of science discussed in this chapter give aid and comfort to the Hindu right wing by providing philosophical arguments for the equality of modern science with other ways of knowing. By denying the reality and the possibility of progress in our knowledge of the natural world through the methodological and institutional innovations in modern science, these theories defend the worldview that *all* systematic attempts to learn about nature are equally scientific. By denying the fact that the modern scientific view of the world has put all metaphysics in question, these theories disarm the struggle for Enlightenment and secularization.

### Conclusion

In this chapter I have presented a critical exposition of the family of social constructivist schools of thought. I have tried to present as unbiased a description of these ideas as I possibly can. In each case, I have shown that despite strenuous denials, all varieties of social constructivism end up opening the door to a serious epistemological relativism. In each case, I have also shown that despite their honorable political intentions, all varieties of social constructivism end up giving aid and comfort to the forces of fascism in India.

I will close this chapter with two final clarifications. Some have argued that by laboring to establish this link with the Hindu right wing, I am censoring any attempt to reveal the role of social context/interest in science and therefore, indirectly, encouraging a return to a discredited "positivism." I believe that disclosing the social structuring of knowledge is a worthy enterprise, but it need not take a relativist turn. I believe that the undeniable fact of scientific progress in providing more adequate and trans-culturally true accounts of natural phenomena can—and should have—served as a pragmatist-realist check on social constructivist enthusiasm for empowering culture over nature. I also believe that the total and universal denunciations of the Enlightenment are just bad social theory, lacking any sense of real history of non-Western societies. Those critics of science who uncritically embrace the ideals of non-differentiated, "amodern" ways of knowing could have easily asked: How have been the lives of real men and women in societies under holist epistemologies and worldviews? Are individuation and objectivity really opposed to the interests of the oppressed? The social history of the subalterns in India could have shown that the subalterns have in fact struggled *against* the undifferentiated holism of Hindu cosmology and epistemology (see chapter 7).

This brings me to the second clarification. It is true that the postmodernist and social constructivist critiques started out with the goal of deconstructing the hegemonic position of science in the West. The fate of science in the rest of the world was not their *primary* focus, even though there was always some concern with issues of science's contribution to imperialism. One cannot therefore, some argue, hold these critics responsible for how their theories are used in the world outside the West.

This defense will not wash. As I have shown at length above, the evidence of alternative rationalities of non-Western peoples has been an *integral* part of social constructivist arguments for denuding modern science of its special and universal stature. It is ironic that with all the emphasis on the social context of modern science, social constructivist theories, for the most part, completely ignored the social context of alternative sciences in other parts of the world. Even a cursory familiarity with the social history of local knowledges in postcolonial societies could have shown the limits (both social and institutional) of these alternatives. Even a modicum of critical engagement with the "other" could have shown that the aspects of ethno-knowledges they admire have quite often been a part of the dominating, hegemonic culture in non-Western cultures. This knowledge is easily available. That it was not sought, or played down, is a result of the political biases of the postmodernist critics. It is simply bad scholarship.

My motivation for this engagement is to remind us of a simple truth: ideas have consequences. Those of us who trade in ideas have a responsibility to ensure that our ideas should do no harm. In the face of the rising threat of reactionary populism in India and many other parts of the developing world, it is high time critics of reason and Enlightenment asked themselves if they are fulfilling their responsibility.

# Six

# We are All Hybrids Now!

## Paths to Reactionary Modernism

*The Hindu genius is to blur issues,*
*not to confront them.*
—Girilal Jain, The Hindu Phenomenon

*Is there an Indian way of thinking? . . . The Indian way*
*of thinking is sensitivity to the context.*
—A. K. Ramanujam, "Is There an Indian Way of Thinking?"

*Borderland epistemology is a knowledge collage [that*
*helps us decide] . . . when to do our vitamins or other*
*"folk science" therapies.*
—Sandra Harding, "Science Is Good to Think With"

"The Hindus" are the Original Cyborgs. Their way of thinking was post-modern even before anyone else was even modern. And it is a good thing, too, for now they can help heal the split soul of the modern West. They can help the modern West learn—all over again—how to think contextually, without dualisms and without confrontation with the wisdom of the ancestors hallowed by traditions.

This is a conclusion that finds support from both the right-wing and the left-wing postcolonials. Hindu nationalists carry it embossed on their saffron banners, as it were. Most left-wing academic postmoderns do not state it in such obviously ethnocentric words, even though the conclusion follows as a corollary of their ideal of science as a bricolage or hybrid.

As we saw in the last chapter, both sides concur that removing knowledge of nature from culture, denuding facts from values, is impossible, and dangerous, if attempted at all. Only a bridging of the gap, a fudging of the boundaries between nature and culture, facts and values, can save humanity form the perils of objectifying nature with its supposedly value-free science. Over the last three decades or more, the left-wing postcolonialists, feminists, and other cultural critics of modernity have looked toward non-Western, premodern cultures as "trustees" of the values of hybridity and non-dualism.

There is yet another dimension of the fudging of the boundaries, this time between local knowledges and modern science, that I want to examine in this chapter. I will argue that the contradiction-ridden hybridization of local knowledges and modern science that the assorted postmodern critics favor as "resistance" to mental colonialism, is the very same mechanism that creates the reactionary modernist mind-set which provides recruits for religio-political movements.

Having denied that in modern science the world has found a truly universal corpus of knowledge which can cut across cultural boundaries, the assorted constructivist critics find themselves in a quandary. How can *the fact* of the universal reach of science be explained? If modern science does not carry rational weight universally, then is the rest of the world simply being brainwashed? If, furthermore, Western knowledge serves to consolidate Western power, why *shouldn't* the recovering colonies around the world simply say "No!" and cultivate their own traditions?

The postmodern solution to this quandary has been to extend the idea of hybridization, collage, or borderland epistemology which invites non-Western peoples to take elements of Western sciences and technology, interpret them in a way that fits them into their own local cultures. The same populism that leads the critics of science to exhort the lay people in the West to defy the experts' criteria for demarcating between what is science and non-science is at work in their advice to non-Western peoples; they, too, should use "their own" cultural meanings to selectively use those elements of modern science which are suitable to their own cultural values and material needs. The need to confront the many contradictions between the new and the old, the scientific and the ethnoscientific, is seen as a source of "epistemic violence" and the intolerance of modernity which cannot live with ambiguities and differences.

In this chapter I will examine how this hybridization comes full circle back to the premodernist way of opportunistic bricolage-making which ignores the need for logical consistency and lack of contradictions. The co-habitation of contradictions, beaten into place by the "strategic needs" of non-Western cultures (read: cultural re-assertion) bears uncanny similarities with the elite-Hindu way of traditionalizing innovations, making the new old, and finding a place for it in the hierarchy of heterogeneous ideas decided by the Brahminical concepts of what is rational and good. I have already examined this mechanism at work in the theoretical defense of Vedic science (chapters 3 and 4). Here I want to look at how this hybridization works at the level of popular consciousness. New social movements, inspired by anti-Enlightenment philosophies, have encouraged this hybridity in the name of "resistance" to mental colonization. Yet, this split consciousness in which modern technologies are hybridized with a supernatural and hierarchical worldview is serving as the seedbed of reactionary modernity in India today.

Recall from chapter 1 that we have defined reactionary modernism as a contradictory, half-way house characteristic of those societies where the process of secularization has not had a chance to take root. These societies are marked by an uneasy and relatively superficial mix of modern technologies and modern institutions of government, coupled with an active opposition to the values of liberal modernity and the Enlightenment. I have argued so far that Hindu nationalism is a variety of reactionary modernism which propagates itself by mobilizing the masses by invoking their traditional religiosity and hitching it to high-tech militarism on the one hand, and retrograde and intolerant social policies on the other. Hindu nationalists, like fundamentalists in other societies, use the tools and the institutions of modernity to subvert the values of modernity.

Using anthropological studies from India, I will show that technological modernization is indeed creating a hybrid consciousness. Ordinary people, including farmers in villages, have become adept at using modern material inputs, while affirming and celebrating the worldview of their forefathers. Hindu nationalists are able to invoke this mixed-up, contradictory consciousness, promising ever more modern technologies, from consumer goods to nuclear weapons, but circumscribed within the cosmology and social values sanctioned by the Great Traditions of Sanskritic Hinduism. The rush to declare Hinduism "scientific" is an attempt to create an updated Hinduism for the new consumers of modern goods and technologies. This updated Hinduism, as we have seen, does not question the outdated and irrational worldview of Brahminical Hinduism. It only drapes it in the language of science.

In the next two sections, we will examine two distinct but related solutions to the demise of the "universality ideal," to use Sandra Harding's formulation. The first comes from cultural studies of science and feminist epistemologists, the second from postcolonial theorists from India. In the third section, we will examine the evidence of hybrid consciousness from recent anthropological studies of the Green Revolution. The final section will examine the resonances between this popular hybrid consciousness and Hindu nationalism.

### Post-Universal Science I: Multicultural Bricolage

The humanist-Enlightenment thought has understood modern science as the common heritage of all humankind. No one part of the world was supposed to have a perpetual patent on it. Modern natural science, in the evocative imagery of Joseph Needham (1969), was the sea into which all the rivers of local sciences flowed and which, in turn, brought new ideas and innovations from around the world to all shores alike. Indeed, the distinguishing mark of modern science was that like an ocean, it was not confined within narrow embankments of local cultures. While all medieval, pre-Galilean sciences, whether from Europe, Asia, or Africa, explained nature through anthropomorphic metaphors peculiar to their time and place, modern science alone managed to break free from time and place. According to Needham's historiography, it was only in late Renaissance Europe, at the time of Galileo, that important innovations—mainly mathematization, experimental methods, the distinction between primary and secondary qualities, and the acceptance of mechanical philosophy (ibid., 15)—were introduced. These innovations provided a common vocabulary which cut across culture-specific languages of other sciences. Over time, modern science became the lingua franca for natural philosophers/scientists in different civilizations.

This view of universalism rested upon a faith in the rational unity of humankind. Needham, one of the most eloquent spokesmen of rational unity, saw the "ecumenism" of science as an affirmation that "in the investigation of nature, all men are potentially equal, all can comprehensively speak the universal language of science, . . . the ancient and medieval sciences . . . were concerned with the same natural world and could therefore be subsumed into the same ecumenical natural philosophy" (1970, 417–418).

Progress in science could be accepted on rational grounds by even those who did not share the cultural background of the scientific innovators. As Charles Taylor puts it, it is possible for a member of a pre-Galilean culture to appreciate that Galileo had made an important advance over Aristotelian physics. Because the pre-Galilean and the Galilean cultures, despite their difference, share an interest in understanding and controlling nature, the "pre-Galilean can recognize the significance of Galilean science's massive leap forward" (Taylor 1993, 221). Faith in the rational unity of humankind thus attempted to preserve the intuition that there has been real progress in our ability as a species to understand the material world better. It pointed toward a conclusion that would be considered highly politically incorrect today, namely, while all peoples were equally capable of scientific thinking, all traditions of science were not. Whatever was worthwhile in premodern sciences could be explained by modern science relevant to the domain, and whatever could not stand the critical scrutiny of science, was to be either discarded altogether, or retained as a myth, or as prehistory of science. The hope, of course, was that growth in understanding will lead to human freedom from fear of the unknown and all the attendant social practices this fear breeds.

This view of universalism and progress in science has fallen out of favor in the era of postmodernism, both epistemologically and politically.

Epistemologically, *science as a progressive and ecumenical knowledge cannot survive the demand for symmetry between premodern and modern sciences.* If all sciences are at par in being equally local, then modern science loses its distinctive power to contain whatever is valid in premodern science and go beyond to explore dimensions of nature hidden from them. The history of science ceases to be a universal history of one single enterprise at different stages. It turns into a history of different ways of knowing, which can potentially produce alternative, contradictory answers in the same domain of nature, with no universally acceptable criteria for judging better answers from the worse ones. We have already examined in detail the arguments for epistemic parity offered by major schools of science studies, cultural studies, and feminist and postcolonial studies (see chapter 5).

Politically, too, the ecumenical picture has come under suspicion. Unity of science, which was celebrated as a source of universal enlightenment, has been condemned as a cover for "epistemic violence" against those who did not fit into the worldview of modern science. As we saw in the previous chapter, modern science is condemned as a "rude intrusion" into the lives of premodern and/or women's cultures. The universal spread of science is ascribed to the "free ride" it hitched with colonialism (Harding 1998). With the political power of colonial masters and mentally colonized postcolonial governments behind it, modern science is able to mobilize more "actants" (scientific institutions, scientists, technological projects, and "nature" itself). This assemblage trains more "natives" who, not surprisingly, hold modern science to be self-evidently truer than their own local knowledges (see Turnbull 1997). All in all, critics agree with Paul Feyerabend's judgment: *"Today science prevails not because of its comparative merits, but because the show has been rigged in its favor"* (1978, 102, emphasis in the original).

Yet, the question remains: What does one do with modern science? The universal spread of science and technology is an indisputable fact of the world we live in. Rigged or not, modern science has succeeded in engaging the imagination of even the most reactionary opponents of the modern age. Obviously, simply saying "No!" is not an option. No one but the most extreme fringe of traditionalists proposes complete autarky in matters of science.

The assorted debunkers of universalism that we have been considering are not naysayers. They do not object to non-Western people using science and technology when it serves their own self-defined purposes. What they do object to is non-Western people feeling obliged to revise their inherited cosmologies and social norms informed by these cosmologies. On these accounts, scientific theories, new methodologies, and technological products can be put in the existing "toolboxes" of other cultures, but as just one among the other existing tools: definitely no better, and perhaps a little worse. (This follows from the belief in the epistemic parity of all sciences as equally local. See chapter 5.) It is left to the non-Western peoples to fit the new into the old in ways that are conducive to the preservation of their self-identities.

Thus, Sandra Harding argues for a multicultural mélange, or "knowledge collage," at the "borderlands" of local and modern sciences. This collage grants equal value, without prejudice, to *all* distinctive understandings of nature that different cultures have produced. It does not disbar "Northern-Eurocentric science," but neither does it give it the special role it enjoyed in the past to reorganize the remaining pieces of the collage. What the members of cultures living at the intersection of universal and local sciences are asked to do is to learn when to use one science and when to use another. For example, when they want a boost in productivity in agriculture, they can (although they should not) use modern agriculture, and when they want to conserve the environment, they can fall back on local traditional knowledge; they can use vitamin pills for some ailments and traditional therapies for others (1996, 16, 22). This kind of opportunism, Harding argues, is no different from the predatory eclecticism of modern European sciences. On Harding's reading of the history of modern science, none of the conventional objectivity-enhancing features of modern scientific practices or institutions is responsible for the universal persuasive power of modern science. Rather, modern science became so powerful because, thanks to the power of colonialism, it could freely "forage in other cultures for elements of those cultures' ethno-sciences to incorporate into European science" (1998, 179). Modern science raided the rest of the world for useful knowledge that could be fitted into the European conceptual framework. Now it is the turn of the rest of the world to raid the West and (re)-take elements of modern science that fit *their* conceptual worlds. It is the special responsibility of women and other subjugated groups to take a lead in constructing these borderland epistemologies, for their own standpoints contain valuable correctives to European-masculine knowledge.

David Hess, another prominent spokesman of multicultural sciences, also argues for creative collage-making. As we saw in the previous chapter, Hess views modern science as a "technototem" of powerful White males in Western capitalist societies. In this account, Western science grows by an ad hoc opportunistic process of building alliances and networks of objects, institutions, and ideas. Scientists and engineers, Hess argues, are bricoleurs, in that they "take the versions [of natural sciences] of other communities and reconstruct them so that the elements are recombined to fit with their own local culture." In a multicultural world, all cultures should enjoy the same privilege to reconstruct modern science and technologies by "positing alternatives consciously linked to their social identity," that is, create their own technototems (Hess 1995, 40–41). In a multicultural world, moreover, it should not be the certified scientific experts alone who decide what criteria to use to demarcate science from non-science. Other cultures should have the same rights to use their accepted beliefs to redraw the demarcating line between what they see as scientific (ibid., chapters 6 and 7). Hess gives the example of Spiritists in Brazil who reconstruct modern biology and physics to include the presence of spiritual elements in nature. Spiritists, in other words, draw the demarcating line between science and superstition to include the spirit world in the scientific. (This case bears a striking similarity with Vedic science.) They blur the conventional boundaries. They draw new connections and equivalences between local and Western beliefs about nature. They, in other words, create new cyborgs by investing the objects of nature identified by modern science with their own cultural meanings.

Hess chides those who would object to this cultural reconstruction on the grounds that there are glaring contradictions between Spiritism and modern biophysical sciences. On the cultural studies view of science, this kind of debunking is no longer justified epistemologically or politically. Epistemologically, modern science is no less of a hybrid enterprise; it, too, creates "nature" out of Western metaphysical beliefs and cultural metaphors. There is thus no rational obligation to accept its accounts of reality as truer. Politically, too, debunkers are on the wrong side, for they fail to see in the spirit-doctors, the "agency" of ordinary people to "remake science and medicine in a way that makes sense to [those] who live in a culture in which spirits are as much a part of reality as are televisions and trees" (ibid., 172). The task of new social movements of feminists, environmentalists, and other social justice groups in toady's world in which the "universality ideal" (Sandra Harding's phrase) is dead and gone, is to empathetically enter the world of the Spiritists, Vedic science, and other popular reconstructions. This is supposed to be good for the Western scientific establishment as well, for it can reveal the latter's blind spots and suggest new questions for research.

This kind of contradiction-ridden hybridity is elevated to the level of the collective subject of worldwide "oppositional movements" by Donna Haraway. As we saw in the previous chapter, Haraway's celebrated essays present all sciences alike as conversations between historically located cyborgs, or nature-culture hybrids; the accepted community beliefs shape the background assumptions of scientists, who, in turn, ascribe those properties to natural objects which make sense from within their background assumptions. If all science, in the end, is cultural construction, then whose culture leads to a "better," "no-nonsense" account of nature? Haraway accepts the general Hegelian/feminist principle that those from the bottom can see more of the world and with fewer biases, but she rejects that there is any such universal class—women, the proletariat, non-Western people. The solution is a universal cyborg, so to speak, a web of "partial knowledges" of different cultures and different social groups all united in an "oppositional consciousness" to global networks of technoscience and capitalism. A global collage of so many different cultural reconstructions of dominant sciences, with no view claiming total truth, is Haraway's answer to the feminist and Third Worldist angst over who is better equipped to lead to alternative sciences.

To summarize, those critics who reject the ecumenical and progressive view of science are confronted with this question: What should non-Western cultures do with modern science? All the three influential answers to this question that we have considered here offer very similar advice: reinterpret science to bring it in accord with the existing belief system; hybridize the global with the local in a way that the local retains its identity and its creative tension with the global.

All three answers, however, suffer from the same problem: What does one do with the obvious contradictions in such a hybridization? For example, in the kind of reconstruction recommended by Hess, the Spiritist and the materialist account of nature cannot be equally true. In Harding's recommendation for mixing up modern biomedicine and "other folk therapies," there are obvious contradictions between the two systems. How can the differences be resolved? What elements of the hybrid should be altered? What elements must be given up as false?

All three answers, and others in this genre, deal with this problem by downplaying the need for consistency. They simply exhort the new social movements to live with contradictions. Indeed, they make a virtue out of contradictions by turning them into openings for different voices and different viewpoints. Modern science's ideal of advancement of knowledge by putting contradictory propositions to controlled tests is replaced by a seemingly tolerant acceptance of all "truths" that are self-evident to different peoples.

This multi-culturalist tolerance to alternative sciences has already become the first principle of new social movements. Intellectuals and activists tend to show extreme indulgence toward even those local beliefs which they themselves might find repugnant. Any attempt at debunking the local understanding of nature, however false, outdated, and harmful it may be, is seen as insensitive and authoritarian. In the absence of an organized attempt to correct popular myths about nature—which shape popular social practices, many of which are highly inegalitarian and intolerant—most exuberant and contradictory hybrids are growing freely in modernizing societies. We will examine some evidence of this in what follows. But before we get to the evidence on the ground, I want to examine how hybridity has been theorized by academic postcolonial theorists who formally work outside science studies but are allied with it.

### Post-Universal Science II: Critical
### Traditionalism and "Hybridity Studies"

Secular and mainstream academics from India, some of them with international name recognition, have been major participants in debating the fate of modern science after the purported demise of the "universalist ideal." Indian voices, especially those of the neo-Gandhian critics led by Ashis Nandy and his group at the Center for Study of Developing Societies in Delhi (also referred to as the "Delhi School" of science studies) were beginning to sound out critiques of universalism of science in the early 1980s, much before the postmodernist storm gathered full force in the West. By the end of the 1980s, inspired largely by Edward Said's *Orientalism*, a new crop of world-renowned Indian historians, associated with the famous Subaltern Studies group, literary critics, feminists, environmentalists, and other scholars had gained full proficiency in postmodernist thought. Since then, Indian scholars have played a central role in crafting ever more sophisticated theories of hybrid consciousness. Subaltern historians, including Partha Chatterjee, Gyan Prakash, Dipesh Chakrabarty; literary critics, such as Gayatri Spivak and Homi Bhabha; feminists Chandra Tolpade Mohanty, Lata Mani, Gauri Vishwanath; and environmentalists like Vandana Shiva have become top-ranking names in the new wave of social theory. Their work has been instrumental in sensitizing feminist and science studies scholars to the issue of difference and alternative sciences.

Indian critics of modernity took a page from their patron saint, Mahatma Gandhi, to argue for "critical traditionalism." What is critical traditionalism? It is both an intellectual argument and a political program adopted by numerous Gandhian, small-is-beautiful, alternative science/post-development movements in India. The gist of critical traditionalism lies in accepting the need to update inherited traditions with carefully chosen foreign inputs from science and technology, as long as they can be fitted into "India's unique gestalt . . . its view of man and universe" (Nandy 1983, 73). The concern is that "even in defeat, [the colonized cultures] should retain their authenticity" (Nandy 1987, 124).

Why this obsession with authenticity? The nation states that emerged after the overthrow of colonialism, on this account, won their political economic freedom at the cost of losing their authenticity—their unique gestalts. Modern India (which stands in for all non-Western postcolonial societies) is a "derivative discourse of colonialism," because it accepts the "sovereignty of science" (Chatterjee 1986, 169). As long as postcolonial India tries to remake itself in the name of science, reason, and modernity, it cannot be truly free. Thus, *all* nationalist ideologies, whether led by secularist-socialists like Nehru, who would open the tradition to a rational critique, or cultural nationalists like the neo-Hindus who claim to find tradition itself as rational, are considered equally colonized, because they accept the normativity of modern scientific reason. Opposition to colonialism must extend to opposition to the "thematic" of the colonizers (see chapter 5).

Only a movement that rejects the borrowed worldview of modern science and modernity can *truly* set India free. Such a movement, which first commits itself to the recovery of India's suppressed and silenced gestalt, can be allowed to borrow, very selectively, those Western sciences and technologies that blend in and support the Indian structure of thinking. As Bhiku Parekh, another Gandhian theorist of critical traditionalism put it, India could and should, "borrow from the West only that which is consistent with its own traditions, temperament and circumstances" (1995, 24).

Where is this authentic Indian gestalt to be found? In the thought of Gandhi, critical traditionalists answer in unison. Gandhi alone, it seems, accepted the non-modern Hindus as they were, without trying to find modernity in their traditions, or without returning to some authentic pure origins of Hinduism. Gandhi alone, on this account, escaped all the allures of the Enlightenment. He alone "articulated the consciousness which has remained untamed by the British" (Nandy 1983, 100). All of the numerous European romantic influences on Gandhi—from the Bible, to Ruskin and Tolstoy—are conveniently forgotten and Gandhi is made into a genuine son of the soil.[1]

Finally, what is the content of this authentic Indian gestalt? Here, we return to the staples of postmodernism and classical Sanksritic Hinduism. The Indian way of thinking treats truth about nature as a moral truth. Indians treat nature as enveloped in cultural values of right and wrong, and conversely, culture is enclosed in nature, so that they cannot tell the difference (Ramanujam 1990, 50). Thus, Gandhi's judgment of a devastating earthquake in Bihar in 1934 as "divine chastisement" for the sin of untouchablity is presented as a unified (or non-binary) conception of truth (Chatterjee 1986, 96–97), completely overlooking that through most of its over two thousand years of history, this conception of truth has worked to instill fear, meekness, and obedience among ordinary people. Second, the authentic Indian gestalt holds truth as a felt experience of certainty and changelessness. Again, the social conservatism of such a conception is passed over in complete silence.

But critical traditionalism is not the final word. It has been overtaken by more avant-garde theories of "hybridity" which retain the anti-Enlightenment bias of neo-Gandhians, but reject their essentialism. There is plenty in these newer theories which is unexceptionable and useful, even though mixed up with fashionable cynicism against science. Theories of hybridity argue, correctly, that in the era of globalization, there is no purely indigenous knowledge left anywhere. The condition of postcoloniality is the condition of hybridity that disrupts the binaries between pure modernity and pure tradition. The colonized don't discard their traditions in toto when they become modern. Rather, traditions are a condition of becoming modern, and will always be with us. This hybridity, moreover, is not a sign of defeat or a loss of authenticity. It signifies "resistance," because by mimicking the West while holding on to his cultural universe, the postcolonial subject refuses to become the inferior, unchanging, traditional "other" of the West. In other words, by absorbing the West into its own tradition, the non-West shows it can change, and that it has some control over what and how it will change.[2] This is the bare-bones gist of "hybridity studies," some of the most abstrusely written works in all of postmarked scholarship, mostly associated with the work of Homi Bhabha.

Notwithstanding their differences, critical traditionalists and hybridity theorists both rule out a modernist, debunking stance toward traditional knowledge. The "critical" in critical traditionalism is targeted at modern science. Likewise, in the jargon-filled hybridity studies, it would be hard to find the word science without the sneer quotes that have become the trademark of all postmarked writings.

### "Hybrid Times, Hybrid People"

The Indian countryside makes an excellent setting for testing whether ordinary, poor to middle-class farmers are indeed showing the signs of a hybrid consciousness which is not-quite-modern and not-quite-traditional.[3] Since the middle of the twentieth century, India's villages have been flooded with modern high-yielding seeds, chemical fertilizers, and all the other new technologies and practices associated with the Green Revolution. The Green Revolution has transformed the class/caste dynamic in the countryside. It has produced a whole new class of mostly backward-caste (lowest in the fourfold varna order, but higher than the untouchables) capitalist farmers, who cultivate the land they own with wage labor (mostly marginal farmers and landless untouchables), and with purchased inputs made available at subsidized rates by the Indian government. The areas of Green Revolution agriculture thus provide a natural laboratory to test the claims of the various theories of cultural reconstruction/hybridity we have examined. After four decades of using new seeds and new bioscientific technologies, how have Indian farmers reconstructed the science behind the new technology? If there is a hybrid consciousness, what does it look like? Is it "emancipatory?" Is it any kinder to women? Does the component of tradition make Indian farmers more ecologically responsible? What kind of "oppositional consciousness" are these cyborgs displaying? What is being opposed, and in the name of what?

While the Green Revolution has been studied exhaustively by political economists, there are not too many qualitative, ethnographic studies of cultural changes brought about by new technology and new relations of production. In recent years, however, anthropologists influenced by developments in science studies and postcolonial theory have begun to examine the cultural aspects of the Green Revolution. In what follows, I will use some of these recent studies that specifically set out to address the issues of hybridity among the beneficiaries of the new biosciences-based technologies in India's villages. My major sources of information are Akhil Gupta's *Postcolonial Developments* (1998), a book-length study of Alipur, a village in western Uttar Pradesh, A. R. Vasavi's 1999 book, *Harbingers of Rain*, set in the southern state of Karnataka, and Ann Grodzins Gold's work on the moral ecology of villagers in Rajasthan, a northwestern state. I have taken the liberty of using these case studies to illustrate the potential and real dangers of hybrid consciousness. I do not present an exhaustive analysis of these cases—the reader will have to read these ethnographies himself or herself. I also do not claim to have made a complete survey of the literature, or carried out any fieldwork myself. These are just three case studies. While they do not represent the whole of India, I believe that they are not very atypical either.

There is no doubt that the theories of knowledge-collage, critical traditionalism, and hybridity are amply borne out. The Green Revolution capitalist farmers are indeed reconstructing new science and technology according to their own meaning systems. Traditions are not being evacuated from their lives; they take their traditions with them as they adapt the new technologies to their context. They are no longer peasants, nor are they turning into carbon copies of their Western counterparts. All this is true. The question is, are these kinds of culture-preserving, opportunistic knowledge-collages really conducive to the creation of a more egalitarian and humane culture that the critics of modernity hope for? I will let the evidence speak for itself.

First, the evidence puts to rest some common misconceptions about the Green Revolution. For all the academic critiques of modernization as violence, there is very little nostalgia for the past. Even when villagers reminiscence how good food tasted in those good old days, and how green was their village, they still agree that life is better today (Gold 1998, 169). Even when the farmers were critical of high-input practices, they did not look back to the older farming practices as more desirable, either ecologically or economically. In his study spanning the 1980s and the 1990s Akhil Gupta could find only one farmer who chose to grow non-hybrid seeds. Others had voted with their pocketbooks for the new technology. Indian farmers behaved like any other rational profit maximizers in their choices. Second, Gupta's study demonstrates that it is a myth that modern varieties require inorganic chemical fertilizers to give high yields: farmers are happily mixing old-fashioned organic manure with the new seeds. Moreover, Indian research labs have been able to offer wheat seeds that give high yields even without high inputs of fertilizers of any kind. Overall, farmers are trying out various combinations of their traditional, time-tested understanding of their soils with new seeds and new machines.

So far so good. The story gets trickier as we move from *using* the new technology, to *understanding* why and how the new technology works. While the farmers have become adept at using modern seeds, fertilizers, and pesticides—all products of modern biochemistry and genetics—their understanding of why these inputs work partakes very little of modern scientific understanding of plant growth. Instead, they explain the why and how of high yields of new seeds in terms of a humoral understanding of nature, which they combine with an enchanted and morally coded cosmology in which inanimate objects have needs, motives, and agency. Indian farmers, in other words, are doing exactly what feminists, cultural studies and postcolonial theorists recommend they should do, that is, they are reconstructing modern science according to cultural meanings derived from their own traditions and their own cosmologies.

The farmers in Gupta's and Vasavi's ethnographies explain the workings of the new technology in terms of the balance of four humors: hot, cold, dry, and wet. (While these agronomical humors are different from those used in the Ayurvedic understanding of human health, there is a close parallel in how the health of both people and plants is understood in terms of the right balance of humors). In humoral agronomy all elements of nature have their own innate "natures" which are both material and moral. Thus, soils, seeds, fertilizers, and the resulting food crops can be hot (angry, strong); cold (peaceable, quiet, weak); wet (malleable, young, fertile), dry (brittle, old, weak). All the inputs pass on their morally coded properties—unchanged—to the plants, and plants in turn, pass them on to human consumers. You *become* what you eat, not just in biochemical terms, but also in terms of character attributes. Because chemical inputs are seen as impurities, "weakening" the food, all three ethnographies report the widespread belief that eating hybrid grains grown on chemical fertilizers, which release too much "heat," makes people rude, hot-tempered, and weak in character. Vasavi reports that farmers never use hybrid seeds in sacred rituals, reserving only organically grown traditional varieties for that purpose. But on the other hand, they have no trouble extending the traditional fertility rites to new mechanical devices: tube-wells and pump-sets are ritually worshipped (1999, 117–118). This illustrates the cultural schizophrenia the new technology has engendered: while the tools of production are accepted as powerful enough to be accepted and ritually propitiated, the product (i.e., the grain) itself is seen as a source of pollution and weakness of character. There is enthusiasm for the new technology, mixed with an inchoate resentment and irrational fear about what it is doing to their lives.

Gupta provides interesting examples of how farmers adjust their choice of soil, depth of tilling, the amount of water and fertilizer based upon their understanding of how much "heat" or "wetness" is needed. A couple of examples will suffice. Alipur's farmers judge the suitability and/or amounts of fertilizers in terms of the amount of "heat" they provide which, they believe, pushes the plant upward. Estimates of "heat" can and do lead to overuse of fertilizers (1998, 188, 211, and passim). Likewise, the suitability of different soils for particular crops is judged by the "disposition" of the soil to "catch" the plant roots. As Gupta puts it, "the land was not merely the object of labor or the medium through which the agency of the farmer could be exercised. It had its own volition, its own character, its own disposition, its own ability to act on plants" (ibid., 187). Gupta hints that this view of nature with agency stems from the Alipur's peasants' belief in the interconnectedness of all beings derived from their belief in the transmigration of souls: "peasants in Alipur thought that all living matter was connected, because it was interchangeable. Such a view of connectedness rests on a belief in the transmigration of souls. . . . A human life could become any other form of life in the next birth. It was not a question of equality of all forms of life, but of their transmutability" (ibid., 88).

Vasavi's study of Bijapur, a Green Revolution village in the southern state of Karnataka, and Gold's ethnography of Ghatiyali, a village in Rajasthan, add another dimension. Their studies show that, consistent with the humoral agronomy, a moral causation of natural disasters like drought, deforestation, and other misfortunes remains prevalent. Vasavi's investigations, carried out in the closing decade of the twentieth century, show a widespread acceptance of the belief in droughts and other natural disasters as caused by divine anger brought on by the collective bad karma of the village, to be remedied by religious rituals, including fasting and praying (1999, 54–56, 62–64). Here we see the lack of differentiation ("dualism") between nature and the cultural meanings of good and evil in full swing; forces of nature are understood to be influenced by human actions and divine will. The ethical life of the community, as Vasavi points out, has a bearing on the cycles of nature. Likewise, Gold reports villagers in Rajasthan see the lack of rainfall and declining forest cover as divine punishment for the various sins of the people common in the degenerate age (*Kali Yuga*) we live in. A similar nature-human-divine continuum is also invoked among the Buddhists and Hindu Tamils in Sri Lanka (Weeratunge 2000).

Anthropologists are often content to simply describe local theories of climate change and natural disasters, regardless of their accuracy and validity (Gold 1998, 174). But for those who have to live with theories which are far removed from the real causes operating in nature, there is a price to pay for ignorance. The price in the Indian countryside comes in the continuing hold of the moral and natural rightness of a hierarchical social order. A telling case is Gold's finding that some villagers in Rajasthan hold the decline of the caste order and greater freedoms for women as a sign of the breakdown of the cosmic order for which the gods are punishing the village by withholding rains. In response to a question from the anthropologist about why there is not much rain now as compared to the past, this is how a villager, a Brahmin by caste, responded:

> There used to be much dharma. People used to feed Brahmins, and people used to do fire oblations for the goddesses and gods, and spread fodder for the cows. But now the Degenerate Age (Kali Yuga) has come. . . . Now dharma is completely suppressed. . . . It is my dharma that I won't drink water from anybody's hands. . . . But today, the sweepers, the leatherworkers—we have all become one. It used to be that sweepers and leatherworkers could not come near me, but now . . . it is as if God has forgotten us as we have forgotten him. (ibid., 180–181)

The belief that laxness of moral law brings about changes in natural laws, and vice versa, is widespread. Vasavi (1999, 56) cites a 1988 poll in Karnataka that showed 43 percent of those polled professing a belief in sin as a cause of drought. While there are no detailed studies of what exactly constitutes sin, or laxness of moral law, the breakdown of caste and gender roles is seen by many as prime symptoms of the coming of the degenerate age or *Kali Yuga*. Uppity untouchables and immodest women are a sign that something has gone wrong with the entire fabric of the universe; even rains don't come on time any more. Deep down, core metaphysical beliefs which are the foundations of tacit assumptions about nature and society, have not changed sufficiently to make room for the idea of equality and individual freedom. Neither has the connection between natural law and moral order been broken.

This overlay of modern technology on the deeply unsecular common sense creates an electorate responsive to reactionary modernist politics. Secular reformers, in this context, come to be looked upon as harbingers of *Kali Yuga*, the degenerate age, when people forget how to live by the natural-moral order or dharma. Parties like the Bharatiya Janata Party that can satisfy the material demand for modern technology while promising the restoration of dharma, the eternal moral-cosmic law, enjoy a great advantage among ordinary people. In Akhil Gupta's ethnography of Alipur, for example, upper-caste, landowning farmers consistently blamed the welfarist, pro-lower caste policies of Indira Gandhi for the breakdown of the social order. Indeed, they did not feel that they owed the untouchable castes even the legally mandated minimum wages for they saw them as being coddled by the state. The older, upper-caste farmers were clearly nostalgic for the good old times when everyone knew his place. Ideas of male honor and lineage remain strong in local politics. The landless laborers, on the other hand, welcomed the developmentalist policies of the Indian state, including the populist but authoritarian policies of Indira Gandhi.

Such examples can be multiplied manifold. Rural people are not the only ones to display a schizophrenic culture. Even the most educated and wealthy urbanites show this kind of split consciousness in which modern technologies and all other insignias of modernity (Western attire, fast food, English language) are put on full display for the outside world, but modern ideals of reason and freedom are not allowed to enter the inner sanctum of family life and personal relations.

Anthropologists and psychoanalysts have called this pan-Indian characteristic "compartmentalization" (Singer 1972; Roland 1988). A split consciousness, with a great tolerance for contradictions and ambiguities, has been largely accepted as the Indian way of dealing with change. Indians modernize by first keeping the modern and the traditional separate, and then slowly accept the modern innovations by turning them into simply another form of tradition. India, it was declared, has a unique "cultural metabolism" which demands that "for a foreign import to enter the realm of traditions, it must become old, it must conform to the customary or the scriptural norms" (Singer 1972, 397). The modern, the innovative, the new, has to lose its novelty, its difference. It has to be cast in the old mold before it can be accepted. The deeply conservative nature of India's unique cultural metabolism is self-evident.

This downgrading of consistency in favor of compartmentalization and contextuality, already accepted as India's peculiar cultural style, got a boost from the relativist turn in social theory that we have been examining in this and the previous chapter. Intellectuals, whose job it is to agitate and educate on behalf of universal and humane values, began to see the protection of traditions from the onslaught of modernity as more important than combating the tyranny of traditions on social relations. Non-modern worldviews were indiscriminately declared to be "innocent" because of their victimization by the West. The problem is that those who appear as "victims" from a global anti-Enlightenment vantage point are actually the *beneficiaries* of traditional cultural legitimations, or are willing to upgrade their status within the traditional order ("Sanskritization") in order to have someone below them to look down upon.

### The Route to Reactionary Modernism

Why is this hybrid consciousness problematic? The simple answer is that it opens the way for false prophets, charlatans, and demagogues to manipulate popular consciousness. A couple of examples will clarify what I mean.

First, take the "progressive" case for the political uses of a morally coded understanding of nature, like the one we have examined in the preceding sections. The best example is provided by none other than Mohandas (aka Mahatma) Gandhi. There was a terrible earthquake in the northeastern state of Bihar in 1934. Gandhi declared it to be a "divine chastisement" for the sin of untouchablity. Gandhi's view was not unusual at all for his time, and as his respectful admirer, Partha Chatterjee points out, "could have been adopted by any number of the traditional intelligentsia . . . who shared the modes of thought of a large precapitalist agrarian society" (1986, 100). In any case, Gandhi was challenged by Rabindarnath Tagore who was appalled by his view of God who could inflict such indiscriminate punishment. Gandhi held on to his belief, admittedly derived from faith and instinct, that "physical phenomena produce results both physical and spiritual. The converse I hold to be equally true." Gandhi then went on to offer a political defense of his faith. He claimed in his response to Tagore that by making the sin of untouchablity the cause of the earthquake, he was interpreting the event in a manner that would aid self-purification and introspection among ordinary people about the practice of untouchablity. (Gandhi was opposed to the excesses of the caste system. But he was an ardent supporter of the basic fourfold division.) There are obvious contradictions in Gandhi' argument, for he completely overlooks that a moral coding of nature is partly responsible for condoning the caste system and untouchablity in the first place.[4]

Gandhi was arguing, in effect, that it did not really matter if a belief was objectively true or not, as long as it served a good cause. It did not matter if a belief contravened everything we know about the way the world actually works, as long as it was morally uplifting. Truly significant truths were not found through scientific experiments in the lab, Gandhi insisted, but through experiments in moral living.

It should be obvious by now that Gandhi's position was not very dissimilar from the cultural studies critics who favor interpretive flexibility toward the content and methods of science for the sake of building an oppositional consciousness against the dominating ideas of the modern age. Like Gandhi, they give the advancement of a political cause preference over the objective truth of the belief. Like Gandhi, they give the subjective, lived experience a crucial role in deciding what is true. They assume that *their* political cause, because it advances the poor and the oppressed, leads to a "stronger objectivity" anyway.

The problem is that interpretive flexibility which disregards norms of inquiry and the objective truth of beliefs is an all-purpose, equal-opportunity political tool of manipulation. With the Hindu nationalist parties in power, this process has gained a new momentum, especially in the so-called "tribal areas" where Hindu parties are trying to bring the population into the Hindu fold, or to "reconvert" them from Christianity to Hinduism. As the following two examples show, the traditional belief in the causal connection between human morality and natural order is serving as one of the vehicles for social engineering. The relevance of these examples to the hybrid consciousness that we have examined above will become obvious shortly.

Both examples come from northwestern India. The first is from a western Himalayan village, Singtur, described by William Sax (2000). Singtur is a village with the usual caste hierarchy, but with a predominantly non-Brahmin culture, which permits meat and alcohol consumption. Villagers believe in the connection between sin and natural phenomena and routinely appeal to their patron god, Karna, for relief from droughts, as well as from excessive rains. Karna answers their prayers by possessing his Brahmin oracle. Speaking as the voice of Karna, the oracle demands changes in social customs in return for relief. Invariably, the reforms demand behavior more in keeping with upper-caste Brahminical values. The oracle, for example, demands that villagers give up the practice of brideprice and instead adopt the upper-caste norm of "gift of a virgin" for which the bride has to bring a dowry. Drinking liquor and eating meat are condemned as sinful. A false understanding of nature, thus, serves to act as an agent of social engineering by exploiting the fear of calamity in people's mind.

The second example comes from a village of indigenous people (officially called the "scheduled tribes") called Alirajpur in the Narmada valley, in the state of Madhya Pradesh. In her illuminating ethnography of this area, Amita Baviskar (1997) describes how the "sins" of the villagers bring the goddess (Māta, or mother) sweeping through the area, passing from one village to the next, demanding "reforms." The goddess demands, through her many mediums including women, that the villagers adopt the social customs of "clean" castes. The villagers are exhorted to give up their marriage customs, which are actually more liberal for women than those of upper castes, and take on arranged marriages with dowry. Animal sacrifices in their worship ceremonies, meat-eating and drinking are all proscribed, while fasting, vegetarianism, and cleanliness are stressed. Under these conditions, can one depend upon the standpoint epistemology of the tribal people to serve as a sufficient defense against upper-caste pressures? It is not as simple as that, for the indigenous people have already imbibed many ideas from the surrounding culture so that Hinduization resonates with their existing culture (Baviskar 1997, 101).

Even though Hinduization is not the only demand the goddess makes, it is the most significant in the current political atmosphere. Hindu parties have been active in these areas to claim the indigenous people for Hinduism, or to "reconvert" the Christians among them to Hinduism. They use various means of inducement, from setting up Vedic schools and hostels, to plain coercion. These attempts are bearing fruit, as was clear in the riots in Gujarat in 2002 in which tribals and dalits took an active part in the anti-Muslim violence. It is in this context that the Māta phenomenon, which exploits a moral understanding of nature, takes on a dangerous meaning.

These two examples reveal, in pure form, the mechanism of how a moral conception of natural order is turned into an instrument of control. This kind of control reaches deep into the consciousness of the people, as it touches their faith.

The indigenous people are not the only group at risk for such manipulation. We have seen above that even technologically modern, prosperous farmers in the heart of Green Revolution areas have retained the traditional understanding of homology between natural phenomena and human affairs. Indeed, anthropologists studying an area of intense competition between indigenous people and caste-Hindus (in the eastern state of Bihar) find that it is the caste-Hindus (the Bhumihars) who host rain-making prayers every year with most pomp and show meant to assert their social dominance, as well as to propitiate the gods (Jackson and Chattopadhyay 2000, 153–154). Even those who have made substantial gains by adopting the new technology, still think and speak in the grammar supplied by an enchanted, alive, and holistic view of the world, linked however indirectly to the Great Traditions of Hinduism. The bioscientific worldview has not made much of an impact at the level of common sense of these beneficiaries of modern technology.

This kind of overlay of modern technology on a traditional mind-set is ripe for takeover by religious zealots. It is not a surprise that the Hindu nationalist virus has been slowly spreading into villages and small towns, even though it started as an urban middle-class disease. Of course, many of the underlying causes are material and economic, often having to do with land grab, business competition between religious communities, or quarrels over wages. But leaders—increasingly religious leaders and village priests—who can translate these demands into a religious idiom get an enthusiastic hearing. It is no secret that the Rāma temple agitation swelled the rural votes for the Bharatiya Janata Party. A party like the Bharatiya Janata Party is ideally suited for the new rural and provincial middle classes, for it promises outward signs of modernity, while also promising to restore the moral order (dharma) of the society. Indeed, the whitewashing of such obscurantist ideas as "Vedic science" is tailor-made for these hybrid farmers and other beneficiaries of economic modernization: it gives them the satisfaction of being modern and "scientific", without questioning their superstitions.

### Hybridity and the Fate of Contradictions

Theories of hybridity do capture the reality in the Indian countryside. The farmers are indeed hybrids. They treat modern technology purely contextually, as one more tool in their cultural toolbox, just as Harding and her numerous followers had hoped for. They retain the non-modern gestalt, even when they borrow the modern technology, just like critical traditionalists had hoped for. They think like cyborgs, for they do not separate nature from culture or culture from nature: sin brings about rain, and a life of dharma will ensure rains. And they do display an oppositional consciousness to Western ideas of modernity (breakdown of caste norms, the incipient signs of individualism, feminism, for starters), even while they use Western technology. Indeed, the critics needs not have worried, for there are not many signs of a "colonized mind" here: Indian farmers are happily taking what they need and fitting it into their non-modern gestalt.

The problem with postmodernist theorists of hybridity is obviously not their failure to capture the reality on the ground. The problem is *their celebratory stance toward hybridity as the politics of "emancipation."* Both politically and epistemologically, they have bent the growing evidence of reactionary modernity to fit into their own animosity against modernity and Enlightenment. They have allowed their own deep disillusionment with the West to turn every problem in modernizing societies into a pathology of the "colonized mind." They have allowed their own ideal of "emancipation" to celebrate even obscurantist ideas as progressive. Meanwhile, the religious right has learnt how to exploit this hybrid mind-set for its own purposes.

Paternalism toward non-Western cultures is matched by a breathtaking indulgence toward contradictions in epistemology. The factual understanding of nature upon which many of the non-modern values are constructed, continues to coexist with contradictory facts on which modern technologies are based. The Green Revolution farmers, as we have seen above, continue to believe in the defunct humoral-moral theory of nature, while using technologies derived from a modern bioscientific understanding of nature. There is no cultural agency—neither the schools, nor other institutions of civil society including the temples, religious leaders, the intellectuals, the media, non-governmental groups including new social movements for the environment, etc.—that seeks to educate the public opinion in a manner that brings the bioscientific worldview to challenge and falsify the humoral worldview. Of course, the mere presence of contradictory worldviews is bound to create some degree of intellectual dissonance, which is bound to lead to some spontaneous re-thinking and revision among ordinary people. But that this revision will happen cannot be taken as an automatic by-product of modernization. On the contrary, modernization of production relations creates a need for holding on to familiar signposts.

This is where the intellectuals come in. In a culture where modernist values have legitimacy among the intellectuals, there will be an attempt to question and remove the contradictions between the traditional and scientific understanding of nature. As we will see in the next chapter, India *does* have a tradition of this kind of modernist hybridity in which non-Vedic rationalist traditions are invoked to challenge and eliminate contradictions. Most ironically, the truly subjugated groups, who postmodernists are so concerned with, have been the most committed supporters of this kind of modernist epistemology in India.

The postmodernist movement in science has, on the contrary, served to actively *delegitimize* modernist values among intellectuals and social movements. Postmodernist intellectuals display a supreme unconcern with the coexistence of contradictions and even treat them as an act of "transgression" and "resistance." A case in point is Akhil Gupta, whose ethnography of the Green Revolution farmers we have encountered in the preceding section. Speaking from a postcolonial perspective derived from Homi Bhabha's impenetrable musings on hybridity, and Gayatri Spivak's advocacy of "strategic indigenousness," Akhil Gupta offers a celebration of hybridity. After describing the incorporation of modern technologies into a humoral worldview, and duly describing the deeply hierarchical and patriarchal values of the farmers, Gupta goes on to rejoice in the "multivalent" and "alternative modernity" of his farmer-subjects where "incommensurable conceptions and ways of life implode into each other, scattering, rather than fusing, into strangely contradictory yet eminently 'sensible' hybridities" (1998, 238). Leaving aside the vexed question of how incommensurable conceptions can "implode" into each other, there is a sense here of reveling in contradictions. The refusal to accept bioscientific explanations of the world as any more valid than their own humoral explanations, the farmers in Alipur, according to Gupta, are depicted as "destabilizing" the Western bipolar episteme and therefore, challenging the colonial power. Gupta rhapsodizes about the farmers' hybrid-ness in these words: "neither occupying a pure opposition to the modern, nor assimilable to a homogenizing Western episteme, farmers in Alipur constantly destabilized the oppositions that have framed explanations of the subaltern, Third World people" (ibid., 186). In Gupta's account such willful "mistranslations" and "alternative constructions of modernity" are a necessary aspect of how postcolonial people assert their difference and their agency in the face of dominance and inequality.

Gupta expresses in clear and exuberant language, the sentiments of his fellow travelers. In postmodernist discourse about cross-cultural encounters of science, the need for questioning contradictions and removing falsified conceptions is considered as oppressive and imperialistic. Obviously, one coherent story built upon the foundations of fundamental scientific laws would see non-modern, local knowledges as pre-scientific and in need for a critical updating. This is unacceptable to those critics of the West who see this updating as the source of the epistemic violence of colonialism.

The result has been an exuberant, unchecked growth of dangerous and uncontrollable hybrids. These hybrids are a part of the worldwide oppositional consciousness toward Western forms of modernity. But contrary to postmodernist hopes, this opposition is hardly conducive to a humane and tolerant future. And this is precisely where the dangers of postmodernism lie.

### Conclusion

The condition of hybridity is real. But contrary to postmodernist hopes of an "emancipatory" politics, in India at least, the hybrid consciousness has prepared the grounds for the onslaught of religious right-wing forces.

There is an alternative hybridity, a modernist appropriation of rationalist and naturalistic traditions. This hybridity is intolerant of contradictions and intolerant of objectively false, supernatural holism. It is this alternative hybridity that we turn to in the next chapter.

*Seven*

# A Dalit Defense of the Deweyan-Buddhist View of Science

*Be your own guide.*
*Take refuge in reason.*
*Take refuge in truth.*
—*The Buddha*

*The central problem of philosophy is the relation that*
*exists between the beliefs about the nature of things*
*due to natural science to beliefs about values.*
—*John Dewey,* Quest for Certainty

*If you meet the Buddha, kill the Buddha.*
—*A Zen Buddhist commandment*

Intellectuals have a choice to make, Max Weber wrote in his well-known essay "Science as Vocation": they can either become demagogues and treat ideas as "swords against enemies," or they can be teachers and use ideas as "plowshares to loosen the soil of contemplative thought" (Weber 1946, 145).

As the last two chapters have demonstrated, social constructivists and allied critics of science have lost hope in the Enlightenment faith that science can serve as a "plowshare to loosen the soil of contemplative thought." They see modern science more as a sword that the powerful wield against the powerless. They invite the powerless to wage a war against the hegemony of modern science by cultivating "their own" sciences.

To recapitulate some of the highlights, Sandra Harding, a leading feminist epistemologist, has argued that it is merely a sign of "civilizational Eurocentrism" on the part of the West to pretend that modern science is rational, objective, and universal. In reality, Harding claims, the vulnerable groups in non-Western societies actually experience the purported value-freedom and objectivity of science as a "rude and brutal cultural intrusion," because they "do not value [value]-neutrality; they value their own Confucian, or indigenous American, or Islamic or Maori, . . . or Judaic or Christian values" (1998, 61). Helen Longino has argued that feminists must join postcolonial peoples, especially the more oppressed among them, to develop an "oppositional stance" toward the dominant ideology coded into the "background assumptions, language, models and arguments and theories" of modern science (Longino 1997, 117). Well-known postcolonial critic, Ashis Nandy echoes this sentiment, arguing that "there must be skepticism against science [because] modern science is the basic model of domination of our times and is the ultimate justification for all institutionalized violence" (Nandy 1988, 121–122). Indeed, as I have tried to show in the last two chapters, the repudiation of modern science as an advance over other ways of knowing has become the first principle of postcolonial and post-development studies. For these "new cynics," to borrow Susan Haack's (1998) apt epithet, modern science has ceased to be a source of organized skepticism against dogma, but has become a new dogma that requires a radically skeptical scrutiny from the standpoint of its "victims."

What is truly amazing about this neo-cynical agenda is that it fails to even acknowledge—let alone allow any substantive role in understanding the historical role of modern science in the non-Western world—this one simple but inconvenient historical fact. There are many instances in the non-Western world when those who suffered the worst indignities and injustices heaped upon them by "their own" cultural-religious values, were the first to *embrace* modern science. One such example comes from the victims of India's hierarchical caste order—the untouchables, or dalits (literally, the "broken" or "crushed" people).[1] Dalit intellectuals have been among the most ardent advocates of a de-sacralized understanding of the natural world. Far from experiencing the objectivity and value-freedom of science as a "rude and brutal intrusion," important dalit intellectuals have *celebrated* the contents and the methods of modern natural science as a source of the demystification of the elite Hindu understanding of nature as permeated with Brahman, the divine spirit. Since Hinduism justifies untouchablity and caste hierarchy as being in accordance with the order of nature itself, the content and the method of modern science have obvious attractions for the victims of caste prejudices.

In this chapter, I hope to bring to light a remarkable confluence between dalits' aspirations for freedom, and the American pragmatist call for subjecting all inherited values to the test of scientific method. I will describe how Bhimrao Ramji Ambedkar (1891–1956), one of the most important dalit intellectuals of the twentieth century, was able to "hybridize" John Dewey's (1859–1952) conception of scientific temper with the teachings of the Buddha (563–483 BCE), and use the reinterpreted Buddhist tradition to challenge Hindu metaphysics and the ethics of natural inequality it sanctions.

Very briefly, Ambedkar spent three years in America as a graduate student at Columbia University (1913–1916), where he came under the influence of John

Dewey, then a professor of philosophy at Columbia and the nation's best-known public intellectual. Young Ambedkar attended Dewey's lectures, read his books, and took his ideas back to India with him. After many years of civil disobedience against caste discrimination, and after a long, frustrating fight against Mahatma Gandhi's paternalistic reformism which never renounced the principle of the fourfold division of human beings into castes, Ambedkar, along with nearly half-a-million of his fellow untouchables, publicly renounced Hinduism and converted to Buddhism. Ambedkar's call for making local cosmologies and sciences answerable to the findings and methods of modern science still reverberates in the neo-Buddhist and other segments of the contemporary dalit movement.

In this chapter, I will argue that how dalits understand science, and why they find it empowering, holds important lessons for contemporary critics. Ambedkar followed Dewey in finding in modern science an attitude, a temperament, that had the potential to challenge unexamined tradition and prejudices by cultivating a collective, democratic "will to inquire, to examine, to discriminate, to draw conclusions only on the basis of evidence after taking pains to gather all available evidence . . . to treat all ideas as working hypotheses to be tested by consequences they produce" (Dewey 1955b, 31). What is more, Ambedkar followed Dewey in believing that the *content* of modern scientific theories demanded rational acceptance by *all* people, universally, because these theories are the products of the most systematic practice of the scientific attitude. He believed that with modern science, a new kind of knowledge was born that could replace the supernatural, metaphysical knowledge, accessible only to the pure and the wise, with the fallible, testable experience of reality available to all. Ambedkar, again like Dewey, believed that the most important task facing intellectuals was to reconstruct the inherited cultural values and social ethics by bringing the spirit and the content of science to bear upon them. Both Dewey and Ambedkar would have actively resisted the project of "alternative sciences" as turning the clock back on the hard-won progress modern science has made in learning how to learn.

I will offer a reading of Ambedkar's *The Buddha and His Dhamma* (1992), the bible of dalit Buddhists, to show that Ambedkar turned Dewey's call for reconstructing philosophy and society in the light of scientific inquiry into the central message of the life of the Buddha. I will argue that Dewey's ideas helped Ambedkar make the historic rebellion of Siddharth Gautama relevant for his own quest for a civic religion of "equality, liberty, and fraternity" in India. Dewey was by no means Ambedkar's only inspiration: powerful nineteenth-century anti-caste movements in his own province were important influences, as were the histories of numerous heterodox, anti-Vedic, materialist sects/schools that have always existed on the fringes of Hinduism. But I contend that Dewey, and his American experience more generally, served for Ambedkar as a bridge between the past dalit traditions of protest, and a self-consciously liberal and secular worldview. By emphasizing scientific temper as the central message of the Buddha, Ambedkar made respect for systematic inquiry a part of the religious obligations of dalit neo-Buddhists. Ambedkar saw it as a creative reworking of the Buddhist tradition to serve the goals of human freedom in the modern age.

While Dewey's influence on the Chinese Enlightenment, the May 4th Movement, is very well documented (Westbrook 1991), as is his continuing influence in China today (Youzhong 1999), his indirect connection with the aborted *Indian* Enlightenment is hardly known outside the small circle of scholars. Unfortunately, even these scholars tend to treat Ambedkar's American experience and his great regard for Dewey as just one more biographical detail, "counting for very little" (Zelliot 1992, 85).[2] In the last few years, however, there has been a beginning of the long overdue recognition of the formative influence of Dewey and the liberal Protestantism of early twentieth-century America on Ambedkar's thought (Queen 1996). But overall, the Dewey-Ambedkar connection is very little known, the pragmatic reading of the Buddha is not much appreciated, and neither is its significance for the current quarrels over modernity, science, and Enlightenment recognized. Even more problematic is the relativist talk of 'a different voice' for dalits that is making its appearance in a segment of dalit scholars, albeit not without strong protest from others.[3]

This chapter is a contribution toward recovering the Dewey-Ambedkar-Buddha connection. The chief aim is not so much to add to the rich intellectual history of American pragmatism, although that would be a wonderful bonus. The motivation is to retrieve the ideas of Dewey—the original Dewey as Ambedkar understood him, and not the 'hypothetical' postmodernist Dewey made popular by Richard Rorty[4]—in order to dispel the cultural despair that has befallen Indian intellectuals and their postmodern allies in Western academia. I hope to ride the rising tide of pragmatism in North America and Europe to bring back a Deweyan respect for scientific temper to the continuing *Kulturkampf* in India against elements of Brahminical Hindu cosmology that proclaim permanent inequalities to be built into the very nature of some categories of people, natural objects, foods, occupations, and even gods themselves. Even more importantly, I hope to bring out the relevance of the Ambedkar-Dewey synthesis for the creation of a secular and humanist civic culture which can, hopefully, combat the rising religious nationalism in India today.

I will use Ambedkar's Deweyan Buddha as an ideal-type of a *modernist hybrid* which refuses to accommodate contradictory truths on a contextual basis. In contrast to the opportunistic and contradictory eclecticism of postmodernist and Hindu nationalists, Ambedkar represents an Enlightenment-style confrontation with traditions which seeks to challenge, falsify, and reject those elements of traditions that fail the test of scientific reason. Ambedkar and other Indian modernist, humanist intellectuals (including Nehru and many members of the

organized left) who held the advances in science as relevant for a growth of secular culture in India, sought to retrieve those traditions from India's history
which were conducive to a naturalistic (as opposed to enchanted) ontology and
critical (as opposed to mystical) thinking. Ambedkar's reading of the Buddha remains one of the most brilliant attempts to revivify India's critical and naturalistic
traditions so that the orthodox Vedic cosmology could be seriously interrogated
and the values it supports questioned. This was clearly the path *not* taken by the
mainstream of contemporary Indian intellectuals who are more comfortable with
the opportunistic eclecticism of "critical traditionalism" and/or postmodernism.
(Their attraction to postmodernism is perhaps related to the fact that postmodernism is very close to the orthodox Hindu way of drawing equivalences between different and contradictory beliefs.) Ambedkar's Deweyan Buddha shows
that the development of a modern, secular culture does not demand discarding
*all* traditions (which is impossible). What modernism requires is a revitalization
of those traditions which can provide a home to the temper of modern science.

In what follows, I will continue with my engagement with the critics of science
who we encountered in the previous chapters. I will start with juxtaposing the
main ideas of John Dewey's philosophy with the feminist and postcolonial theorists we encountered in chapters 5 and 6. The rest of the chapter will examine
how Ambedkar understood Dewey's ideas and reinterpreted them in a Buddhist
idiom. I will conclude with a consideration of the relevance of Deweyan Buddhism for our own times.

### John Dewey: Science and the Revaluation of Values

One wonders what Dewey—and more crucially, Ambedkar, whose entire project was motivated by the most vivid experiences of oppression—would
make of the feminist and postcolonial valorization of the subjective experience of
oppression as a source of more objective knowledge.

There is plenty in Dewey's writings to suggest that he would have seen such
valorization of experience as an "immunizing strategy by which the rationales of
oppression in other cultures can be protected from criticism" (Putnam 1992, 185).
Epistemologies that treat values derived from oppression as truth-enhancing can
end up shielding the very sources of oppression from a critical examination.
Dewey would see the subjective experience of the underdogs neither as a matter
of "epistemic privilege," nor as an indictment of the existing corpus of knowledge. He would instead treat the experience of oppression as a call for *enriching*
the subjective experience of the oppressed with intelligence and controlled infer-
ence. As Richard Bernstein points out, for Dewey, the proper contrast is "not between experience and reason, but between experience that is funded by the
procedures and results of intelligent activity, and experience that is not" (1971,
50). Far from being inimical, Dewey finds modern science to be indispensable for

enlarging and enriching the experience of ordinary people by enabling them to draw better-tested and more reliable causal connections between their experience and the underlying reality.

Dewey's idea of experience is as situated and embodied as the best that feminist theorists have developed. In his account, the "knowledge-experience" is never separated from the non-cognitive, affect-, habit-, and tradition-laden spheres of social life. He shared the feminist antipathy to a positivist conception of experience as a passive mirror of nature, and stressed the active and selective character of human experience. For Dewey, the data of science are not "the given" but rather "the taken," that is, they are selected from the totality of experiences by socially embedded human beings, with an express purpose of finding clues to the solution of the problem at hand (Dewey 1929, 178).

But no experience, and no values that we derive from reflection on that experience, can stand apart from what we already know about the world through science. To borrow a metaphor from Susan Haack (1993), science is a massive crossword puzzle we humans have been collectively trying to solve through all of known history. Our background assumptions—our guesses, worldviews, and biases, derived from habit, cultural traditions, or explicit political commitments— must face and adjust to the already completed entries in the crossword, just as the latter must remain open to revision in the light of the new clues our background assumptions lead us to discover. Given the crossword-like, multi-directional, and mutual checks-and-balances between values and the already known facts, determination of how good some evidence is is simply too deeply embedded in a whole web of other meanings and claims to be self-evident to *any* particular group of inquirers, be they distinctive by their class, gender, or nationality. Yet, social interests and values play a crucial role in influencing whether or not the warranted evidence will be *accepted* by a group of inquirers, and what *significance* and meanings would be given to a finding once it is accepted. Social interests do not serve as reasons for beliefs, but as causes for accepting or rejecting beliefs.

The hope that with the advancement of science, more and more of our accepted beliefs will *also* be the most warranted, animates the entire corpus of Dewey's writings. Dewey insists upon and welcomes "a certain purification of traditional beliefs," by making them face the tribunal of science (1955a, 30). To bring our metaphysical assumptions about the world in consonance with what and how we learn about the world was the whole point of Dewey's naturalistic humanism. Human values and purposes need not, any longer, be dictated to us by any external power—be it the Church, the state, the dictatorship of the proletariat, or custom and tradition. Rather, the success of science shows that human beings are capable of creating their own regulative standards by subjecting socially embedded experience to a collective, democratically conducted inquiry. As we shall see, it is this prospect of "purification" of traditional beliefs and a reconstruction of cultural values in a more rational and secular direction that Ambedkar retains from Dewey and finds a cultural ground for in the teachings of the Buddha.

For Dewey there is nothing imperialistic or scientistic about interrogating the experiential knowledge of lay people and non-Western cultures against the findings of modern science. A pragmatist in his philosophy of science, Dewey judged the validity of science not by antecedents or origins but by consequences, judged by success in predicting and controlling the course of nature. A naturalist in conviction, Dewey saw modern science as a continuation of the natural rationality of human beings, in all historical epochs and in all cultures, with the difference that modern science had learned not to stop inquiry prematurely, but to make productive use of doubt. Science had institutionalized doubt by converting it into a positive ethic of inquiry—what he calls "scientific attitude" or "scientific temper" defined as "the will to inquire, to examine, to discriminate, to draw conclusions only on the basis of evidence after taking pains to collect all available evidence" (1955b, 31).[5] This ethos was not limited to within the walls of the laboratory but could be used in all situations that become problematic in the course of human intercourse with nature and with each other. There was nothing scientistic about scientific temper in Dewey's view, for he, like his fellow pragmatists, saw it only as a more systematized version of an attitude shared by *all* people when they engage in inquiry.

Rather than take the experience of the subaltern, whether defined in terms of gender, class, or nationality, as a vantage point for creating new rules of inquiry, Dewey would call for extending the hypothetical stance of science to *all* "primary experiences" of *all* social groups alike, so that they are transmuted into "secondary experiences," that are "purified" and "enriched" by a process of systematic doubt and "regulated, reflective inquiry" by a democratic community of inquirers (Dewey 1929). Rather than treat everyday experiences, material practices, and emotions as knowledge-experiences, or as at least relevant to belief formation, as feminist epistemologists do, Dewey insisted that these primary experiences need the aid of systematic inquiry in order to understand the connections and interrelations between the various objects of our experience. The objects and phenomena we encounter in our primary experiences are "gross, crude and experienced as a result of minimum of incidental reflection" which are turned into "more refined, derived objects" of secondary experience which "*explain* the primary objects, enable us to grasp them with *understanding* instead of just having a sense of contact with them" (ibid., 5, emphasis in the original).

Scientific inquiry can have this purifying and enriching result because it routinely does what feminist and social constructivist critics claim cannot be done; that is, it helps inquirers separate and differentiate between their beliefs caused by "things being so," and other beliefs caused by "habit, weight of authority, imitation, prestige, instruction and the unconscious effect of language" (ibid., 14). Science, Dewey continues in his *Experience and Nature* (1925), helps to "depersonalize and de-socialize objects" of nature and in the process, emancipates the human imagination from the weight of social conventions. In stark contrast to the critics we encountered in the previous chapters who see scientific facts to be inseparable from social conventions, Dewey views the growing separation between social conventions and scientifically warranted facts of nature as the hallmark of scientific inquiry.

This purification of experience in scientific inquiry works by the same experimental logic of evaluation that Dewey proposes for ethics and morality. On a Deweyan understanding of science, we bring values we learn from our varied experiences in all aspects of our lives into the construction of scientific facts, but *our values themselves can be, and must be, warranted as any other judgment of fact.* Or as Hilary and Ruth Anna Putnam (1990, 410) put it, "any valuing can be evaluated," using the same laws of logic whether we are reasoning about an ethical question, or about a question in physics, or history, or any other field. Values, on Dewey's account, are not given to us by gods, nor are they mere individual whims or social conventions. Rather, *values are ideas that guide conduct,* and in that capacity, they are the means to solving a problem. They can therefore be rationally assessed in terms of their success or failure in solving the problem adequately. Thus, it should be an empirical question whether or not the values women, non-Western people, and others derive from their experience of marginalization do indeed make for more rational means of conducting scientific inquiry.

The very idea that values can be rationally assessed highlights one crucial difference with the critics we encountered in the previous section. For feminists and others influenced by social constructivism, knowledge of facts presupposes social and cultural values. Dewey and other pragmatists accept this feminist insight, but add that *knowledge of values simultaneously presupposes facts of nature.* Dewey is a thoroughgoing naturalist because he treats both epistemic values (i.e., norms of inquiry) and meta-epistemic values (i.e., goals of inquiry) as amenable to rational evaluation using the established methods of natural science. This Deweyan naturalism is elaborated upon by Larry Laudan (1996) who treats epistemic values as "hypothetical imperatives;" they hypothesize effective means for realizing cherished ends.[6] Given that underneath the great cultural variety of beliefs and ways of knowing all cultures do cherish and try to maximize the reliability of their beliefs, modern science provides methods that have proven their reliability and can be used to assess the reliability of other methods favored by other cultures.

As we will find in the following sections, Ambedkar clearly shared Dewey's faith that scientific knowledge about the natural world cannot remain limited only to specialized, technical fields but must influence the values and purposes of social life. He was drawn to science precisely because of his Deweyan belief in the possibility of rational evaluation and reconstruction of moral and ethical values. The stakes were much higher for Ambedkar. As a member of a community that had endured grave injustices legitimized by an objectively false understanding of nature's laws, Ambedkar saw cultural demystification as the first priority for any progressive social change in India.

### Ambedkar's Confrontation with Tradition

October 14, 1956 holds a special significance for the dalit community in India. On that day, Bhimrao Ambedkar publicly renounced Hinduism and converted to Buddhism. He was not alone: half-a-million other untouchables converted with him.[7] Ambedkar died shortly afterwards. He is reported to have spent his last hours putting finishing touches to *The Buddha and His Dhamma*, published posthumously, translated into Marathi and Hindi, and accepted as a sacred book by neo-Buddhists in India.

In this section, we will follow Ambedkar through his quest that led him to the Buddha, the quintessential Eastern teacher, in whose teachings he found a distinct echo of John Dewey, the quintessential Yankee pragmatist.

The biographical details are well documented. Ambedkar was born in 1891, in the western state of Maharashtra in a Mahar family. Mahars were considered among the lowest of the low in the local caste order.[8] His father was employed by the British army.[9] The army connection secured him an education in English, even though in segregated settings. With financial help from an enlightened local royal, he was able to go to college, first in Bombay, and later in the United States of America and in England.

Ambedkar spent three years (1913–1916) in New York City at Columbia University, where he worked for his Ph.D. in economics. He seems to have availed himself of the courses offered by "as many top ranking professors at Columbia as he could, whatever their field," including John Dewey, Edwin Seligman, James Harvey Robinson, and Alexander Goldenweiser who gave him a "broad and deep exposure to an optimistic, expansive and pragmatic body of knowledge" (Zelliot 1992). Chris Queen (1996) has speculated that Ambedkar could not but have been influenced by Dewey and also by the overall progressivist and modernist *zeitgeist* that prevailed at that time in America. Ambedkar's years in New York City coincided with the peak of the Social Gospel Movement in liberal Protestant circles that dominated Columbia University, the Union Theological Seminary, and other institutions in Morningside Heights, where Ambedkar lived. Queen believes that the beginnings of Ambedkar's liberation theology can be traced back to the religious modernism of liberal Protestant movements.

But it seems that Dewey was the closest to a guru Ambedkar had. Ambedkar not only followed his ideas all his life but, according to his wife, Savita Ambedkar, "happily imitated John Dewey's distinctive class room mannerism—thirty years after he sat in his classes" (Zelliot 1992, 84). (Columbia University conferred an honorary doctorate on him in 1952. Last year, a bust of Ambedkar was installed on the campus of the university.) It is not known, however, if Dewey was aware of the influence he had on Ambedkar. As far as I have been able to ascertain, the two were not in any direct communication, although there is some evidence (Raina and Habib 1996) that Dewey maintained a professional association with nationalist Indian intellectuals associated with the so-called Bengal Renaissance. Ambedkar went on to obtain a D.Sc. from the London School of Economics and to pass the bar exam, returning to India for good in 1923.

For more than a decade after his return, Ambedkar remained optimistic that political and economic changes—access to education, right to vote, etc.—would suffice to integrate the lower castes into the national mainstream. But the bitter struggles of untouchables to exercise their right to drink water from village wells (the famous civil disobedience at Mahad), their right to enter Hindu temples (the temple-entry movements of Pune and Nasik), and his bruising debate with Mahatma Gandhi over the question of separate voting rights for outcastes (the famous Poona Pact of 1932 in which Gandhi prevailed) led him to realize that advancement of the untouchables was impossible without a prior reform of the core values of Hinduism. His disillusionment with Hinduism, and with the largely upper-caste nationalism of the Congress party which put political emancipation from the British above any urgency for internal social reform, was complete by 1935 when he first declared his intent to renounce Hinduism. "I was born a Hindu, but I will not die a Hindu," he is reported to have told Mahatma Gandhi. His quest for a new faith that could anchor his values of "liberty, equality, and fraternity" ended, 20 years later, with his conversion to Buddhism.

What does this religious conversion have to do with questions regarding the place of science, scientific temper, and critique of inherited cultural values? The short answer is everything, because Ambedkar's Buddha was reason and scientific method sacralized. And it was Dewey who gave Ambedkar a mold of scientific reason into which he cast his image of Buddha. In order to appreciate the Deweyan elements of Ambedkar's Buddha, it is important to understand the philosophical source of his disillusionment with Hinduism.

In his short and bitterly angry book, *Annihilation of Caste* (1936), Ambedkar asks why upper-caste Hindus tend to treat their fellow beings with aversion, refusing participation in the "associated activities" of everyday life—eating together, living together, working together, praying together, marrying into each other's families? His answer: they shun social intercourse with fellow human beings not because they are "inhuman or wrong headed . . . but because they are deeply religious" (ibid., 111). The myriad hierarchies and taboos of caste have the "sanctity of the *shastras* [Hindu scriptures] . . . people will not change their conduct unless they have ceased to believe in the sanctity of the *shastras*" (ibid., 112). The real enemy, Ambedkar declares, "is not the people who observe caste, but the *shastras* that teach them this religion of caste" (ibid., 111).

Ambedkar launched a bitter attack on Gandhi's reformist attempts to just say no to untouchablity, while publicly praising the divinely sanctioned *chaturvarna* (the four basic hereditary occupational groups or varnas) as an antidote to Western-style individualism. Ambedkar argues that untouchablity is not an aberration of Hinduism that can be rooted out by good works, moral exhortations, or new legal codes. Rather, untouchablity is a logical corollary of the Hindu understanding of nature, God, and human agency, bound together by the law of karma.

Here Ambedkar came face-to-face with the famous non-dualism and holism of classical Hinduism, celebrated by feminist, postcolonial, and neo-Hindu intellectuals. For Hinduism, karma serves as a theory of cause and effect, occupying the same cultural space as natural laws of physics and genetics. Because the famously non-dualistic Hindu ontology does not separate matter or physical nature from spirit or the moral realm, karma acts to transfer causation from the realm of consciousness/spirit to the realm of nature/matter, and vice versa. Thus, immoral actions in the present or past lives ("karmic crimes") activate different proportions of the five elements of nature (*gunas*) in each person that make him/her innately more or less pure. Treating nature as sacred or as inseparable from the supernatural has implications not just for how we treat nature, but also for how we treat fellow human beings, something Ambedkar was all too painfully aware of. The peculiar nature of Hindu metaphysics which rationalizes injustices and misfortunes as the inevitable consequence of the workings of the laws of nature led Ambedkar to an appreciation of scientifically justified laws of nature which *separate* the physical from the cultural conceptions of the divine immanence. He insisted that the removal of untouchablity will require a "notional change . . . a change in the state of mind" (ibid., 111).

There are two aspects of Ambedkar's call for the annihilation of caste that are of great relevance to the feminist debates. One, in a complete refutation of those who condemn science and the Enlightenment as Eurocentric and colonial, Ambedkar was making a classic case for an Enlightenment-style critique of religious reason in India. He expressly and repeatedly invoked the ideals of the French Revolution—"Liberty, Equality, and Fraternity"—and like the philosophers of the French Enlightenment, gave primacy to scientific reason as the new standard for a "constant revision and revolution of old values" (ibid., 132). Ambedkar realized, more clearly than any other early modern Indian reform movement and without even a trace of compromise with the myth of India's "golden age," that the inherited cosmologies and epistemologies could not—indeed *should not*—survive an encounter with modern science and still retain their legitimacy as explanations of nature.

In this he was not alone. Two other contemporary anti-caste movements—one led by Jotirao Phule (1826–1890), a "touchable" but of backward caste, in Maharashtra, and the "self-respect" movement, led by E. V. Ramasami "Periyar" (1879–1973), in the southern state of Tamil Nadu—had already emerged as nuclei for an alternative form of nationalism. These non-Brahmin nationalists, unlike the Congress, demanded not just freedom from the British, but also freedom from internal, homegrown oppressions. Phule's high regard for natural philosophy and its use by radicals like Thomas Paine to challenge Christianity is well documented (O'Hanlon 1985), as is Periyar's radical empiricism and his avowed atheism (Geetha and Rajadurai 1998). Far from non-Western cultures valuing their own values over the value-freedom of science, as Harding claims, all three of these major movements of the oppressed were unequivocal in their preference for the purely naturalistic and verifiable worldview of modern science of Galileo, Newton, and Darwin. The turn towards Buddha (Periyar, too, held the Buddha in great regard) and ancient non-Brahmin philosophies was in part an attempt to find a cultural homologue for materialist and skeptical traditions in India's minority non-Vedic traditions.

Second, at no point in his writings did Ambedkar romanticize the experience of his fellow caste members as a source of superior knowledge. His entire project was motivated by a great empathy for his long-suffering community, but this love never turned into a romance. The "monster of caste" crosses *everyone's* path alike, every which way you may turn: "you cannot have political reform, you cannot have economic reform, unless you kill the monster [of caste]" (Ambedkar 1936, 5). Unlike some who claim that the dalits' experience of oppression has made them egalitarian and non-patriarchal (see Deliege 1999 for a review), Ambedkar recognized that the experience of oppression has *also* left them deformed and in need of repair: "tolerance of insults and tyranny . . . has killed the sense of retort and revolt. Vigor and ambition have completely vanished from you. All of you have become helpless, unenergetic and pale. Everywhere there is an atmosphere of defeatism and pessimism . . ." (Ahir 1997, 17). While he acknowledged that centuries of caste oppression gives the untouchable community a greater objective interest in annihilating rather than modernizing caste, he never forgot that they had also internalized the Hindu worldview that naturalizes hierarchy and that they also recognized the distinctions of caste. Indeed, that was one of the reasons why he felt that socialism in India was not possible without a prior secularization of the social consciousness of the working classes (Ambedkar 1936, 74), a position which put him at odds with both the Marxists and the Gandhians.

Ambedkar answered the classic question facing all revolutionaries "What can

be done?" with a bold call for the annihilation of the worldview that allows and justifies caste. To accomplish this task he constructed a Deweyan Buddha.

### Ambedkar's Deweyan Buddha

Ambedkar's Buddha teaches how to bridge the gap between facts and
values, between how we know about the world we live in, and how we treat our
fellow beings. Ambedkar understands the Buddha as teaching *prajna* (understanding) in order to create bonds of *karuna* (love) and *samata* (equality). Of the
three, *prajna* is central, for without it the other two can falter: "the path of all passion and all virtue . . . must be subject to test of prajna or intelligence, . . . because
without intelligence, generosity may end up demoralizing and love may end up
supporting evil" (Ambedkar 1992, 30). It is in how Ambedkar understands *prajna*
and the centrality he assigns to it for democratic change that Dewey's presence
is most palpable, and also where Ambedkar stands in stark opposition to all contemporary advocates' standpoint epistemologies of the oppressed.

Briefly, in *Annihilation* he urges his fellow Indians to forego the quest for certain and absolute knowledge of the ultimate Truth of Being, the kind of metaphysical knowledge idealized by Brahminical Hinduism. In words that distinctly
echo Dewey's, he proposes a new ideal of knowledge which embraces change,
and which will learn to constantly revise all that is taken as settled: "the Hindus
must consider whether time has not come for them to recognize that there is
nothing fixed, nothing eternal, nothing *sanatan* [Sanskrit for eternal]; that everything is changing, that change is the law of life for individuals as well as for society. In a changing society, there must be a constant revolution of old values and
the Hindus must realize that *if there must be standards to measure the acts and
men, there must also be a readiness to revise those standards*" (1936, 132, emphasis
added).

Ambedkar bases his call for the transvaluation of values on two long quota-
tions from Dewey (without citing the source) to the effect that it is our duty "*not*
to conserve and transmit the whole of our past achievements, but only as much
as makes for a better future society" and that we should not make "the past a rival of the present, and the present a more or less imitation of the past." The contrast with underdog epistemologies is clear: inherited values are to be critically
examined, not "privileged" as sources of better truths.

What will break the spell of the *sanatan* or the eternal is reflective thought,
which Ambedkar understands in a classic Deweyan manner and connects to the
teachings of the Buddha. Most of our life is unreflective and habitual, he says.
Only a situation that presents a dilemma forces us to reconsider our habits and
the philosophical assumptions that support those habits. He cites the case of
caste-Hindus traveling in railway trains where it is impossible to maintain the customary caste distinctions as an example. There are two ways, Ambedkar says, to

deal with this crisis that modernity has engendered. One way is to follow what the Hindu sacred books commend, that is, to consult first the Vedas (the revealed knowledge), then the *smriti*s (the law books) and only then *sadachar* (customary morality). This traditional "solution" only legitimates a schizophrenic life in which "a Hindu" accepts the modern technological conveniences like train travel, but then comes home and undergoes a *prayaschit*, or repentance, for breaking caste prohibitions. Following traditional values in a modern world will force "a Hindu . . . to break caste at one step and to observe it at the next, without raising any questions." Ambedkar is highly prescient here. As we have seen in the previous chapter, keeping the modern world limited only to technological gadgetry, but isolated from the values that guide the inner life and social ethics is a major mechanism for the remarkable survival of traditional values in India. Although most Indians no longer come home and do penance for breaking caste rules, the compartmentalization of things scientific for the outside life of work, and traditional virtues for the domestic life of marriages, friendships, life cycle rituals remains widespread.

Ambedkar's preferred solution bears the stamp of Deweyan thinking which calls for breaking down the compartments between the instrumental and ethical implications of science. Ambedkar argues for actively using the same scientific revolution that replaced the bullock-cart with a train to reshape the Indian society's understanding of natural laws and its preferred modes of fixing beliefs. He argues for the need to develop new principles of validating facts, and then using these principles to judge if the traditional facts about the natural and social order are warranted. In his magnum opus, *The Buddha and His Dhamma*, he accomplishes just such a task.

The bridge concept is *prajna*, or understanding. Ambedkar presents the Buddha as giving permission to ordinary men and women, regardless of their station, to trust their experience over the authority of the learned Brahmins encoded in the Vedas. But at the same time, the Buddha encourages them not to treat their own experience, at any time, as infallible and exempt from revision. As against the unchanging cosmic order of the Vedas and the Upanishads, the Buddha taught that *everything* is always changing and there is no continuous and coherent self that experiences an unchanging reality. To cling to the idea of permanence is the source of suffering, while cultivating an attitude of "mindful contemplation" of the ever-changing reality is the way to master and overcome suffering. Ambedkar interprets mindfulness as Deweyan scientific temper: "everything must be open to re-examination and reconsideration, whenever grounds for re-examination and reconsideration arise." A re-examination, backed by "logic and proof," and conducted with the spirit of "freedom of thought," will itself change what the inquirer will value: not the certain knowledge of ultimate reality, but reliable knowledge of here and now. Ambedkar reads the Buddha, like Dewey, as offering a method, not a doctrine, as the source of enlightenment.

Ambedkar presents the Buddha's own renunciation and enlightenment as an exercise in *prajna*, with nothing pre-ordained or divine about them. In Ambedkar's retelling, Siddhartha Gautama turns his back on his family and on his tribe not as a fulfillment of a pre-ordained fate, as the Buddhist lore would have it, but as a conscientious objector to war and violence that his caste duties demanded of him. Ambedkar depicts him leaving home not in the dead of the night, stealthily, but openly, with a public affirmation of his pacifism, and in full consultation with his wife and his family. Not finding any of the existing philosophies helpful in explaining the cause of social conflict and suffering that he finds all around him, the young Siddhartha resolves to "examine everything for himself," to hold nothing as infallible and permanent, including the Vedas. His Enlightenment, in Ambedkar's retelling, amounts to the discovery of a method by which to conquer the ignorance, the cravings, and the hatreds that hold humanity in thrall. By making the spirit of inquiry the essence of Buddha's spiritual journey, Ambedkar is making inquiry itself a part of the religious duties of (the mostly) dalit neo-Buddhists.

In his first sermon at Sarnath, Ambedkar has the Buddha proclaim a rough paraphrase of the pragmatic maxim as a centerpiece of his Middle Path: "you may ask, ye Parivrajatkas, why are these principles [of the Path of Purity] worthy of recognition as a standard of life," the Buddha says. "The answer to this question you will find for yourself if you ask, 'are these principles good for the individual? Do they promote social good?'" This leads him to fashion a crude pragmatic maxim: if it makes no difference to our experience, it is meaningless. Ambedkar has the Buddha apply this rough and ready pragmatism to deny the existence of God, the immortal individual soul (atman), the universal soul (Brahma), and all supernatural forces (book III, part IV). Repeatedly, the Buddha refuses to answer any question about metaphysical matters, finding them "not tending to edification." But the answers Ambedkar's Buddha does favor—those amenable to logic and proof, and those that also promote human well-being—help him to radically redefine all the major conceptual categories of Hinduism of his time; purity simply becomes "good conduct," nirvana becomes "control over passions through knowledge and understanding," dharma becomes "energetic action," soul, "consciousness emerging from matter."

Ambedkar's most touching—and to some most controversial—reinterpretation remains that of the doctrine of karma. Buddhism, as is well-known, rejects the idea of immortality of the soul, but accepts the idea of rebirth according to the laws of karma. In the traditional Buddhist writings, the soul is simply replaced by unidentified immaterial constituents that carry over the traces of karma into the next birth. Thus, "even though Buddhism rejects the existence of the soul, this makes little difference in practice, and the more popular literature of Buddhism, such as the Birth Stories (Jatkas), takes for granted the existence of a quasi soul which endures indefinitely" (de Bary 1958, 92).

But Ambedkar seized upon this contradiction and asked: How can there be rebirth if there is no soul? Ambedkar applies the Buddha's own pragmatic maxim to this contradiction in Buddha's teachings. He argues that "if there is anything that can be said with confidence [about the historical Buddha] it is this: He was nothing if not rational, if not logical. Anything, therefore, which is rational and logical, other things being equal, may be taken to be the word of the Buddha." Going by this interpretive principle, Ambedkar assumes that the Buddha would, if he could, agree with the findings of science. Thus we find Ambedkar committing the worst kind of presentism: he brings in the laws of Mendelian genetics and the laws of conservation of matter and energy to argue that karma cannot be inherited in the absence of the soul, and that all that is reborn is matter. To deny the immortality of the soul, Ambedkar interprets the Buddha as saying, you have to deny karma and rebirth, the latter decided by the karmic account of the soul's journey through all the past lives. Denuded of its metaphysics, and contained strictly within one lifetime, karma simply becomes another name for humanism in that "moral order rests on man's own actions and not on anyone else."

*Prajna* held a deeply political meaning for Ambedkar. He saw it as the necessary first step toward the creation of a "religion of principles" which free, equal, and self-respecting people could practice, as against the "religion of rules" which divides people into castes, makes them cower in fear of unseen powers and robs them of an "associated life" in the public sphere. Like Dewey in *A Common Faith* (1934), Ambedkar sought to separate the religious attitude (the religion of principles) from its institutional trappings (the religion of rules). *Prajna* allows him to equate "the cleaning of the mind as the essence of religion." A genuine religious attitude becomes simply to act mindfully, to act with consciousness and responsibility and not obey any rules laid out in advance.

In his interpretation of the other aspects of Buddhism, Ambedkar foregrounded the social message of the Buddha. As Chris Queen has argued (1996, 59–63) Ambedkar offers a contemporaneous interpretation of the Four Noble Truths based upon three modernist hermeneutic principles, namely, reason, social benefits, and certainty. For example, he interprets the Buddha as teaching that it is class struggle, ultimately caused by human passions, that is the source of *dukkha*, or sorrow. Similarly, he interprets the Eightfold path as the way to "remove injustice that man does to man." Traditional Buddhists have reacted with dismay at this "corruption" or secularization of Buddhism. But Ambedkar was doing something that Buddhism has always permitted, namely, a fresh reading of Budhha's Dhamma according to the needs of the time. Ambedkar was offering a fresh reading of Dhamma for this suffering community.

## *The Social Significance of Ambedkar's Buddha*

Ambedkar's Buddha is not meant for dalits alone. Given the close relationship between caste and gender, Ambedkar's intervention is of great significance for women as well. But his true significance is much broader. Ambedkar poses a challenge to the cultural common sense of the entire Indian society. In turning to the Buddha, Ambedkar is attempting to "extend the reach of reason" into the moral sentiments of the Indian society at large, an Enlightenment-style *kulturkamph* that Amartya Sen (2000) has recently argued for. (I, too, have argued a case for re-igniting the Enlightenment process in India, Nanda 2002).

There is no doubt that a turn to Buddhism has given the ex-untouchables, especially those from Ambedkar's own Mahar community, a new self-confidence and a new culture, complete with less superstitious and simpler life–cycle rituals. Many autobiographical accounts and sociological studies attest to the changes: replacing old idols, fasts, and ceremonies with readings from *The Buddha and His Dhamma* conducted by *anyone* with a short training as a Buddhist priest, an explosion of creative expression of new rationalist, humanist themes through song and poetry, and above all, a growing sense of self-worth and pride (see Moon 2001; Narain and Ahir 1994). According to Timothy Fitzgerald, a scholar of Japanese and Indian Buddhism: "Buddhists have achieved a new identity, symbolized by their change of name. The new optimism is strongly connected to ideas about self-reliance, rejection of the old Hindu subservience and a rational approach to their own self-development" (1994, 20).

But all is not well. The ex-untouchables have found the Buddha, but they have not yet killed the Buddha. As Fitzgerald's field studies (1994), and also the important study by Burra (1996) clearly show, the change in identity is not necessarily accompanied by a change in worldview, especially among the lower-income, rural neo-Buddhists. Their "village Buddhism" (Fitzgerald's term) tends to make new gods out of the Buddha and Ambedkar and fit them into the Hindu pantheon. Old Hindu forms of idol worship, ancestor worship, and even the ideas of purity and pollution, directed at castes "lower" than themselves remain widespread. Fitzgerald finds that it is the more educated, politically mobilized minority among neo-Buddhists who take the scientific temper of the Buddhist teachings into their lives.

Ambedkar was the first to link caste with gender inequities. He located the connection between oppression of women and the perpetuation of caste hierarchy in the morbid fear of miscegenation, expressed in the strict rules of caste endogamy that are still the norm in India. Whereas all class societies impose restrictions on women's sexuality in order to ensure patrilineal succession, Hinduism imposes especially stringent restrictions on upper-caste women in order maintain the purity of castes. Purity of women is central to Indian patriarchy, because the purity of castes is contingent upon it.[10]

The brunt of this obsession with caste purity was, and still is, borne by the women of upper castes. Controlling their sexuality and life choices was the easiest way of making sure that they were not accessible to men of lower orders. The peculiarly harsh rules of upper-caste patriarchy—child marriages, veiling, and withdrawal from productive work outside the home, compulsory ascetic widowhood and even sati, and the entire ideology of *pativrata* (worshiping the husband like a god)—become explicable when seen as so many devices to ensure endogamy.

Ambedkar made these connections early on in his Columbia days. He came back later, in his capacity as India's first law minister, to argue for a reform of Hindu personal laws regarding marriage (outlawing polygamy, legalizing intercaste marriages), divorce (ensuring women the right to divorce), and inheritance. He realized that the constitutional provision of caste equality in the public sphere would be meaningless without social provisions for women's equality in the private sphere.

While dalit communities and movements are not free from their own forms of patriarchy, overall, organized protest movements of dalits have historically displayed an exceptional degree of solidarity with women's struggle for greater autonomy. One milestone of mutual support among dalits and women was the support an earlier lower-caste movement led by Phule (see the preceding section) gave to Pandita Ramabai, one of the pioneers of Indian feminism. Ramabai (1858–1922) was a remarkable woman. The daughter of a poor but progressive Brahmin, orphaned and widowed at an early age, she was unique among women for her knowledge of Sanskrit. Given her command of Sanskrit, she was expected by male reformers in Bengal and her native state of Maharashtra to preach a reformed, monist version of Hinduism to women. But reading the sacred texts in the original, combined with her first-hand experience of the "social death" that accompanies widowhood, Rambai rebelled. She converted to Christianity, went to England, and later to the Unirted States of America, where she wrote a book, *The High-caste Hindu Woman*, and lectured to raise funds to start a home for young widows in India. Her independence of mind, and especially her conversion to Christianity, brought her nothing but denunciations from well-known Indian male reformers. Only Phule and his anti-caste movement defended Ramabai, and did so publicly. Phule, like Ambedkar later, was able to connect Ramabai's critique of the religious ideas and rituals that justified women's oppression to the scriptural justifications for caste hierarchy (Chakravarti 1998). The third major protest movement of backward castes—the self-respect movement of Periyar— was an early and passionate campaigner for women's equality in the public and the domestic sphere, and tried to institutionalize new rituals for weddings which respected women's choice of partners across castes, allowed widow remarriage and did away with vows derogatory to women (Geetha and Rajadurai 1998).

What makes these movements a class apart is their uncompromising defense of reason and science as weapons for resistance and reform, as the Weberian

plowshares of the mind. These movements did not "valorize the subjective experience" of the underdogs. Their approach was thoroughly Deweyan-Buddhist. All
myths, taboos and commandments were to be examined against the accepted
standards of modern science at that time. Second, it was "their own" homegrown
values justifying oppressions that were their chief targets. Secularization of the
imagination was their central goal. While Ambedkar and Periyar were unequivocal in their opposition to the exploitative economic policies of the Western imperialists, they held Western political and natural philosophy in the greatest esteem.
Unlike today's radicals, they never made the mistake of conflating the content and
the method of science with the political-economic power of the West.

   This original Buddhist-Deweyan spirit has taken quite a beating in India recently.
Some dalit and "backward"-caste intellectuals have taken an anti-modernist view
and romanticized the knowledge-traditions of dalits. Scholars sympathetic to
ecofeminism (Datar 1999) and Gandhian traditionalism (Nigam 2000) have argued
that dalits should treat the disembedding from traditional communities caused by
modern technology and capitalism, and not traditions themselves, as their primary
cause of concern. These critics accept the postcolonial argument that the modern
secular worldview is silencing the experiential knowledge of dalits and women.

   But these identitarian tendencies are kept in check by those who grant a very
limited epistemological privilege to the raw experience of oppression. They admit that dalits and dalit women "talk differently," but only insofar as their experiences give them access to problems that may not register as problematic to
non-dalits. As Sharmila Rege (1998) has argued, given the preponderance of urban, upper-class/caste intellectuals in India's social movements, the concerns
specific to dalits—the everyday indignities, violence, and social apartheid—often
get subsumed under the standard rhetoric about the working class, the sister-
hood, the environment, etc. It is this silence that dalit intellectuals are trying to
break. But by and large, dalit scholars show a salutary weariness with identity-
based epistemologies, for they fear that, "to privilege knowledge-claims on the
basis of direct experience, on claims of authenticity, may lead to a narrow identity
politics" (Rege 1998, WS44). They have asked, if "experiential knowledge of dalits adds up to a 'knowledge system' [can it provide] tests of verifiability and vali-
dation?" (Guru and Geetha 2000, 133). To the extent these dalit modernists do
privilege dalit culture, they privilege the "proto-scientific" aspects of it which are
rooted in manipulation and control of nature in the process of production, which
was the province of the laboring castes (Ilaiah 1998).

### Ambedkar's Epistemological Revolution

   The relevance of Ambedkar extends beyond dalits and women. His true
relevance lies in his insistence upon an *epistemological revolution as a precondition for the creation of democratic and secular habits-of-the-heart.*

In the seamless meshing of John Dewey and Gautam Buddha, Ambedkar points the way for launching an epistemological revolution in India that draws upon those traditions that share a philosophical continuity with modern science. He found these traditions in the pragmatic and naturalistic ideas of the historical Buddha and early Buddhism that lie buried under the mystical idealism of the later Buddhism and Brahminism.

Because the pragmatism Ambedkar finds in Buddha's teachings bears an uncanny resemblance with American pragmatism's anti-metaphysical bent, it may create a suspicion that it was smuggled in by the America-educated, Dewey-influenced Ambedkar. But there is good evidence that the historical Buddha himself was deeply influenced by anti-Vedic naturalist philosophies prevalent in pre-Buddhist India, especially those of Lokāyata (literally, "prevalent among the people") and Sānkhya (literally "reflection"). The non-Brahmin, lower-caste Lokāyata philosophers were famous (or infamous, among the priestly castes) for insisting upon putting the teachings of the Vedas, including all the rituals and spells meant to bring about a desired result, to a "test of practice" in everyday life. For the Lokāyata "that alone is true which proves itself to be so in practical life" (Chattopadhyaya 1976, 235). Like their contemporaries in Athens before Socrates, the Lokāyatas denied any notion of a self (i.e., consciousness or atman) over and above the material body, saw all consciousness as an attribute of matter itself, and tried to explain all observed phenomena by natural laws. They admitted no God, no soul, no survival after death. It has been established that references to Lokāyatas, both in positive and in critical vein, abound in early Buddhist texts (Chattopadhyaya 1959, 47).

Likewise, there is evidence that Siddharth Gautama, before his enlightenment, had studied with Sānkhya philosophers in his own hometown.[11] The *original* Sānkhya philosophy accepts non-sentient matter (*prakriti*) as the only and the first cause of all of nature, including sentience or consciousness (*purusha*) (ibid., 381). Like the Lokāyata, Sānkhya too holds that the regularities and laws of *prakriti* can be understood through evidence of experience in here and now. This position is in stark contrast to that of the idealistic monism of the Upanishads, which treats the *purusha*, or the spirit, as the first cause of all of nature, the latter having a status of mere illusion. True knowledge, the only kind of knowledge worth pursuing, according to these Brahminical doctrines, is that of *purusha,* which being a part of the Absolute Divine Consciousness, is declared to be beyond the grasp of all mundane experiences of ordinary mortals.[12]

Thus, the pragmatism that Ambedkar highlights in the Buddha was always there in the Indian intellectual history. *But—and this is crucial—it was always there as the repressed and the despised "other" of the true, transcendental knowledge of the Brahmins.* Time and time again, the mystic idealism of Brahminism has ridiculed, absorbed, and in other ways demoted the empirical, experimental understanding of nature as *avidya* or false knowledge before the transcendental knowledge of the spirit: "Gods love the mystic," "Gods are fond of the obscure and detest direct knowledge," the *Satapatha Brahmana* taught. The available historical texts tell the grim tale of the victory of Brahminism; all that we know of Lokāyata comes from the scorn and ridicule that was heaped upon it by the exponents of Vedānta; Sānkhya was forced to compromise with Brahminism by gradually making consciousness (*purusha*) the cause of nature (*prakriti*) and by re-introducing thoroughly un-Buddhist metaphysical questions into the later Buddhism.

What Ambedkar was trying to do was to revivify these suppressed proto-scientific elements of Indian traditions by presenting them as the core of the Buddha's teachings, and thereby turning them into the sacred duty of his fellow Buddhists. He believed, rightly, that the philosophy of nature and the norms of reasoning of science were fully commensurable with *these*, rather than with the Vedic monist traditions. Unlike the contemporary critics who see science as oppressive to those at the bottom, Ambedkar wanted to put science in the hands of the oppressed to challenge the traditional sources of oppression. Unlike those who condemn science in the name of the oppressed, he celebrated science as the weapon of the weak.

What is more, Ambedkar placed the need for this epistemological revolution in Hinduism at the center of the cultural revolution needed for India to become truly democratic and secular.

Creation of new cultural habits is the most daunting task facing the Indian democracy. Study after study shows that the constitutional promise of equal rights as citizens, regardless of caste, creed, and gender remains unfulfilled in practice. There is no doubt at all that the right to vote and the demands of the modern economy have largely destroyed the political and economic basis of the social order based upon caste, even though enormous disparities still remain. Ideology, too, has followed a secularizing trend. Although a nostalgic, communitarian/Gandhian defense of the traditional four varnas as a source of harmony is commonplace, only a minority of die-hard ideologues defend the karma-dharma basis of the caste and gender order. Therefore, one often hears of the irrelevance of the Ambedkar critique of religious reason. There are calls to "move on" from all quarters, including neo-Gandhians (Sheth 1999), feminists (Kishwar 2000), and even dalits themselves (Teltumbde 1997).

An unabashed, full-throated defense of the rituals and the morality of hierarchy has indeed become rare in the public arena. But the actual practice of caste continues. Take for example, the recent Amnesty International and Human Rights Watch reports on the rape of untouchable women as a means of punishing the "uppity" untouchables, the brutal "honor killings" involving upper-caste women marrying lower-caste men, the wide caste gap in literacy, employment, and in ownership of assets, the continuation of deeply humiliating practices of segregating dishes and even throwing food at untouchable servants even in middle-class urban homes, and the continuing restrictions on commensality, especially in rural areas.[13]

This gap between the rhetoric and the reality stems from the fact that the idea of equality has not become a habit-of-the-heart, a truth that Indian people hold as self-evident. Respect for equal rights of individual citizens may be *seen* as a *legal* imperative, and but it is not yet *felt* as a *moral* imperative. While modernity has truncated the many linkages between ritual status and unearned material privileges, the fundamental, metaphysical assumptions about nature and man's place in it that sanctified hierarchy still enjoy the status of self-evident truths, upheld by countless rituals and sacred texts.

As we have documented in the first part of this book, anti-colonial sentiments in the nineteenth and twentieth centuries led Indian nationalists to defend and even glorify as "scientific" the metaphysical doctrines of karma, dharma and the immanence of the divine in nature. This trend of rendering Hindu holy books into "Vedic science" is continuing unabated, encouraged by the postmodernist erosion of all demarcation between science and myth. Thus, an insidious, powerful doublespeak, perpetuated by politicians, clergy, and secular anti-colonial intellectuals alike, continues to dominate Indian society where proclamations for equality are accompanied with obeisance to the worldview and the metaphysics that make inequality appear natural, just, and moral.

Ambedkar's challenge to Indian intellectuals is to put an end to this doublespeak by persuading the Indian people of the validity of a rational and naturalistic metaphysics. Ambedkar, like Dewey, recognized that values presuppose certain facts about nature, and that in order to change the values—to reset the moral compass of the nation, so to speak—Indian masses will have to be introduced to a new understanding of the workings of nature. Ambedkar realized, again like Dewey, that the method and institutional arrangements for arriving at the new knowledge of nature will have to be made the basis for a new civil society in which norms are not dictated from above, but arrived at through a rational, open, and public inquiry. His turn to Buddha was an attempt to resurrect the cultural traditions of skepticism and reason that have for so long been silenced by the mystic idealism of Brahminism.

### Conclusion

Ambedkar is an exemplar of what I call *a prophet facing forward*. Like the philosophers of the French Enlightenment who turned to the humanistic heritage of the Greeks in order to break with institutionalized Christianity, Ambedkar turned to India's Buddhist heritage in order to break with institutionalized religions and the worldview they sanctioned. In Buddha he could find those traditions which satisfied the demands of reason and naturalism *as they have evolved in modern science*. Ambedkar accepted the universal legitimacy of science. He understood that modern science had made a break with the sacred sciences of the past. As a victim of Hinduism's sacred cosmology, he welcomed this break and sought to institutionalize scientific reason for the neo-Buddhists by interpreting it as the essence of Buddha's teachings.

In Buddha he found the beginnings of a kinder and gentler view of the human condition, a view which set aside the old Brahminical quest for unanswerable metaphysical certainties (of encountering the Absolute Spirit face-to-face, for example) and accepted the fallible reason of mortal human beings as the only guide to action. In Buddha, Ambedkar found the still viable remains of ancient India's naturalist and anti-metaphysical traditions which could be legitimately linked to the contemporary spirit of scientific inquiry, understood in the skeptical spirit of the pragmatists like Dewey.

Even more importantly, Ambedkar, unlike the Marxist critics of religion, never dismissed the need for the sacred in everyday activities. In Buddha he found a way to combine a view of the sacred that did not offend the dictates of reason. Of course, as we have seen, the defenders of Vedic science *also* claim to find resources for scientific reason in the Vedas and the Vedānta. The problem with the latter is that they *redefine the norms* of reason themselves as those conforming to the Vedic norms (e.g., associative logic, reasoning by homologies), and *then* claim to find homologies between them and selective interpretations of modern physics and biology. Ambedkar, on the other hand, was trying to reform these very same Vedic norms themselves, to conform to what nature and reason have come to mean after the Scientific Revolution.

It is obvious that the postmodern skepticism regarding modern science completely disarms and delegitimizes Ambedkar's project. Postmodernists, like the Vedic science proponents, demand a radical and culturally relative redefinition of the norms of reason and conceptions of nature.

One can safely conclude that the only option for the friends of the oppressed in the postcolonial world is for them to recognize that the interest of the oppressed in secularization and demystification of traditional ideologies is best served by the naturalism and skepticism of modern science. It would be fair to say that modern science *is* the standpoint of the oppressed.

*Part III*

# Postmodernism and New Social Movements in India

*Eight*

# The Battle for Scientific Temper in India's New Social Movements

*Scientific temper . . . is the temper of a free man.*
—*Jawaharlal Nehru,* Discovery of India

*The ultimate logic of scientific temper is the vulgar
contempt for the common man it exudes.*
—*Ashis Nandy, "A Counter-statement on Humanistic Temper"*

I write this chapter with much trepidation, for I am afraid of running into ghosts from times past and places left behind.

The setting was New Delhi in the early 1980s. I, then a young microbiologist, had freshly arrived in the metropolis from a provincial town for my doctorate at the prestigious Indian Institute of Technology. It is there that I read in print a scathing attack on all that I held (and still do) holy, what used to be called "scientific temper" in India those days. That essay by someone named Ashis Nandy, was followed by an equally venomous attack by Vandana Shiva, both unfamiliar names to me then. Many other attacks followed, in turn followed by many counter-attacks from assorted scientists, historians, and others who disagreed with the critics. The debate went on for a few months in the pages of a left-wing weekly called *Mainstream*. It was the Indian version of the science wars, what I call "Science Wars I," preceding by nearly two decades a similar showdown in the United States, the "Science Wars II," brought on by the outrageous parody of Alan Sokal. The difference, of course, is that Science War I was won by the naysayers.

I did not know then, even though I was very disturbed by the critics, that the famous "scientific temper debate" was the beginning of the end of the intellectual consensus in India over the ideals of a secular modernity. I did not know then, even though I tried to put in my two-cents worth of defense at that time, that I would write a book answering the critics after so many years.

I feel compelled to revisit the scientific temper debate for two reasons. First, it marks the beginning of the kind of cultural-relativist critique of scientific rationality in India that later connected with the anti-Enlightenment strands in science studies. Given the intimate relations the Indian critics of modernity have with science studies and postcolonial studies, the scientific temper debate rightfully belongs to the prehistory of these fields of inquiry. Second, this debate marks the fateful retreat of Indian intellectuals from any serious engagement with the religion question. With their demands of "authenticity" and "mental-decolonization" in matters of science, the critics were playing on the natural terrain of religious-cultural nationalists. It is not a surprise that when the Hindu zealots regrouped to fill in the vacuum left behind by the collapse of the Congress system in the fateful decade of the 1980s, there was simply no effective, organized opposition to the *ideas* of these religious nationalists. (There is, of course, plenty of opposition to the *politics* of the right.) The debate over scientific temper is important, therefore, as the beginning of the grand betrayal of the clerks in India.

I examine, below, the debate, its context, and its reverberations through the burgeoning new social movements in India.

### "Scientific Temper": Its Nehruvian Supporters . . .

If it were not for the raw nerve it touched off among the detractors, the "Statement on Scientific Temper," a five-page document in a little-known left-wing journal, *Mainstream*, would have met the same fate that awaits most such statements of good intentions, namely, oblivion. For that is what the scientific-temper document was—a statement of good intentions of bringing reason to bear upon the intellectual and moral climate of Indian society at large. The signatories of the document had no power to implement any policy change, nor were they backed by the government. The "Statement" was put together at the end of a junket organized by the Nehru Centre, an organization that propagates the ideas of Jawaharlal Nehru, India's first prime minister. The Nehru Centre had invited some 30 scientists, historians, and public intellectuals of all stripes for a seminar in October 1980. These men and women had enjoyed four days of good conversation and fresh air at a verdant holiday spot, at the end of which they brought forth this statement arguing for more scientific temper and less obscurantism. All this is typical academic conference-circuit stuff.

The "Statement" was a sobering stocktaking. More than 30 years after independence, it noted, India had made significant strides in developing institutions for scientific education and research. Already in the 1980s, India boasted of the world's third-largest scientifically trained workforce. The country was also striving mightily for self-sufficiency in industrial capacity. But India, as the signatories very correctly pointed out, was becoming a land of stark contrasts. Sophisticated high technology coexisted with grinding poverty, world-renowned institutions for higher education coexisted with huge pockets of complete illiteracy, the great abundance of scientific expertise coexisted with a very high degree of ad hocism and personality cults in the government. Even where schools existed, they specialized in "stupefaction of the young." Even in the institutions of scientific learning, authoritarianism and superstitions reigned. There was a "retreat from reason," an atmosphere of "conformity, non-questioning and obedience to authority. Quoting authority . . . was substituting for enquiry, questioning and thought."

The signatories looked to Jawaharlal Nehru's Enlightenment faith to find a cure for this cultural malaise. They zeroed in on the concept of "scientific temper" culled from Nehru's writings. Following Nehru, the "Statement" described scientific temper as an attitude of mind which is universally shared by all and applicable to all aspects of life, including the evaluation of values. It called for cultivating science education and scientific thinking as a social force for "changing the intellectual climate of our people." Glimpses of Nehru's paean to science in his *Discovery of India*, to "the adventurous and yet critical temper of science, the search for truth and new knowledge, the refusal to accept anything without testing and trial, the capacity to change previous conclusions in the face of new evidence, the reliance on observed fact and not on pre-conceived theory, the hard discipline of the mind" (Nehru 1988, 36), are all too evident in the "Statement of Scientific Temper."

As must be evident in light of our encounter with Ambedkar's reading of Dewey, there are striking similarities between the two views of scientific temper and its relevance for creating a liberal political culture. This is understandable in view of the so-called "positivist" or anti-metaphysical conception of the world that Nehru shared with Ambedkar and Dewey and indeed with all the leading thinkers of that era, including the Marxists. Even though Marxists obviously did not see secularization as a prerequisite of a just society in India, as Ambedkar did, and as Nehru hinted at (although he mixed it up with a large dose of Hindu romanticism), they were at the forefront of rationalist and secular movements. Many Marxists led exemplary lives, challenging traditional taboos, especially of caste. The conception of science as a progressive, secular (i.e., anti-metaphysical) and universal method of inquiry was the leading conception around the world, before it was declared to be a conservative, paradigm/culture-bound way of knowing in the post-Kuhnian philosophies of science. The "Statement on Scientific Temper" was the last hurrah of this optimistic, progressive conception of science in India.

Like Nehru and Ambedkar (although the statement does not refer to Ambedkar at all), the signatories of the statement wanted this attitude to become a "way of life, a method of acting and associating with our fellow men." They recommended that this attitude be taught in schools, promoted in public institutions, and put to work in all spheres of social life *in order to serve the ends of justice*. The signatories were not flying the flag of reason in the service of some abstract notion of truth, but in their belief that the search for truth serves the ends of justice.

They wanted a critical, scientific conception of the world to enable citizens to understand better the anatomy of religiously sanctified stratification and social barriers. They wanted citizens neither to reject out of hand, nor accept without questioning, the traditional answer to questions about nature and society. They wanted ordinary people to ask: What exactly did the metaphysicians, the philosophers, the priests mean? Did their answers measure up to what was known about the world through the empirical inquiry conducted according to the methods of science? As a consequence of this inquiry, the Nehruvian advocates of scientific temper hoped for a secularization of consciousness to emerge: "Scientific temper . . . leads to the realization that events occur as a result of interplay of understandable and describable natural and social forces and not because someone, however great, so ordained them" ("Statement of Scientific Temper" 1981, 8).

Scientific reason in the service of social justice and secularization: this, in brief, was the content of the "Statement."

A word about those ghosts I mentioned earlier. News of this debate reached me in the Indian Institute of Technology, in New Delhi, where I was working for my doctorate. I read the "Statement" and followed the debate with great interest. I attended the many seminars that were organized under the leadership of the Centre for the Study of Developing Societies where the opponents of scientific temper used to gather. (Considering that the Centre for the Study of Developing Societies and the Indian Institute of Technology are the opposite ends of the vast expanse of Delhi, attending these seminars was not without its hardships!) I took the debate seriously—perhaps too seriously—as it was questioning the value of one of the most formative experiences of my own life, namely, an education as a biologist. I had experienced my encounter with natural sciences, especially molecular biology, as extremely enlightening and liberating. My training in science had set me free from many of the fears and taboos that came with my traditional middle-class upbringing. It is the questioning of authority and demystification of the world that I valued most in science. I eventually gave up on a career as a research biologist. But it was because I thought that science was too valuable to be constrained within the walls of the laboratory. My tragedy was that precisely at the time when I gave up the lab for science popularization through journalism and popular movements, the tide began to turn. The modernist, secularizing impulse started to fade, even among the popular science movements I held in the highest esteem.

### . . . And Gandhian Opponents

It did not take long at all for critics to launch an attack. It started with the publication of a "Counter-statement on Humanistic Temper" by Ashis Nandy (1981), which recycled many themes he had developed in a lecture he delivered in 1980.[1] The implied conflict between "scientific" and "humanistic" set the tone for what was to come.

In the time-honored tradition of all romantics, Nandy lost no time in bringing out the "P" word. He declared the authors of the "Statement," and by implication Nehru as well, to be "fourth-rate pamphleteers for modern science and maudlin ultra-*positivists*." Like Nehru, the signatories of the statement want this attitude to become a "way of life, a method of acting and associating with our fellow men" (Nandy 1981, 17).

Calling one's opponents "positivist" is one of the most effective ways to silence them, for then all the epistemological shortcomings of logical positivists—which are many—are easily mobilized to eclipse the revolutionary anti-metaphysical program of the positivists. It is easily forgotten that the Vienna Circle included socialists like Otto Neurath and many liberal internationalists and staunch anti-fascists like Rudolf Carnap and Moritz Shclick.[2] Indian humanists and intellectuals have shown an inordinate eagerness to use "positivist" and "scientistic" as four-letter words to be flung without discretion at any one who dares to suggest that developments in modern natural science demand a critical clarification of traditional cognitive and moral commitments. Even before the onslaught of radical social constructivism, Nandy and associates had honed the art of invoking Western critics of science to silence the domestic advocates of science—all in the name of "decolonization of the mind!"

Nandy and associates argued that because the Western philosophers and social critics themselves were rejecting modern science as epistemologically privileged, Indians who went around signing statements of scientific temper were dupes and fools. The gist of the argument was this: Post-positivist critics, among whom Herbert Marcuse, Paul Feyerabend, and Thomas Kuhn came in for special mention, have already "proved" that science does not and cannot confront dogmas, because science *needs* dogmas (i.e., paradigms) in order to do its work. In light of the recent developments in the "radical" understanding of science, those who still view scientific temper as a source of cultural critique are behind the times. They are promoting a culture that accepts the hegemony of science, when what is needed is a consciousness that "accepts science as only one of the many imperfect traditions of humankind and which allows the peripheries of the world to reclaim their human dignity and reaffirm those aspects of their life on which the dignity is based." A society like India, on the periphery of the world, needed to protect itself from, rather than cultivate, the hegemony of the "obscene and amoral logic of science" (Nandy 1981, 17).

The twin themes of epistemic parity (modern science as only one among other ways of knowing, no better and often much worse), and populism (the right to live by one's own traditions), became the refrain of the critics of scientific temper. These themes have continued to serve as guiding lights for the new social movements engaged in environmental, ecofeminist, post-development mobilizations. A whole generation of social activists and science studies scholars has emerged in India that takes these two themes as axiomatic. Even those who have doubts in private, dare not challenge the consensus in public for fear of being labeled as "positivists" and elitist.

Claiming support from "seventy-five years of work in history, philosophy and sociology of science," the "Counter-statement" declared that science itself had shown limits of science. There was no need for the younger generation of scientists and social activists in India to follow the outdated ideas of Nehru to treat science as a universally valid method. Nandy assured the younger generation, who he explicitly tried to address, that *science is no less determined by culture and society than any other human effort*" (emphasis in the original) and there is no need to pay it any special attention. We encounter here the beginnings of the social constructivist discourse in India, which only grew more abstruse in writing, but shriller in tone, as it gained more prominence in the West.

Spurious claims of universalism and superior truths, it was argued over and over again during the debate and in the books and anthologies that followed, serve to enforce the authoritarianism of science, for they deprive other traditions from challenging modern science. (This theme later flowered into postcolonial theory once it connected with Foucault and Said.) The universalism of science converts the plurality of traditions, each with its own methods of acquiring and affirming beliefs, into a hierarchy of methods with science at the top. The critics accused the Nehruvian promoters of science of disrespecting and insulting the ordinary citizens by daring to suggest that they need to revise their ways of knowing. True equality demands that ordinary people should have as much right to question science from *their* perspective. The most common example cited was that of astrology. Inspired by Feyerabend, the self-proclaimed defenders of the common man argued that astrology was the myth of the weak, as much as science was the myth of the strong. Scientists and those who dare criticize astrology as a superstition must take the empirical experience of the ordinary people as evidence, and rethink *their* opposition to it. Astrology stood for local knowledges in general. The modern West was exhorted to learn from the epistemologies of the weak, the downtrodden, and the victims of modernity.

The "Counter-statement" was the beginning of a new movement with roots in the alliance between the so-called "alternative West" and Gandhian populism. In other words, Western critics of the West, especially those in the new humanities, joined forces with the "victims" of the West in ex-colonial societies in search of a new model of modernity and science that embodied a more compassionate view of humans and nature. Right on the heels of the scientific temper debate, there was a tremendous rush of energy throughout the 1980s which gave birth to what has been called the "alternative science movement" in India (Guha 1988). The major figures of alternative science, Ashis Nandy, Vandana Shiva, Claude Alvares, Shiv Visvanathan, and in his own way, Rajni Kothari, the mentor and father-figure to these critics, were all associated with the Centre for the Study of Developing Societies and its activist affiliate, Lokayan, in Delhi. (Since its inception in the mid-1960s by Rajni Kothari, CSDS has been involved in developing indigenist social sciences to break the hegemony of Marxist scholarship. Its turn to Gandhian thinking was a part of the general disenchantment with the course of modernization after the Emergency. For more on the larger context, see the next section.)[3] The "Delhi school of science studies," as I think of this elite group, joined forces with another like-minded group, the so-called "Patriotic and People's Science and Technology," or PPST. The members of Patriotic and People's Science and Technology are inspired by the works of Gandhian historians of indigenous sciences. They sought to demonstrate the continued relevance, if not superiority, of traditional knowledge systems. Sympathetic intellectuals from a variety of appropriate technology, rural development, and anti-imperialist movements from India, Pakistan, Malaysia, Sri Lanka also joined in. Major environmental movements like Chipko and the movement against the dam on Narmada, are closely associated with the alternative science movement.

A number of seminars were organized which produced an abundance of anthologies with titles announcing "requiems" for the "death," "end," and "twilight" of modernity. Some of these books gained a worldwide following. Notable titles include: *Science Hegemony and Violence* (1988), *Traditions, Tyranny and Utopias* (1987), both edited by Nandy, *Science, Development and Violence* by Claude Alvares (1992), *The Revenge of Athena* by Ziauddin Sardar (1988), and *Staying Alive* by Vandana Shiva (1988). These ideas became the signature tune of Third World Network and other advocacy groups leading to the *Penang Declaration on Science and Technology* (Sardar 1988). Organizing nationwide conferences promoting traditional science and technologies has been another major activity of alternative science movement groups. Gradually, the agenda of many alternative science movement groups, especially those pushing for "patriotic science" has become indistinguishable from the Swadeshi Science Movement of Hindutva sympathizers. We have already discussed in Part II how these ideas connected with the social constructivist and feminist critiques of modernity in the West.

In practical terms, alternative science enthusiasts work at two levels: producing theoretical defenses of traditional sciences (especially in medicine and agriculture) as viable scientific alternatives, and in practice, encouraging and agitating for the preservation and spread of traditional sciences at the grassroots level. The limitations, and the misconceptions, the lack of sound empirical testing, the objective falsity of the assumptions about nature that traditional medical practices like Ayurveda, or many agricultural practices involve is of no great concern.

To hold these practices to critical scrutiny would, in their view, amount to insulting the ordinary people. And yet, surprisingly, these groups bristle at being called anti-science. According to a recent overview of science movements by V. V. Krishna (1997, 398), alternative science movement groups see themselves not opposed to science, but "only" opposed to its "social, political, and cultural hegemony." The problem is that there is not much of "hegemony" of science to begin with. Of course, there is much attention to cultivating technology and some useful science (especially for warfare), but these are geared to instrumental uses of science. Science is far from being hegemonic in the basic assumptions about the world, neither does scientific rationality provide much of a model for public discourse. Under these conditions, it is not the hegemony, but the very legitimacy of science that gets attacked.

In the next two sections I want to turn to two fairly obvious questions. One, why did the alternative science movement appear when it did? Two, why should we care?

### The Context of the Debate

The "Statement of Scientific Temper" was just so many words on paper. Why did it provoke so much debate? Why were the attacks so vicious? And why were so many well-meaning and intelligent men and women, deeply committed to the cause of social improvement, ready to rally around Nandy's "Counter-statement?" There is no doubt that the debate had touched a raw nerve all around. Intellectuals in India, especially among the non-Marxist left (which in India means the Gandhian left, as there is not much of a self-consciously liberal left) were ripe for the populist message of Nandy and other critics of scientific temper. So wide and powerful has been the impact of alternative science movements that even those among the Marxist left, who disagree with the backward-looking romanticism of the movement, have felt compelled to compromise with them (next section). The enthusiastic reception of the very idea of alternative or indigenous sciences demands an explanation.

Interestingly enough, the fault lines in Indian politics that gave birth to the indigenist tendencies in the new social movements of the left, also gave a new lease of life to the Hindu nationalists on the right. *The alternative science movements and the Hindu nationalists are twins*, born of the same events and carrying forth the same traditionalist social agenda, with the crucial difference that the former lack the latter's penchant for strong-arm tactics and its venomous hatred of religious minorities.

The birth of these twin movements dates back to the call for "total revolution" by "JP" in 1974, which in part, precipitated the Emergency that Indira Gandhi imposed in 1975. Jayaprakash Narayan, or JP as he was called by one and all, was a veteran Gandhian leader who had come out of retirement to lead a student movement in Bihar, later Gujarat, and eventually in all of India against the regime of Indira Gandhi.

The rise and fall of Indira Gandhi is well documented.[4] What is relevant for our purpose is the combination of populism and authoritarianism of her regime that provoked a backlash from a right-left coalition. This coalition, led by Jayaprakash Narain, turned opposition to Indira Gandhi's policies into a referendum against the kind of modernization the Congress party had pursued since India's independence in 1947. Jayaprakash Narain's call for "total revolution" was a calculated call to discredit the "imported" ideals of industrial development, liberal democracy, and secularism. "Total revolution" was a call for returning the country to its own traditions, its own native genius, which supposedly was best represented by Mahatma Gandhi. Gandhi's essentially upper-caste Hindu traditionalism was to be the guidepost of India's future. There was enough in this vision to attract both the Hindu nationalists from the right, and the cultural nationalist, anti-imperialist intellectuals, and activists from the left.

Indira Gandhi, as is well-known, had fought her way to the top by sidelining and/or co-opting the party bosses of the Congress. She sensed, correctly, that universal suffrage and modernization had fueled great expectations among ordinary people. To ensure Congress rule, she co-opted the pro-poor policies of the communists and other regional parties. Under the banner of *Garibi Hatao* ("End Poverty") she built the famous coalition of lower castes and religious minorities to face the traditional, middle-caste oligarchs in the countryside. But unable to meet these promises, and keen on maintaining her personal dominance in the party, her regime became increasingly authoritarian and corrupt. Shortages of food and fuel, spiraling prices, and unemployment added to the sense of a crisis.

University students in the states of Gujarat and Bihar were the first to take to the streets to protest against her. Starting in late 1973–early 1974, students organized street demonstrations against bread-and-butter issues like unemployment and high prices. They also demanded political changes ranging from clean government, decentralization, and "Bharatiya education." These movements were spearheaded by the student wing of the Rashtriya Swayamsewak Sangh, which linked up with the political activists of Jana Sangh, the predecessor of the Bharatiya Janata Party, and the political arm of the Rashtriya Swayamsewak Sangh. But many socialists and non-communist left parties also participated in these movements. (The communist parties resolutely stayed away, branding the emerging coalition as fascistic.) The Rashtriya Swayamsewak Sangh and Jana Sangh saw this movement as their opportunity to become "respectable" again, after years of being shunned as the party that killed Mahatma Gandhi. Hindu nationalists approached Jayaprakash Narain, a Gandhian with most excellent credentials, to lead the student movement. Jayaprakash Narain's leadership brought the Rashtriya Swayamsewak Sangh and affiliates back into the mainstream of

Indian politics. After many months of extra-parliamentary agitation to remove the state government in Bihar, Jayaprakash Narain decided to nationalize his total revolution. This is when Indira Gandhi clamped down on all dissent by imposing the infamous Emergency in June 1975.

To understand the agitation over scientific temper, one has to understand the ideological content of Jayaprakash Narain's movement. As the detailed account by Christophe Jaffrelot (1996) makes amply clear, devout Gandhians like Jayaprakash Narain had long felt betrayed by the industry-heavy modernization policies of the "Nehru dynasty" which they saw as marginalizing the social message of Gandhi. According to Jaffrelot, "if in March 1974, at the age of 72, JP agreed to return to the political scene, it was because he had seen in the student movements in Gujarat and Bihar a chance to rehabilitate certain Gandhian priorities" (ibid., 260). What were these "certain Gandhian priorities" that Jayaprakash Narain sought to rehabilitate? These priorities included, above all, a "purification" of the government and the political system by countering state power from above by reasserting the primacy of society over the state, as Hindu customs and history prescribed. Jayaprakash Narain's "total revolution" visualized rehabilitation of village-based cottage industries, utilizing local knowledge and skills, and managing their affairs through village assemblies or *panchayat*s (ibid., 262–266). Decentralization of power, patterned after traditional social arrangements, was to be the answer to modernization which had been promoted by Nehru and then by his daughter. Anti-statism, coupled with a bias for the "genuinely indigenous," was the minimal basic agenda of all those who united under Jayaprakash Narain's leadership.

It is my contention that the coalition of the left-wing, non-Marxist socialists and the right-wing Rashtriya Swayamsewak Sangh, Jana Sangha, and their student wings that coalesced around Jayaprakash Narain's Gandhian program of Total Revolution never really broke up, even after the Jayaprakash Narain movement itself died out: the left and the right Gandhians have pursued a very similar, religion- and tradition-infused agenda in the three decades since the Jayaprakash Narain movement. The left-wing Gandhians gravitated toward new social movements for "alternative" and "patriotic" science/development. The leading lights of these alternative science movements gained international recognition, as they linked up with the Western anti-Enlightenment intellectual currents. The right-wing Gandhians are none other than the present day Bharatiya Janata Party, Rashtriya Swayamsewak Sangh, and their various affiliates. What unites the left and the right Gandhians is "Gandhian socialism," also known as "Vedic socialism" or "Integral Humanism."

The left–Gandhians of alternative science movements that we are discussing in this chapter have resolutely tried to differentiate themselves from the right-wing Gandhians by insisting that the Hindu right wing has nothing in common with the "real" message of Gandhi which they, the alternative science/post-development movements, represent. Indeed, it is almost a truism among the Indian left intellectuals that Gandhian thought has nothing whatsoever in common with Hindutva, and that the Bharatiya Janata Party and the Rashtriya Swayamsewak Sangh are making an opportunistic use of Gandhi. They make much of the fact that Nathuram Godse, Gandhi's assassin, was a member of the Rashtriya Swayamsewak Sangh (see Noorani 1997, for a good example of the left-wing view of these issues.)

There is no doubt at all that Gandhi himself had no sympathy for the vicious anti-Muslim and anti-Christian sentiments that mark the muscular nationalism of Hindu nationalists. Hindu nationalists themselves recognize this difference, declaring Gandhi to be "too saintly" to fully comprehend the "evil" of Islam (Agarwal 1999). But when it comes to the larger vision of a good society, the Hindu right wing's relation to Gandhian philosophy is far from opportunistic. Indeed, the official philosophy of the Bharatiya Janata Party, the "integral humanism" of Deendayal Upadhyaya, is almost an exact paraphrase of Gandhi's vision of a future India. Both seek a distinctive path for India, both reject the materialism of socialism and capitalism alike, both reject the individualism of modern society in favor of a holistic, varna-dharma based community, both insist upon an infusion of religious and moral values in politics, and both seek a culturally authentic mode of modernization that preserves Hindu values (see Fox 1987).

The left/postmodernist Gandhians have refused to recognize the vast agreement between the Hindu right wing and the holistic socialism of Gandhi. Using the discourse theory of Foucault and Said, they have argued that Hindu nationalists are corrupted by Oriental stereotypes of India's spirituality and contaminated by the allures of Western theories of nationalism and modernity. They have used Gandhi as their mascot of an authentic Indian alternative, totally uncontaminated by Enlightenment ideologies of reason, progress, and secularism. (This thesis of Gandhi being uncontaminated by Western ideas of India has been refuted ably by Richard Fox [1987] who shows in great detail how Gandhi's understanding of India was shaped by European Romantics. Be that as it may, it is clear that in their zeal to oppose Western power/knowledge, the left/postmodernist Gandhians have closed their eyes to the historical connection between neo-Hinduism and Gandhian thought. They have in the meantime, continued to propagate, in theory and in practice (of new social movements), the Gandhian ideas of cultural authenticity, anti-secularism, and "community." Their insistence upon cultivating an "alternative" and "patriotic" science belongs to this overall project.

### Impact on New Social Movements

India is the land of a million mutinies. The Jayaprakash Narain movement set an example of a non-communist, extra-parliamentary route to social change. It was followed by a flourishing of numerous new social movements of radical activists, working in small autonomous groups, trying to address every imaginable social cause including popular science, education programs, health, agriculture, forestry, environment, women's rights, civil liberties, tribal welfare, housing for the poor . . . the list goes on. In addition to these movements, there is a burgeoning non-governmental voluntary sector, with literally thousands of "action-groups," big or small, pouring their energies into social improvement.

The dominant tendency of new social movements is to look for a vanguard that is outside modern society (as if that were possible). Indeed, a sympathetic critic asks the question whether India's new social movements are "fundamentally anti-modern, drawing upon the traditions and world-view of non-modern cultures in their opposition . . . to the conventional models of development, both capitalist and socialist" (Guha 1989, 14). Guha suggests that these movements' attitude to modernity has been shaped by a historic compromise between distinctly anti-modern "crusading Gandhians" and the anti-traditional "ecological Marxists," with the former playing a hegemonic role. According to Gail Omvedt, one of the most astute participant-observers of these movements, this neo-Gandhian vocabulary expresses itself in "anti-industrialism, hostility to the . . . modernizing interventions of the centralized state . . . expressed in "traditional" or "populist" symbols drawn form Indian traditions" (1993, 300). All of these are accepted as a "point of departure" for new social movements: it is what makes them "new" as compared to the "old" social movements that did not question the legitimacy of the modern age (Wignaraja 1993, 6).

One feature of the "historic compromise" between "ecological Marxism" and neo-Gandhism mentioned above is that the Marxist elements have had to mute their Enlightenment heritage, especially the belief in the universal and culturally progressive nature of science. Furthermore, even those people's science movement intellectuals and activists who do not accept the civilizational incompatibility of modern and traditional knowledge, end up admitting the "anti-imperialist" uses of critiquing science. Gail Omvedt, for instance, while admitting that the critics of science "idealize the traditional society," nevertheless praises their work for focusing attention on the "imperialistically determined modern science and technology" as the main barrier to "genuine scientific and sustainable development" (1993, 145). Whatever doubts there may be in the new social movement discourse regarding equating science with the West, these doubts get muted in favor of creating a broad front against imperialism. Philosophical critiques of science as Western knowledge resonate well with the traditional politics of anti-imperialist nationalism, which has a wide appeal in most postcolonial countries. As a result of this mutual affirmation, it is fair to suggest that epistemological critiques of science have come to constitute the common sense of new social movements in India.

Of all social movements in India, the largest and most active are the people's science movements (PSMs) which are engaged in "bringing science to the people," especially in villages and small towns, through lectures, demonstrations, street plays, booklets, and magazines explaining scientific ideas in simple language while making them relevant to the local context, providing translations of scientific works in local languages, etc. The oldest and largest of these movements, the Kerela Sastra Sahitya Parishad (KSSP, which translates roughly into Kerela Science Literary Forum) is active in the southern state of Kerela. The Kerela Sastra Sahitya Parishad serves as a model for other science popularization societies in other parts of the country. The people's science movements boast a large and growing membership. The Kerela Sastra Sahitya Parishad alone has some 50,000 members, the people's science movement in West Bengal has over 25,000 members, 16,000 in Tamil Nadu, 12,000 in Andhra Pradesh, 5,000 in Karnataka. All major cities, Bombay, Delhi, Hyderabad, Ahemdabad boast of significant people's science movements (see Issac et al. 1997 for more data).

In theory at least, the people's science movements are not to be confused with alternative science movements. People's science movements do not condemn science. Quite the contrary. The Kerela Sastra Sahitya Parishad, in fact, upholds the slogan "science for social revolution" as its motto. Most people's science movements take the potential of science for change of cultural attitudes very seriously. The overall philosophy broadly follows the ideas of Needham and Bernal, who held modern science as a progressive force once it is freed from capitalist interests. One could say that in the philosophy of these groups, the old Nehruvian faith in modern science as a tool of social and economic transformation is alive and well, albeit in a heavily Marxian vocabulary (which the Fabian-Socialist Nehru also shared to a large extent). Even though Ambedkar, Periyar, and dalit rationalists do not find explicit mention in most people's science movements writings, clearly the two positions are not at odds, with the crucial difference that the largely Marxist people's science movements lack the priority Ambedkar gave to religious reformation over class conflict as a source of progressive change in India. (In standard Marxian theory, Ambedkar's call for Hindu reformation as a precondition of socialism would be rejected as idealistic.)

Given this high degree of mobilization in the cause of scientific temper, one could well ask, why worry about the romantic alternative science theories? What does it matter if Nandy and his associates wrote a few essays? What real difference does it make if some well-meaning Gandhians try to reconstruct past traditions of science, or try to build workable alternatives starting from the practices of women in the farms? One could even argue that these kinds of initiatives are necessary to moderate the hubris of science popularizes, and give ordinary people confidence in themselves. But the impact of the romantic alternative sciences has been a lot more complicated than that.

To understand the long-term damage the alternative science movements have done, consider the following question. Given such a state of ferment of science movements, both of the "people's" and "alternative" kind, why was the civil society unable to resist the onslaught of religious nationalists? No doubt, the combination of political and ideological factors (especially the collapse of the Nehruvian consensus of Congress at home and the collapse of the socialist alternatives around the world) probably made the tide of religious reaction hard to hold back. I am not suggesting that *any* social movement, by a sheer act of will, could have altered the course of events. But why wasn't there a more aggressive response to the ideas of the religious right? Why did the civil society fall like ninepins, notwithstanding pockets of resistance in some state or in some social group? Why have even the dalits and tribal people heeded the message of Hindutva?

There is obviously no single cause for the turn India has taken toward religious politics. No one group, movement, or personality can be singled out. Social change is a lot messier than that. Yet, ideas have the ability to set eddies and currents in motion which enable a myriad social institutions to either go along with the flow, or to resist it. I believe that alternative science movements put the relatively more secularizing people's science movements on a defensive, making them ever more timid to question and critique the common sense of the society. More than two decades of combined assault on secular thought and scientific reason by those seeking alternatives to modernity created currents and eddies that pushed the civil society along in the direction of the religious right.

The people's science movements had been sitting on the fence on the religion question for a long time. While a critique of religious common sense was built into their program of popularizing science, for the most part, they took a non-confrontational stance toward religion. But under the influence of the alternative science movement philosophy, defense of tradition, which is inextricably mixed with religion in India, became almost obligatory. Gradually, and imperceptibly, people's science movements diluted their secularizing agenda and became more like any other "sustainable development" movement, content with condemning the big capitalists and the state for despoiling the environment and exploiting the workers, issues on which both people's science movements and alternative science movements and even the Hindu nationalists could agree.

The case of the Kerela Sastra Sahitya Parishad illustrates the Enlightenment aspirations held back by Marxian economism on one hand, and a fear of offending the religious sensibilities of the people on the other. The Kerela Sastra Sahitya Parishad proclaims "science for social revolution" as its operating principle. A secularizing spirit is built into this program because the Kerela Sastra Sahitya Parishad tries to bring in the scientific worldview in order to demystify the religious legitimations of caste, patriarchy, and other sources of discrimination based upon concepts of purity, etc. A secularizing and humanistic spirit is also built into the program of providing alternative, demonstrable, and rationally defensible explanations for diseases and physical disasters commonly ascribed to supernatural causes. In the context of India, where the supernatural has not yet lost its hold on rationalizing and explaining the natural and the social, science teaching cannot but have a secularizing impact, whether intended or not. The defenders of orthodoxy in all the major religions of India recognize the potentially subversive impact of science. The spokesmen of all major organized religions, from the youth wing of the Rashtriya Swayamsewak Sangh, the Muslim League, and Catholic Bishops, have condemned the Kerela Sastra Sahitya Parishad for "promoting atheism" (Kanan 1990).

For all the accusations of corrupting the faithful, the Kerela Sastra Sahitya Parishad officially maintains a neutral stance toward religious issues. Officially, the Kerela Sastra Sahitya Parishad maintains that "God is outside the realm of science [and] refuses to be drawn into a debate on God and religion" (Issac et al. 1997, 36).[5] This is perhaps a tactical stance designed to avoid confrontation. But it displays an amazing lack of historical understanding of the relationship between science and religion: it is only *after* the hegemony of science has been established that God is removed from the realm of nature, and therefore from the preview of science. That has not yet happened in India. More than that, the reluctance of the Kerela Sastra Sahitya Parishad and like-minded groups to systematically engage with religious sources of cultural attitudes is a part of the economism inherited from their Marxist leanings. The general belief is that it is more important to change the social conditions that encourage dependence upon religion and supernaturalism, than to confront religion itself (see Kanan 1990). There is a naïve faith that a change in social relations will necessarily change ideas in people's heads.

Indian people's science movements, in other words, were already pre-disposed not to fight the battle openly and directly on cultural-religious grounds, even though they *hoped* that is where their efforts will bear fruit. Even when they talk of science education and critical thinking, they see them as dependent variables that follow structural changes in the distribution of economic resources. The Ambedkarite notion that cultural-religious factors could be impeding economic and social justice is brushed off as "idealism." Despite their disagreements with the romantic, ciliv[z]ational, and relativist views of science that alternative science movements hold, people's science movements have been more than willing to participate in the initiatives of the former in areas of environmental protection, indigenous health-delivery systems, the indigenist ecofeminist critique of modern agriculture, and similar projects. Even when they deeply disagree with the romantic traditionalism of Vandana Shiva, Ashis Nandy, or even with the opponents of the Narmada dam, people's science movements tend to go along with them because they share their anti-capitalist and anti-imperialist views.[6] Close collaboration and compromises over years has eroded whatever little explicitly secular agenda people's science movements had in the first place. Groups like Kerela Sastra Sahitya Parishad have transmogrified into a "sustainable development" group, one among literally thousands of similar groups, fighting on policy matters rather than on fundamentals.[7]

But the influence of alternative science movements goes deeper into how science is understood and presented to the people, even by people's science movements. It has been observed by many that since the scientific temper debate, the modernist, people's science movements have a hard time defending the distinction between modern and traditional sciences. In the post-Kuhnian, social constructivist view of science popularized by proponents of alternative/indigenist sciences, even sections of modernist people's science movements have begun to see all "science as culture" (Raina 1997, 16). We have examined the content of these philosophies in chapter 5. As we have seen, social and cultural constructivist theories accord complete parity between the rationality, within culturally sanctioned assumptions, of all cultural knowledge systems. This parity has a salutary effect insofar as it allays the suspicion of Third World social movements of racist overtones of modernization theories. In the light of the anthropological and social history of sciences around the world, no one can claim that scientific modes of reasoning are a Western cultural peculiarity. This part of social constructivism is welcomed by people's science movements (see Raghunandan 1989). But then, following the post-Kuhnian social constructivist theories and feminist epistemologies, it is not just rationality but the content and method of science that is seen as a social construct. This handicaps the people's science movements in challenging the objective truth of the culturally constructed assumptions within which traditional sciences operate. Moreover, any distinction of the methodology of sciences is not admissible, as it is seen as internal to the culture of the West. This leaves the people's science movements as simply passing on scientific facts and information without really enabling their audiences with any meaningful way to see the method of inquiry that led to these facts and theories. In the absence of any meaningful way to question and challenge the existing common sense, science popularizers simply become conveyers of information. Science becomes just one more system to be taken on faith. It is no wonder that such an effort does not leave any lasting impact on the audience (Zachariah and Sooryamoorthy 1994, 163).

### Conclusion

The scientific temper debate in the early 1980s marks the beginning of the postmodernist currents among Indian intellectuals. This was the first time that newly emerging voices from the secular and left-inclined social movements in India had rejected the very legitimacy of modern science, both as a resource for cultural change and as a part of the state-led modernization. There had been many critics in India who had been agitating on behalf of people-oriented *uses* of science among the left. But the criticism of the very content of science as ethnocentric and a "superstition of the strong" was a new trend among the popular movements.

This Indian postmodernism owed its origin to Western and indigenous influences. Works by Thomas Kuhn and Paul Feyerabend were being read all around the world. Their influence was keenly felt in India. The anti-war movement in the West had also given a new lease of life to Frankfurt-school style critiques of instrumental reason. This genre of critique of instrumental reason, per se, abstracted from the historical context, popularized by the works of Theodore Roszak, Herbert Marcuse, and Ivan Illich had a substantial following among Indian intellectuals. The problem, as I see it, was that in India, scientific or instrumental rationality hardly constituted the hegemonic or dominant position which was crowding out other aspects of a humane society. Quite the contrary: even the most rational and instrumental goals of modernization were constantly compromising with romantic and paternalistic notions of "village community" inherited from Gandhism. Under these circumstances, attacking all science, abstracted from historical context, as authoritarian and one-dimensional, and to top it all, Western in its cultural provenance, showed a lack of understanding of the precarious hold of modernity in India.

But the reasons anti-rationalist, anti-Enlightenment currents could strike a responsive chord among Indian intellectuals were quite local. As I have described in this chapter, the scientific temper debate is set against the background of the "Total Revolution" against Nehru's conception of a modern India. The populist mobilization against Indira Gandhi's populist but authoritarian policies turned into a repudiation of a modernist-developmentalist vision of a future India. The vision itself came under attack. The rebellion against Nehru's ideas led to a revival of a Gandhian/ neo-Hindu brand of indigenism. It is in this context that many Indian intellectuals found the Western critiques of science and technology relevant to their own society. The problem was that the Gandhian vision of a good society contained many elements (e.g., trusteeship of the rich, non-individualistic, non-competitive caste organization of society, respect for traditional village-based institutions for industry and social justice) that also appealed greatly to the Hindu nationalists (which is not surprising, given that Gandhi spoke from deeply religious beliefs). Indeed, as I have tried to show, the postmodernist new social movements and Hindu nationalism are twins born out of the same circumstances (of the agitation for "total revolution" in the early 1970s).

In the long term, the opponents of scientific temper have done much damage to the secularist cause. As this chapter has demonstrated, they have succeeded in putting the secularizing elements of popular science movements on the defensive. For at least the last two decades, popular science movements have behaved more like sustainable development movements, with many overlaps with indigenist green movements. In the absence of adequate support for a secular worldview among the intellectuals and social movements, Hindu chauvinists have had an easier time presenting themselves as the voice of all of India.

# Nine

# The Ecofeminist Critique of the Green Revolution

*In the end, the glorification of the splendid underdogs is nothing other than the glorification of the splendid system that makes them so.*
—*Adorno*, Minima Moralia

The call for "total evolution" and the debate that followed over scientific temper were the harbingers of the wave of neo-traditional populism that was to flood India's political and intellectual landscape for the next two decades. Everywhere, social activists of all political inclinations, in various combinations of red, green, and saffron (the color of Hindutva), went looking for local traditions as alternatives to the industry- and state-led capitalist modernization. Just about every tradition—but curiously enough, mostly those traditions with a pre-Islamic, Hindu lineage—were dug up for revitalization. Even India's peculiar institution, caste, was elebrated as a strategy for the conservation of natural resources (Gadgil and Guha 1992). Almost overnight, panchayats, the traditional village-level assemblies, became all the rage (Anil Agarwal 1994), their role as enforcers of caste and gender hierarchy conveniently forgotten. Environmentalists and feminists also found religion, from sacred groves to various goddesses, as a resource for their agenda. All that was local and traditional suddenly became chic.

One of the earliest and most influential voices in this neo-traditionalist populism was that of Vandana Shiva, a physicist-turned-social activist. Shiva, who had already made a name for herself during the scientific temper debate, came out with a book, *Staying Alive: Women, Ecology and Survival in India* in 1988. That book put her on the international map. Mercifully, by now, the book and its author have lost much of their appeal. But this book was quite a rage in feminist, science studies, and cultural studies in the United States for about a decade or so after its publication. Shiva herself became an international celebrity, well-known for her passionate harangues against modern science and the modern West in general. In India, she became the object of intense devotion and intense criticism. Even those who deeply disagreed with her—as I did (Nanda 1991)—found it impossible to ignore her, for she represented the emerging Weltanschauung of India's rightward-tilting "radical" new social movements.

In this chapter, I examine the ecofeminist criticism of the Green Revolution. I will look at Vandana Shiva's writings on the subject in her *Staying Alive*, and in *Ecofeminism*, a book she co-wrote in 1993 with Maria Mies, a German ecofeminist. ("Green Revolution" is the name given to the project of agricultural modernization that started around the mid-1960s and was aimed at increasing the yield of food grains in the Third World through the introduction of high-yielding varieties of cereal crops, mainly wheat and rice. These high-yielding varieties are not to be confused with genetically engineered seeds, for they are produced by traditional Mendelian crossbreeding and selection.)

Briefly, Shiva and Mies condemn the Green Revolution as a species of "Western patriarchal violence" against women and nature in the "colonies." Why "patriarchal?" And why "violent?" Here Shiva and Mies are arguing from the first principle of ecofeminism, namely, anything that is violent to nature causes violence to women, and vice versa. Because modern science, technology, and capitalism disrupt the organic interconnectedness of nature and women, they "dis-embed" women from nature, family, and the moral economy that prevails under subsistence agriculture. The assumption, of course, is that removing women from subsistence economy is against their interests. They locate the source of this violence against nature and women in the very rationality of science and the underlying "Western cosmology." They wish to counter this rationality with holism and interconnectedness they believe characterizes women's work in subsistence agriculture in India.

I have two goals in this chapter. First, I will counter the romantic and idealistic understanding of the scientific worldview as the source of oppression of women and nature, and conversely, peasant women's worldview as liberatory. This kind of idealism is simply bad social theory. It is more useful to understand women's work as embedded in a complex ensemble of social relations in which class, gender, and caste interact in culturally sanctified ways to appropriate women's unpaid labor. More specifically, I will juxtapose a materialist analysis against the ecofeminist analysis of empirical findings regarding three indices of women's well-being, their claims to life, livelihood, and productive assets, and argue for the superior explanatory power of the former.

Second, I hope to show that the alternative, "standpoint epistemology" of peasant women that ecofeminists celebrate contains exactly those elements that have, in fact, been used to legitimize women's inequality and inferiority in India. Shiva, Mies, and their sympathizers, I will show, have offered an *ecofeminine* support of the status quo, couched as a critique of the instrumental rationality of the Green Revolution. Their "critique" only succeeds in glorifying the "splendid underdogs" and the "splendid system" that produces them. This kind of traditional, feminine romance has dangerous political implications. In the next chapter I will examine how Shiva and her associates' defense of traditions has been picked up by the conservative elements of farmers' movements.

## Three Dogmas of Ecofeminism

What is unique and interesting about ecofeminism is that it stands at the intersection of three different streams of social thought: feminism, philosophy of science, technology and development, and native/indigenous/local worldviews (Warren 1997, 4). In this space, ecofeminist theory has developed a set of dogmas. It will be useful to locate Shiva and Mies's brand of Third World cultural ecofeminism in these dogmas, for it will prepare us for the actual empirical claims they make about the "violence" of the Green Revolution.

### Women-nature relationship

Ecofeminists start with the assumption that women have a special relationship with nature, which gives them a particular stake in conserving the environment. Western industrial civilization, it is argued, was built in opposition to nature. Because women were seen as closer to nature in all patriarchal cultures, women, too, were made into objects of domination. The ultimate source of this shared domination is an anthropocentric humanism which makes a distinction between humanity and the natural world; all that is not a part of human culture is looked down upon as wild and feminine nature which must be tamed. Ecofeminists, along with other radical ecologists, trace the beginning of this anthropocentrism to the Scientific Revolution when nature was transformed from a caring and bountiful mother to a fearsome, wild woman who must be controlled. Since the beginning of modernity, domination and control of nature has been justified in sexist and patriarchal metaphors, and that has had the result of justifying women's oppression as part of the domination of nature as well. How can one end this twin oppression of nature and women? By re-conceptualizing the relationship between human beings and nature as one of continuity, non-separation, and harmony. Once we see ourselves as interconnected and one with nature, ecofeminists argue, we will respect it and cease from dominating it. Because women have a more interconnected sense of the self, they can and should lead the way to an ecological future. Rather than sever their connections with nature in order to pursue a false and dangerous equality with men, women are invited to choose to identify themselves with nature in order to protect it. These are fundamental assumptions of ecofeminist literature and can be found in just about any text (see Braidotti et al. 1994; Zimmerman 1994 for sympathetic reviews).

On the one hand, ecofeminists claim to reject any biological explanation of women's closeness to nature and blame any such connection on the philosophical error of modern ideas that separate nature from culture, material gain from ethical values and so on (more on this shortly). But on the other hand, the actual goals and policies that ecofeminists support seem to assume that any such association of women and nature is not an ideological aberration but is *in fact* so, and that women *in fact* have a special relation with nature.

The core of the ecofeminist position is derived from a handful of a priori philosophical assumptions regarding the essences of "the West," "Capitalism," "White Men," "Third World peasant women." At the broadest level, modern Western social order is understood as constituted through the opposition between incommensurate binaries, each of which is defined as the absence of the other. Maria Mies and Vandana Shiva state in their recent book: "Capitalism is based upon a cosmology that structurally dichotomizes reality: the one always considered superior, always thriving, and progressing *at the expense of the other*. Thus nature is subordinated to man, women to men, consumption to production and the local to the global" (1993, 5).

In order to explain the Green Revolution through this logic of dualism, Shiva and Mies add more binary categories to the list: First World/Third World, urban/rural, Western/indigenous, rationality/irrationality. In all these cases, they associate the Green Revolution as imposing the first and the more powerful category over the second, less powerful, non-Western category in the pair. Thus, the Green Revolution is charged with imposing the interests of the First World over the Third World, the urban over the rural, the Western over the indigenous, culture over nature, and of course, men over women.

For all their critique of dualistic thinking, Shiva and Mies happily fall into a rather crass dualism themselves when they portray the non-West as the binary opposite of what they take to be the West. If the cultural essence of the modern West is dualistic thinking, the non-West is presented as the polar opposite: guided by a cosmology of harmony. For Mies, women's work in the subsistence sector of "the colonies" is the source of such non-dualistic work which integrates production with an ethic of care and non-domination (Mies and Shiva 1993, 298, 303). Mies indeed goes so far as to hold up the impoverished economy of contemporary Cuba as an example of subsistence economy for the rest of the world to emulate.[1] Shiva finds models of non-dualistic thinking in an idealized "Indian cosmology:" "Contemporary western views of nature are fraught with the dichotomy or duality between man and woman, and person and nature. In Indian cosmology, by contrast, person and nature (Purusha [literally, man]-Prakriti) are a duality in unity. They are inseparable complements of one another in nature. . . . Every form of creation bears the sign of this dialectical unity, of diversity within a unifying principle, and this dialectical harmony between the male and female principles and between nature and man becomes the basis of ecological thought and action in India" (1988, 40).

Anyone with any experience of living in India will attest that social life in India is anything but this "dialectical duality in unity." If nature was indeed considered an inseparable complement of human life, why is it that those who work with nature—working the land, handling human and animal waste, coming in contact with bodies—are seen as polluted and untouchables? Or if men, women, and nature were in "dialectical harmony," as Shiva claims, why don't upper-caste Hindus allow menstruating women—when they are at their most "natural"—even to enter the kitchen? That Indian culture is averse to thinking in dualisms is patently false. As Aijaz Ahmad points out (1992, 184), Indians regularly posit a Hindu spirituality against Western materialism and Muslim barbarity; and Hindu holy texts like the *Mahabharata* abound in making the foreigners, the lower orders, and women into dangerous inferior others.

Instead of accepting a priori that Third World women actually have a special connection with nature, I believe it is more fruitful to ask why women may *appear* to be more embedded in nature in some societies. Why do women and not men belonging to the same socio-economic group get this "privileged access to nature," which allows them to do, in Shiva's words, the "invisible work of earthworms" (1988, 109) that goes into keeping the family farm in good health?[2] Work that, ecofeminists forget to mention, is also unpaid and unrecognized as valuable.

In contrast to ecofeminism, my starting assumption in understanding the effects of the Green Revolution on women will be free from any gender bias. I will assume only that *both men and women, in the Third World just as everywhere else, relate to nature in their search for a livelihood and survival*. This struggle for livelihood and survival shapes (and is shaped by) the relationship between the sexes. But the concrete forms the social institutions and gender relations will take in any given historical era cannot simply be deduced from some a priori philosophical assumptions regarding the nature of femaleness, or Western-ness. It is the structural and historically contingent juncture of technologies, social relations of production, and gender relations that decide who will do what and with what rewards.

### Ecofeminism as the "standpoint epistemology" of Third World peasant women

Their obvious hyperbole apart, Shiva and Mies place their work firmly in the contemporary feminist critiques of science (see chapter 5). They simply assume, without argument, that the feminist and other post-Kuhnian critiques of modern science are valid and beyond any doubt. Having assumed the patriarchal and Western ethnocentric nature of science, they present Third World women's everyday knowledge of flora and fauna as their "standpoint epistemology."

Following Carolyn Merchant's well-known but highly biased history of the Scientific Revolution, Shiva holds science (dualistic, objectifying, and patriarchal) to be responsible for women's oppression in India.[3] In Shiva's rendering of Indian history, precolonial India was a harmonious society guided by the Hindu belief in *prakriti* (or nature) "the feminine and the creative principle of the cosmos." With the introduction of modern science and technology through colonialism, this living, loving relationship was replaced by man dominating the earth: "The ecological crisis is, at its roots, the death of the feminine principle" brought about by the forces of industrialization and market relations, leading to a "patriarchal maldevelopment," of which she holds the Green Revolution to be a prime example. Mies and Shiva find in the subsistence work of rural women in India a paradigm for a "non-dualistic science." *Prakriti* is based upon "an ontological continuity between society and nature" which "excludes possibilities of exploitation and domination" (1988, 41) and which brings women and nature together "*not in passivity but in creativity and in the maintenance of life*" (ibid., 47, emphasis in original). Clearly, Shiva's *prakriti* and Mies's "subsistence perspective" both serve as models for ecofeminist resistance to the Green Revolution and Western technology in general.

Shiva seems totally oblivious to the numerous scholarly works that clearly show that *prakriti*, the feminine principle, is not actually understood as superior and more valued than *purusha* (or consciousness), the male element. In the major Hindu tradition of Sānkhya, *prakriti* is the cause of bondage of p*urusha*; the female element (Shakti) is treated as passive and inferior to the active male element (Siva) (Verma 1995, 434). Shiva also shows no awareness of the fact that precolonial India was already a highly hierarchical society with the worst elements of caste-based bonded labor and gender inequities firmly in place long before the Western contact (see Patnaik and Dingwaney 1985).

Apart from the historical inaccuracies, Shiva and Mies make a rather opportunistic use of standpoint epistemology. In keeping with the work of Sandra Harding, they present ecofeminism as the standpoint epistemology of women of the Third World. But they simply equate women's raw experience with their eco*feminist* "standpoint." This completely ignores the many denials by Sandra Harding and other standpoint theorists that experience itself does not constitute a feminist standpoint, but rather emerges through a *critical, feminist reflection* on the experience. Critical reflection on experience is sadly lacking in the ecofeminist texts under consideration here. Shiva and Mies follow this line of reasoning: because Third World women work with nature, they have a special knowledge of their own environment, their farm, their forests, etc. And because of this work that lets them participate in the rhythms and productivity of nature in a certain way, they have a special interest in protecting nature.

### Women's "experience"

Standpoint epistemologists in general, and Shiva and Mies's in particular, take women's experience as a given. Satisfaction of experienced needs is taken as a necessary *and* sufficient criterion for assessing change.

Shiva, for instance, believes that subsistence economies were the "original affluent societies," for they took care of the basic vital needs of their members (1988, 12). But the problem is that she defines the nature of these needs, the level at which they could be assumed to be satisfied, and the social relations of satisfying them as defined by the prevalent cultural norms alone. By so relativizing needs, Shiva is able to gloss over some grim facts behind her much-celebrated "affluence" of non-Western agricultural systems. The fact, for example, that subsistence economies did not (and still do not) supply all the nutritional needs of all their members at a level that is biologically adequate for maintaining basic capabilities. Or the fact that social practices that go into meeting even that minimal level of culturally defined needs include backbreaking work of poor women (who by Shiva's own admission work harder than farm animals) as well as other forms of degrading work like gleaning the farms to collect grain which would otherwise be eaten by birds and pests. Or the fact that the culturally defined needs did not include access to education, personal autonomy, freedom of thought, and a host of other higher-level cultural capabilities. A similar problem crops up with Maria Mies's assertion that Third World women reject Western ideas of self-determination and autonomy for they value their connections with the community (Mies and Shiva 1993, 220). Again, behind this totally amazing assertion lies the impulse to reaffirm local traditions as regulative ideals of social life for women.

Such cultural relativization of needs is challenged by Amartya Sen and Martha Nussbaum in their capabilities approach to human functioning. The capabilities ethics is based upon a "critical universalism" that holds a set of basic human capabilities as *intrinsically* worthwhile for a flourishing human life, while admitting that these capabilities may be expressed differently in different cultures and different historical epochs. Amartya Sen holds that "personal interests and welfare are not just matters of perception; *there are objective aspects of these concepts that command attention, even when the corresponding self-perception does not exist*" (1990a, 126, emphasis added). Two levels of human capabilities serve as the objective criteria for judging whether or not any society can be said to be developing or regressing. The first level sets the ground-floor for a life that is suitable for humans as a species distinct from other animals and includes bodily capacities like avoiding hunger and thirst and includes the distinctively human traits like humor and play and the ability to reason and make moral judgments. The next level describes capabilities that make a human life a *good* life. These capabilities include, for instance, the ability to "imagine, to think and to reason," and the ability for critical reflection and crucially for women, the ability to "live one's own life and nobody else's" (Nussbaum, 1995).[4]

Contrary to Shiva and Mies, Nussbaum and Sen believe that it is both possible and desirable to move *all* people, everywhere, above the threshold for a good life defined in terms of capabilities. This reflects their commitment that "all persons are equal bearers of human claims, no matter where they are starting from in terms of circumstance, special talents, wealth, gender, or race" (ibid., 86). By holding these capabilities to be universally valid (because they are grounded in our species characteristics) Nussbaum and Sen are not seeking to homogenize all cultures, sub-cultures, and traditions. Rather, they have identified components that are fundamental to *any* human life, while granting that these components can and do find expression in culturally and linguistically specific local ways. What Sen and Nussbaum are suggesting is that there is a sufficient overlap between different traditions and cultures so that they can be evaluated on the same metrics of capabilities. If some cultures fail to allow all or some of their members (based on their gender, class, caste, race, or sexual orientation) access to a good life, they cannot hide behind the irrelevance of these capabilities to their culturally sanctioned idea of a good life, because, as Nussbaum puts it, "human capabilities exert a moral claim that they should be developed. Human beings are creatures such that, provided with the right educational and material support, they can become fully capable of major human functions, . . . their very being makes reference to functioning" (ibid., 88).

By making local cultures the final arbiters of what is just and worth preserving, ecofeminists and like-minded critics of development end up justifying the status quo, which has been anything but just or fair to women. In this context, it is important to heed Sen's caution against mistaking the pleasure some deprived people may take in "small mercies" as a sign of contentment:

> It can be a serious error to take the absence of protest and questioning of inequality as evidence of the absence of that inequality or the non-viability of that question. . . . Deprived groups may be habituated to inequality, may be unaware of possibilities of social change, may be hopeless about upliftment of objective circumstances of misery, may be resigned to fate and may well be willing to accept the legitimacy of the established order. . . . But the real deprivations are not just washed away by the mere fact that in the particular utilitarian metrics of happiness and desire fulfillment such a deprived person may not seem particularly disadvantaged. (Sen 1990a, 127)

*How* the deprived get to accept their lot plays an important role in Sen's concept of "entitlements." Entitlements refer to socio-cultural consensus about resource allocation: Who deserves what? Who is worthy of what and how much? Who is perceived to be entitled to a good is decided by a society's larger cultural beliefs about distributional justice. In Sen's account, women end up poorer than their male counterparts in all socioeconomic groups because of the culturally determined consensus about their "worth" that determines their relative access to the available goods and resources, both material (food, health care, etc.) and cultural (education). An important mechanism that ensures a lower entitlement of women to capability-enhancing resources enlists women themselves in their own deprivation, through a process that Hanna Papanek (1990,162) has aptly described as "socialization for inequality."

This description of the theoretical blind spots in Shiva's and Mies's work makes it clear why I believe their work cannot be rightfully called *feminist*. Feminism is a discourse of emancipation, a discourse about the equality of men and women, but Shiva and Mies are willing to sacrifice the emancipation of Third World women at the altar of "Third World difference." So keen are they on establishing the difference in the lives and experiences of women, that they do not stop to investigate the political-economic and cultural genesis of the differences they want to affirm. Without understanding the social relations that give rise to these differences, these critics deprive themselves of any objective bases to distinguish *legitimate diversity from illegitimate inequities*. Once the cover of culture is thrown over the material inequities, preservation of culture begins to take priority over the elimination of inequality.

For a *feminist* understanding of the lives of women in the Third World, it is crucial that we learn to distinguish between legitimate and illegitimate differences. If some differences are born out of unjust relationships, then justice requires that we try to eliminate these differences. This project requires lifting the cover of culture from material relationships and seeing cultural differences in their materiality, that is, in the way they socialize Third World women for the inequality of entitlements.

### Women's Entitlements and Green Revolution Agriculture

I now turn to how ecofeminists understand the impact of changes in agrarian economy brought about the Green Revolution on the lives of peasant women in India.

#### Entitlement to life—or the case of the "missing women"

"More than 100 million women are missing" around the world, as Amartya Sen reminds us (Sen 1990b). Sen goes on to demonstrate, if women in the developing world covering most of Asia and North Africa and to a lesser extent Latin America, were provided with the same level of care in terms of food, medical care, and other basic necessities that men in the same societies receive, nearly 100 million more women would still be alive today.

Despite the biological advantage in survival that women have compared to men, the number of women falls far short of men in Asia and North Africa, though not in sub-Saharan Africa. If we took the European and North American sex ratios as standards (where there are about 106 women for every 100 men), the disparity in sex ratios becomes apparent: there are only 97 women for every 100 men in the Third World as a whole. The deficit in women is most marked in parts of Asia—India, China, Bangladesh—with roughly 94 women for every 100 men in India, and only 90 women for every 100 men in Pakistan (ibid.). China has more than 44 million missing women, while India has 37 million, with the total exceeding 100 million worldwide.[5] There are regional variations inside these countries. In India, Punjab and Haryana, the richest states of the country, which are also the seat of the Green Revolution, have only about 86 women per 100 men while the much poorer southern state of Kerela has a sex ratio (103 women for 100 men) similar to that in industrialized countries. The 1991 and 2001 census reports continue to show a declining sex ratio in India.

If the very survival of female persons is not on a par with males born in the same socioeconomic-cultural space, discrimination against them in access to all other promotive entitlements—to food, health and education—is only to be expected. Countless micro-level studies from India (reviewed most recently by Agarwal 1992 and Barbara Harriss 1992) show that women in India, on an average, fall behind men in life expectancy (though the gap is closing), nutrition (in Punjab, discrimination in food is seen only among the poorest families, while studies in other areas of India indicate protein and caloric intake to be lower for females as compared to males in all socio-economic groups), and health care (with fewer numbers of girls and women seeking medical treatment as compared to boys and men for similar ailments). Not surprisingly, girls lag behind in schooling with the gap wider among the poor agricultural laborers and the lower-caste women who are a century behind caste Hindu women in their schooling.

Why are these women missing? Why did they not get to live and thrive?

Shiva is most explicit in her answer. The Green Revolution killed them: "underlying infanticide is dowry and underlying them both is the Green Revolution" (1988, 119). Shiva reads the fact that the Green Revolution areas are also the areas with the lowest sex ratio as a sign of the "violence" of the Green Revolution. But worse, she believes that the frightening rise in the practice of selective abortions of female fetuses—a practice that is widespread in Punjab and Haryana, as it is in most parts of India—is also caused by the Green Revolution.

Shiva explains these damning charges through her overall critique of modernity, of which the Green Revolution is only an agent. Thus, following the basic ecofeminist tenet of women's closeness with nature, Shiva valorizes "precisely those links in farm operations which involve a partnership with nature and are crucial for maintaining the food cycle." With the Green Revolution, women are "dispossessed" as they are forced out of the "ecological work" of soil-builders into the "economic work" of wage-earners (ibid., 114). Or, in another formulation, Shiva claims that commercialization of labor and farm inputs devalues the survival work women do without wages (ibid., 117). Commerce, wages, and capitalist agriculture in general that are promoted by the Green Revolution are seen as men's work, which stands in total opposition to the non-commercial "survival economy" of women. As we will see in the next section, Shiva's correlation of wage work with devaluation of women is entirely at odds with what the empirical evidence shows. But let us first examine her claim that the green revolution is a cause of the lower sex ratio.

Shiva can only sustain her charge if she ignores the history of patriarchy in India. Female infanticide has a long tradition in northwestern India. British records dating back to 1901 when the British annexed Punjab to the empire show that female infanticide had been widely prevalent among dominant castes in that region. The 1931 census of India, conducted under the British, shows that Punjab had only 831 women for each 1,000 men, even then the lowest ratio in all of India. Most geographers and demographers agree with Bina Agarwal's summary of the extent and causes of female infanticide: "Historically, female infanticide was practiced widely in the northern and western (Gujarat upwards) belts, especially in the states of Rajasthan, Punjab and Haryana, with very few and scattered instances noted elsewhere. It was most common among the upper castes and is attributed to factors such as hypergamy, heavy dowry expenditure, prevention of excessive land fragmentation etc." (1988, 91).

Any understanding of the changes in women's lives as India modernizes will have to start with this grim history. This is especially true of technological change and new innovations which are typically fed through a *preexisting* sexual division of labor and family relations. In Shiva's account, these pre-existing, lethal biases find only a passing reference, and in Mies's celebration of subsistence, none at all. The Green Revolution is made to carry the weight not just of its own inequities, but of these historical injustices as well.

Female infanticide is only the most extreme and brutal signs of the devaluation of female life in northwestern India. Most missing women are casualities of persistent neglect in nutrition, health care, and other entitlements necessary for survival. Here, one meets another dimension of inequality that is completely missing in the ecofeminist romance—the family. Village-level studies have shown beyond doubt the gender discrimination in intra-household food allocation and health care (Agarwal 1992). None of these findings are mentioned in the ecofeminist celebration of peasant life.

These pre-existing gender inequities in survival are not just a result of poverty. Empirical studies show that female infanticide is an upper-caste and upper-class phenomenon (motivated by a son-preference and the cultural pressures to withdraw women from visible work), while female neglect persists across classes and is more acute among the very poor. How a group or society distributes available resources among members reflects not only economic power and authority relations, but also the moral basis of that group, its consensus about distributive justice and its implicit priorities. The long history of gender biases, dating way back into India's most remote history, shows that contrary to the claims of ecofeminists, the so-called "moral economy" of the peasant did not value the work women have always done in the agricultural economy. Despite their contribution to the family's wealth through their work in the farms and forests, and where the culture of seclusion kept them homebound, work in processing food and fuel, the premodern and even the pre-British peasant economy in India did not consider women as *deserving* of resources as were men.

Many cultural discourses, including folk tales and popular proverbs in the northwest part of the country show that devaluation of female lives is not a pathology of modernity but an integral part of the traditional cultural order. Prem Chowdhry (1994) has collected many of these common proverbs and folk discourses:

> The son of an unfortunate dies,
>> The daughter of a fortunate dies.
>
> Who can be satisfied without rain and sons;
>> for cultivation both are necessary.

These folk traditions serve to "socialize women for inequality," in Papanek's (1990) apt phrase. They reflect and simultaneously construct the collective consciousness regarding what girls and boys in a family and community are entitled to or deserve. It is through these discourses mothers come to believe that their daughters do not "need" as much food or as good an education as their sons and it is through these discourses that daughters come to accept their share as fair and just.

From a materialist perspective, the lower sex ratio and the spread of female feticide in Green Revolution areas appear to exist *despite* and not because of the Green Revolution. Given the long and culturally entrenched bias against females that prevails in the Green Revolution enclaves, all one can defensibly claim is that three decades of rapid agricultural modernization has not been able to reverse or even weaken the culture of female sacrifice—at least, not yet. Even though in Punjab women are increasingly participating in wage labor which, as we shall see in the next section, is linked to improvement in female survival and well-being, the practice is largely "distress driven" and therefore limited to women from poor and low-caste families. But because women's withdrawal from fieldwork is associated with higher social status, most small and/or lower-caste peasants tend to follow the norms of their social superiors and withdraw their women from farm work as soon as they can afford to. Confined to home, where they continue to work in the "invisible" farm economy, women are not perceived as making an economic contribution. The general consensus is that Punjab, the state with the lowest sex ratio is also the state with the lowest rural female labor participation rate in the entire country (Chowdhry 1993, 139). Thus *withdrawal* from visible work, rather than an increase in it, as Shiva believes, is the reason why women are seen as burdens. So many women are missing in Green Revolution areas because of culturally enforced withdrawal to the invisibility of work done in the private space of the home, without any remuneration and recognition.

### Access to visible work

A striking feature of women's work in rural India, writes Kalpana Bardhan, an Indian feminist economist, is that "while women's overall workforce participation rate is low, and decreasing over the last two decades, the proportion of wage-laborers among those working is high and increasing" (Bardhan 1989, A33). In other words, more rural women who work in the farms are doing it for wages, and the proportion of all women wage laborers active in farm labor is increasing. In 1971 and in 1981, half of the total women in the job market in India were agricultural workers, as compared to only a fifth of the male workers (ibid.). Bina Agarwal (1992) and Hanumantha Rao (1994) both provide figures supporting this assertion.

But it appears that apart from the situation in Punjab and Haryana where an increase in female wages is drawing some women from small farmer/upper-to-middle-caste households into wage labor (Chowdhry 1994), women's presence in the wage market is "distress driven." This is indicated from the fact that outside of Punjab and Haryana, most women who work for wages are from lower income classes, and/or lower castes and tribal groups. (Upper castes and richer farm households tend to withdraw women from farm labor altogether once their economic status begins to improve.) As a general rule, it appears that those regions which have seen a sustained economic growth, accompanied with an increase in irrigation and multiple crops have also seen a decrease in the disparity between male and female wages. In these areas there has been some improvement in the bargaining power of female workers as a result of the increased demand for their work in transplanting and interculturing, both considered female jobs (Rao 1994, 43; and also Bardhan 1989). But apart from these areas, the wages for female agricultural workers remain much lower than the minimum wage and lower for men's wages for similar work. The lower-caste status of most female farm workers, along with cultural biases against women in general, are the major reasons for their lower wages. Thus while the increase in female proletarianization is a fact supported by census data, its causes and its effects on women's well-being are not as clear cut.

As is to be expected, Shiva and Mies read the increasing proletarianization of women as a cause of women's oppression and hold women's work for wages as a sign of devaluation of their work. Again, in a logic similar to that deployed to link the Green Revolution with female infanticide and neglect, Shiva sees all commodity relations, including work for wages, as leading to the increased devaluation of women. There are two parts of her argument against wage work: first, modernization favors production for profits instead of needs, and this "masculinist equation of economic value and cash flows creates a split between the market economy controlled by men, and the survival economy supported by women. Commercialization leads to increased burdens on women for producing survival and decreased valuation of their work on the market" (1988, 117). Second, Shiva believes that with wage labor, women's work shifts from "ecological work" to "economic work" making women, as she describes, "subsidiary workers and wage earners on an agricultural assembly line" (ibid., 114).

The first part of Shiva's argument hinges on a neat and a near-complete split between a cash-mediated, masculine farming sector and a non-commercial, non-market, not-for-profit female sector: any change that gives more power to the former—as the Green Revolution supposedly does—is considered by definition a loss for women. The problem with this logic is that the male and female spheres of agriculture are nowhere as clearly distinct, not even in Africa on which Shiva seems to be basing her analysis of India. In the case of Africa, it has been argued persuasively by Anne Whitehead (1990) that development scholars from Ester Boserup onwards have overstated the extent of female labor in subsistence farming and underestimated the involvement of women in the modern sector of the economy. The problem is not with markets as such, but with the terms on which people come to the market, which is shaped both by gender and class. The real problem for women is not producing for the market but the lack of independent ownership of assets that make their entry into the market difficult.

The second part of the argument—that women's work ceases to be ecological and becomes economic—is also problematic. It is true that with the Green Revolution, many of the inputs that women provided through their free labor (weeding, making organic manure, etc.) are being replaced by chemicals or hired labor. But what is not clear is whether this shift implies a devaluation of women's work. Shiva seems to assume that the ecological work that women did was valued by their families in the first place. The available evidence suggests that because women's ecological work was naturalized to such an extent—that is, considered simply what women do because that's what is appropriate for their womanly natures—that work was (and still is) rendered invisible. Shiva valorizes the invisibility of women's work for the value it has for the environment—although it is debatable whether only women in traditional societies were involved in ecological work, or whether all of women's work was ecologically beneficial to nature, or for that matter whether the Green Revolution has been as unremittingly disastrous for the environment.[6] What Shiva misses entirely is that *women's subsistence work is not only invisible ecologically, it is also invisible socially.* And this social invisibility did not start with the "ecological disruption rooted in the arrogance of the west and those that ape it" (1988, 44). All available signs from history, mythology, and folk narratives indicate that the lot of all of those who worked with the elements in traditional agriculture was not a happy one.

The ecofeminist denigration of wage labor flies in the face of the evidence accumulated from myriad village-level studies linking women's entitlements and well-being to their participation in wage work. Indeed, a *positive relation between female labor participation and entitlements* is one of the few guiding principles of development studies that enjoy a general consensus. This principle is stated firmly and simply by Barbara Miller (quoted by Papanek 1990, 167): "where female labor force participation (FLP) is high, there will *always* be high preservation of female life, but where FLP is low, female children may *or* may not be preserved" (emphasis in original).

It is important to note the simple fact that women's labor is not sufficient to translate that labor into well-being for women. For labor to lead to "preservation of female life," it has to be *visible*, or what Amartya Sen calls "gainful" labor as compared to unpaid and unhonored work as a part of the family either inside the house or on the family farm. There is no denying that women's ecological work has a substantial economic worth, but given the repertoire of cultural and religious narratives of caste purity and class exclusiveness, and the female propriety required for maintaining both, traditional Indian culture does not recognize this work as valuable. The kind of gainful labor most positively related with improvements in women's entitlements is "working outside the home for a wage or in such productive occupations as farming" (Sen, 1990). Bina Agarwal (1992, 403) has ranked, on the scale of visibility, agricultural work that is physically more visible over home-based work and work that brings in earnings over the "free" collection of fuelwood, fodder, or water. Although women have always worked on their family farms (except where they observe the *purdah*, as in Muslim and some upper-caste communities), because of cultural prohibitions and gender stereotypes, they are assigned to tasks that are not valued as productive.

This positive correlation between visible or gainful employment and lower gender disparities in female survival and well-being is strong enough to explain the variations of sex ratios across different regions of India and indeed, the world. It is well established that the relatively better survival rate and status of women in southern India is related to the greater demand for female labor generated under rice cultivation in these states, than under wheat cultivation that prevails in most of northern India. Similarly, Amartya Sen finds a positive correlation between employment and survival in different parts of the world: countries where cultural norms permitted women to work for wages outside the home were also the countries where fewer women were missing.

It is important to realize that the increase in wage labor in Green Revolution areas is affecting different women differently depending upon their class and caste status. The increase in wage labor may improve the perception of worth of women workers who bring in wages. But these women who work for wages are for the most part drawn from lower classes and lower castes. They face numerous problems (ranging from sexual assault and low wages) and their social location simply does not give them the power needed to set new cultural norms. In fact, as their families move up the economic ladder, they also tend to withdraw these women from wage labor, following the custom of upper castes. Moreover, female farm laborers replace the work the farmers' wives, daughters, and daughters-in-law did gratis—converting these relatively privileged women into economic liabilities and raising the rates of dowry. The preexisting cultural norms which associate upper-caste status with women's withdrawal from visible work tend to dampen other improvements in women's status that are generally supposed to follow from economic growth. This is evident in a recent study from Maharashtra (Vlassoff 1994) which shows that women who have withdrawn from farm work have gained more education and more leisure time, but have actually become more conservative in terms of marriage and dowry. Another study from Haryana (Chowdhry 1994) shows that even though there is an increased involvement of women, even from the middle class and caste status, in wage labor, it has not led to any change in the evaluation of women's work or loosened the controls of the family over their life choices. Indeed, as Chowdhry points out, a new kind of semi-proletarianization seems to be in the making under which men continue to work on their own farms, where they retain full control, while they send women to work for wages.

With her bias for subsistence agriculture where women and nature were in a more intimate and mutually nurturing relationship, Shiva and associates refuse to come to terms with the growing proletaritization of women. All their development efforts are geared to recreating mythic self-provisioning village republics where there would be no need for wage work. Consequently, ecofeminists simply ignore the real issues of poor wages and poor working conditions that women as farm workers face. Additionally, Shiva fails to appreciate how the new relations of production are powerfully shaped by preexisting cultural norms and social relations. The devaluation of women is not "inherent" or built into the "Western patriarchal ways of knowing" that have produced the high-yielding seeds of the Green Revolution. Instead, their effect on women's lives in the Third World is shaped, at every step of the way, by this local cultural mores that define female and caste propriety along with the existing inequities in social and economic power. Shiva makes it appear that the preexisting social relations in traditional peasant families were devoid of an economic rationality but operated only to satisfy need. But as I have tried to show, any attempt to portray traditional subsistence economy as driven by the logic of cooperation and altruism simply does not match up against the historical data.

*Access to land*

The "poverty literature" supports the common sense idea that ownership of land, however small in size, serves as a hedge against absolute poverty and destitution. However, as Bina Agarwal's (1994) detailed study shows, women in India are not allowed by custom and religion to inherit land, plow it, or self manage it. Although, after independence, the Hindu civic law has been revised to allow women the right to individually own, use, and dispose of land, strong cultural sanctions prevent most women from exercising this right. Lack of independent ownership further constrains women's access to credit, extension services, and new technologies.

Women still have the customary entitlement to the forests and the village commons which provide them, especially the poorest among them, with essential items for personal use and sale. But most village commons and forests are shrinking as a result of environmental degradation (chiefly due to deforestation, water logging, and salinity), population pressures, and privatization (caused in part by the government's land redistribution programs) causing disproportionate hardships for women and children who have the primary responsibility for collection of firewood, fodder, and water (Shiva 1988).

While ecofeminists have done a service in highlighting the deterioration of the commons and the resulting hardships for women, they have characteristically turned a blind eye to the issue of women's independent ownership of farm land and other productive assets. Both Shiva and Mies are generous in their praise of "Third World women" who: "appropriate nature, yet their appropriation does not constitute a relationship of dominance or a property relationship. Women are not owners of their own bodies or of the earth, but they cooperate with their bodies and with the earth in order to grow and make grow" (Shiva 1988, 43).

In a similar vein, Maria Mies holds up women's work in subsistence agriculture as a critique of the prevailing capitalist, profit- and growth-oriented development paradigm. Both hold that Third World women "expect nothing from 'development' or the money economy. They want only to preserve their autonomous control over their subsistence base, their common property resources: the land, forests, hills" (Mies and Shiva 1993, 303).

These claims about women's lack of interest in ownership of land or material benefit from development are ideological interpretations ascribed to rural women by the ecofeminists. Contrary to the ecofeminist claim that Third World women organize spontaneously on behalf of the embattled Mother Nature, Indian women have shown an equal zeal in organizing to win the right to own and manage land in their own name (Bina Agarwal 1994 provides many useful examples). Women value ownership of land on the understanding that ownership in their husbands' names may improve the welfare of the family but would not au-

tomatically improve their position in the family. None of these interests of women
find any mention at all in the ecofeminist literature.

Defending the commons against environmental decay and fragmentation
through the government's land redistribution and privatization programs is important, and ecofeminists are right in bringing this issue to the forefront. But so
is women's private ownership of land. One may be ideologically opposed to private ownership and wish for women to lead the way to a new way of living. But the
fact remains that rural women in India labor under a regime of property laws
which is thoroughly privatized. Nearly 86 percent of India's arable land is in private hands, with only a very small and shrinking area serving as village commons
(Bina Agarwal 1994). In this context, to expect women to concern themselves
only with the commons is tantamount to ignoring the real problems they might
be experiencing in their families and communities as a result of their being denied the right to own and inherit land in their own names.

The same problems arise in the traditional taboos against women using—or
even touching—certain technologies of production. Taboos against women plow-
ing hold across all communities in India: men have complete monopoly over the
plow, while women are allowed to use the hoe only. Different communities have
different religious justifications for why women should not use the plow. These
strictures are followed quite strictly, with communal punishments for women
who out of desperation might try to plow the field themselves. Similar strictures
also apply to other technologies, like the potters' wheel or the loom in some other
communities. For anyone trying to understand the relationship between women
and nature in the Third World, these cultural taboos have to be taken into consideration.

### The environment, wage labor, and women

Two sets of objections are most often made against the materialist understanding of the Green Revolution presented here. First, it is claimed by the
critics that a materialist analysis ignores or minimizes the problems the Green
Revolution (and modernization in general) has caused for the environment. Second, it is objected that the analysis offered here valorizes wage relations and
overlooks the plight of wage workers in the capitalist sectors of the economy.

First, the question of the environment. It is true that the perspective offered
here is not ecocentric; that is, it does not treat the stresses introduced by the
Green Revolution as central to understanding its impact on the lives of women
and the rural population at large. While I am not unsympathetic to the problems
of soil erosion, salination, and loss of diversity that ecofeminists have raised, I
don't think these problems are "inherent" in the nature of modern science or
technology, and neither do I believe that these problems have been created, ab

initio, by the new technology. Environmental problems, like anything else, have a history and must be understood historically.

As an illustration, take the ecofeminist charge that the Green Revolution is responsible for deforestation and soil erosion. Critics arrive at this conclusion by faulty reasoning: they see that the forests are diminishing and the soils are decaying, and they also see that farmers are adopting new farming techniques and they quickly conclude that the new techniques have caused the sad state of affairs. But in fact, if one keeps in mind the long history of low productivity, extensive agriculture, and the combined pressures of the subsistence needs of the growing population and the profits of timber industries both under colonialism and today, a different picture emerges. The historically low productivity of Indian agriculture even as compared to its neighbors, China and Japan, is a well-established historical fact (Moore 1966). For a variety of reasons, the traditional agricultural system was primarily an extensive farming system, that is, it could only increase the total output by extension of farming to new areas rather than increasing the yield per unit of land. This feature of the traditional farming system has played a historic role in the environmental crisis India faces today; it has been estimated that out of the total forest lost in this century, nearly half was cleared in order to put it under food production (Rao 1994, 161). Increase in the land area under the plow brought with it a proportionate increase in the livestock population which has had a big role in the erosion of the grazing lands. These were not the only factors; colonial use of timber for the railways played a role as well. But if the history of deforestation and soil erosion is kept firmly in the background, it becomes totally unjustified to hold the Green Revolution alone accountable for India's ecological crisis. In fact, it becomes reasonable to think that the yield-increasing, land-saving nature of the Green Revolution has *reduced* the pressure to put more land under the plow. Indeed, the recent data bear out this interpretation: Indian food grain output has continued to grow at a healthy rate of around 3 percent annually through the last decade (1981–1991) while the land brought under cultivation has actually *decreased* annually by about 0.3 percent (Sawant and Achuthan 1995; Rao 1994).

The ecofeminist complaints against chemicalization of agriculture are similarly overblown and abstracted out of the totality. That fertilizers and pesticides are often misused and overapplied is true, but it is not something that cannot be corrected through proper training and extension services. There is another far more basic problem with the way ecocentric critics of the Green Revolution have wrongly portrayed the Green Revolution seeds as necessarily dependent on chemical inputs. It is simply not true, as charged ad nauseam by Shiva and her sympathizers that the high-yielding varieties of seeds "inherently" need chemical fertilizers and pesticides in order to produce higher yields. Modern seed varieties are designed to produce more grain (and less stem and leaves) by making better use of nitrogen and other nutrients in the soil, *irrespective of the source of the nutrients*. There are many studies showing that high-yielding seeds continue to produce high yields even when supplied with organic manure or other organically derived inputs (Gupta 1998).

Let us turn now to the vexed issue of the role of wage work in women's emancipation. The way I have tried to resolve the issue of wage work in the agricultural economy does give the impression of simply turning the ecofeminist negative assessment of wage work on its head: they decry it, and materialist feminists should celebrate it. This impression is regrettable and unintended. I am by no means suggesting that wage labor in itself will liberate women or that Third World societies must simply embrace capitalism and call it progress. All I am suggesting is that wage labor moves the struggle for gender equality from the family to the public sphere, where it can connect with the larger struggles of other workers, locally and globally, to challenge the dominance of capital and patriarchy. I see increased opportunities for wage labor for women as a point of departure, rather than the end point of Third World feminism.

The role of wage relations in women's lives has to be understood historically and in the context of the ensemble of social relations, all of which set distinct limits on human relations and autonomy. That the Green Revolution is replacing the relationships based on kinship, family, patronage with relationships decided largely by calculations of profits, productivity, and the demand and supply of labor as a commodity is not in dispute.[7] The question is the relative assessment of these changes.

I strongly urge all those who believe that wage relations are more oppressive for Third World women today than the work relationships based on family and kinship, etc., to read a recent account by Martha Chen (1995) of the struggle of poor and widowed women in Bangladesh and India to win the right to work for wages as against the pressures to continue to work as a part of the family economy. The Bangladeshi and Indian women in Chen's account had to struggle against the culturally sanctioned norms of *purdah* and upper-caste seclusion, respectively, so that they could sell their labor in order to survive and raise their families in the absence of the male members of the family. What becomes very clear from Chen's account is that the existing institutions of the civil society are no longer meeting the basic subsistence needs of these women without denuding them of all personal dignity. To expect women to live by the same norms of femininity and domesticity when men and the extended family are no longer able or willing to keep their end of the bargain is simply unjust. Wage work allows women to survive in the face of the loss of the traditional structures of support.

Indeed, now we come to the crux of the matter. The ecofeminist (and postdevelopmentalist) critique of modernization is a lament over the erosion of the classical patriarchal bargain, while the analysis I have offered does not consider the prospect of the erosion of classical patriarchy as something that feminists should grieve about. Thus I contend that Third World women need wage labor, not simply as a matter of strategy to tide over the turbulence caused by modernization but as a *welcome means to break out of the classical patriarchal bargain altogether.*

To say that is obviously not to claim the end of patriarchy or class and caste inequities. As the experience of women in the West shows, wage labor under capitalism changes the private patriarchy regulated by the household and headed by the husband/father to a public patriarchy guided by relations of markets and non-kin men (Walby 1992). The personal liberties and autonomy that this change from private to public patriarchy affords women are limited as compared to the degree of autonomy truly human societies are capable of. But, limited though they might be when seen against an as yet unborn future, *these liberties are by no means insignificant* as compared to the far more oppressive limits the classical, private forms of patriarchy have imposed on women in Third World societies.

It is important to remember both the potential and the limits of emancipation possible under capitalism. By converting the process of extracting surplus value from labor from extra-economic means (of feudal lords, slave owners, or family) into an economic exchange (wages for labor), capitalism devalues extra-economic powers. The extra-economic powers in peasant societies, as Wood (1988) has argued, have a disposition to male dominance. The data presented in this paper regarding female survival and well-being bear out the cost of male dominance in peasant societies for women; it is amply clear that women in the largely peasant society of northwestern India have not fared well at all in terms of even basic survival needs. Modernization can erode the very material bases of the male dominance of peasant societies and for that reason, feminists cannot afford to turn their backs on it.

### Conclusion

This chapter has taken a critical look at the dogmas of ecofeminism and the circulation of these dogmas in the India environmental movement. I have argued that the fundamental ideas of ecofeminism and how they have been used in India, are both problematic. They romanticize those traditions which equate nature with a feminine, nurturing mother, without honestly confronting the many oppressions actual flesh-and-blood women suffered in these traditions.

By extolling the traditional order and making modernization look like a conspiracy on the part of an unchanging, congenitally imperialistic "West," India's best-known ecofeminist, Vandana Shiva, along with Maria Mies, and all those who share their animus against modernity, are asking women in the Third World to forego the opportunity they have to cut loose from the ties that have kept them subservient to men for so long. Mies and Shiva's "solutions" for the Green Revolution's imagined and real sins are exactly what have denied Indian women access to the most basic rights to life, livelihood, and assets. Such "solutions" can only be welcomed by those whose chief interest lies in decrying the West in favor of "their own civilization."

But this is the classic agenda of all cultural nationalists and blood-and-soil populists. *Feminism* should have nothing to do with it.

*Ten*

# The "Hindu Left," Agrarian Populism, and the Hindu Right

*Whoever refuses to discuss capitalism and its symbiotic relationship with the forces of the old order should keep silent about fascism.*
—*Arno Mayer,* Why Did the Heavens Not Darken?

"Whatever happened to the Hindu Left?" This question was recently asked with much anguish by a prominent Indian feminist, Ruth Vanita (2002). Why don't many more Indian intellectuals acknowledge their religious beliefs—like Gandhi did, or like Ashis Nandy does[1]—and still continue to support socially progressive causes? If they did, they will discover that they share quite a bit with those among the religious nationalists who swear by Gandhi and Gandhian socialism. And that, according to Vanita, will be a good thing, for then we will all work together for such good things like socialism, ant-imperialism, anti-pornography, social and sexual equality. It gets even better: the "Hindu left" will be able to extricate Hinduism from right-wing fanatics and redefine it to accord with a progressive social agenda.

I saw a photograph recently that explains why the left in India cannot simply embrace Hinduism *and* continue to be true to its own values of secular democracy. This picture was carried on the front page of the *Organiser,* the biweekly magazine of the Rashtriya Swayamsewak Sangh, the politburo of Hindutva. It shows Vandana Shiva, a shining example of the "Hindu Left," addressing the national convention of the Bharatiya Kisan Sangh, or BKS (which roughly translates into Indian Peasants Organization). The Bharatiya Kisan Sangh is a part of the "Rashtriya Swayamsewak Sangh family," expressly set up by the Rashtriya Swayamsewak Sangh to compete against regional farmers' groups, and has been active in Gujarat and neighboring states since 1986 in bringing peasants into the Hindutva fold. The picture shows Shiva making a speech, with all the bigwigs of the Rashtriya Swayamsewak Sangh looking on from the dais. The accompanying news item ("Kisan Sangh Urges to Counter MNCs") informs the readers, "Dr. Vandana Shiva, the noted environmentalist, formally inaugurated the Convention by lighting the traditional lamp." Shiva's message to the farmers in the audience was no different than the message she always carries to her mostly left-wing anti-globalization audiences in the West. She exhorted the farmers to "save their soil from destruction by refusing to use seeds and fertilizers produced by foreign multinational corporations . . . from the United States, which was practically ruling the country." By refusing to use foreign inputs, the news item implied, the farmers will be serving the interest of the nation as well as that of the environment. This was not an isolated incidence. Shiva has served as an adviser to M. S. Tikait, the leader of the biggest farmers' union in north India, widely recognized as having helped the Ayodhya campaigns and the Bharatiya Janata Party's elections. Shiva also works closely with Swadeshi Jagran Manch, an economic nationalist outfit of the Rashtriya Swayamsewak Sangh and other allied nationalists (Brass 2000, 114).

This picture answers Vanita's question most eloquently. Yes, many in the left *have* tried to combine their religious faith with their politics. Feminists and environmentalists have made several attempts to ground their message in traditional symbols, be they goddesses or sacred groves or both. But once they go down this road, they come to realize that the symbols begin to control their message: the "Hindu" ends up redefining what is "left," the Hindu patriarchal family ends up as a model of "socialism," caste becomes the model of an ecocentric way of life. In the absence of a prior secularization of traditions, traditions carry (not surprisingly!) a traditionalist meaning for the people, which has much that rebels against feminist and democratic ideals. It is no surprise that the religious parties win hands down every time secular left movements have tried to compete with them on the terrain of traditions.

In this chapter, I want to take a closer look at what happened to the ideas put out by our left-wing anti-modernist postmodernists. What impact have their ideas had on Indian politics? Who has been the real beneficiary of their longing for a "total revolution?" Is it "the people" who gain when intellectuals betray their calling for critique? Is it possible to create a good society—a liberal, just and egalitarian society—out of the conservative and deeply hierarchical values of a premodern society?

While my focus will be on India, questions naturally arise regarding the overwhelming support left-wing academics in the West have given to these ideas. People like Shiva, Nandy, Chatterjee and many others in the environmentalist camp got their intellectual stature in India in good measure because of the great accolades bestowed upon them by Western postmodernists and multiculturalists. The Western academics, it appears, were picking and choosing ideas that affirmed their own deep disillusionment with their own societies. Secure in their sinecures, Western academics could afford this kind of "radicalism" without much concern for its impact on other societies. Having had their fill of "differences," these tenured radicals are now ready to close the book on postmarked theories, again without concern for how these ideas are faring in other societies.

Ideas, however, do not die when they stop being fads in the Ivory Tower. The ideas of ecofeminists and postcolonial critics of scientific temper have acquired a life of their own.

This chapter will explore the fate of ecofeminism and alternative science in India. Evidence will be offered that shows that the left-wing critiques of modernity have served as the mobilizing ideology of farmers' movements. These movements are made of the kind of "hybrids" we encountered in chapter 6: comfortable with new technology and wanting more of it, but subordinating the technical and economic modernization to the distinctively patriarchal values of traditional elites. It is this constituency which has emerged as the natural ally of Hindu nationalist parties. The postmodernist populism of the left, I show in the following sections, has gradually become indistinct from the populism of the right.

### Farmers' Movements and the Agrarian Myth

Unlike the West where universal suffrage and parliamentary democracy made an appearance after the Industrial Revolution, independent India became democratic before it became industrialized. As a result, the agricultural sector, where more than 60 percent of India's one billion people still live and work, has come to wield a great influence on Indian politics. The influence of agrarian interests on politics is proportionate to the demographics, but disproportionately large in relation to the smaller and decreasing economic contribution of agriculture to the total gross domestic product (about 35–40 percent). Starting around the 1970s, when the spread of the Green Revolution made the use of purchased (even though at subsidized prices) inputs like fertilizers, pesticide, and seeds necessary, major farmers' agitations broke out in different parts of the country. These organized, non-party farmers' movements demanded more subsidies, cheaper electricity, debt write-offs, and better procurement prices of crops. These movements have become important players, forging alliances with environmental and alternative development movements on the one hand, and with Hindutva on the other.

In the beginning, there was hope that the political influence of these movements will be self-limiting because farmers are divided over regional, caste, and class interests and would not be able to emerge as a unified lobby demanding economic advantage for all the farmers. It was predicted—not without relief, for subsidizing such an enormous agrarian sector would surely bankrupt the state—that multiple identities would fracture the unity of the agrarian sector. Primordial identities of caste, language, and religion were expected to overwhelm the economic interests of the farmers' lobby (Varshney 1993, and 1995, chapter 7).

But such centrist optimism in the self-limiting nature of ruralism had not counted on the possibility of an upsurge of Hindutva parties that can *combine* an appeal to the primordial identities of farmers as Hindus, with a promise of greater emphasis on the economic interest of the rural sector in the name of promoting cultural authenticity. Far from overwhelming the economic demands with issues of identity, the rise of Hindu politics is merging the two demands, creating a powerful rural voting block for Hindutva parties. What is more, the more sanguine interpretations of agrarian populism had failed to take into account the growth of new social movements of the Gandhian, post-modernist left which have brought a progressive gloss to the agrarian myth of the wholesomeness and naturalness of the small-scale, decentralized peasant-owner farming embedded in nature and family. Hindu nationalists, meanwhile, have turned this agrarian myth into a part of India's uniqueness, its cultural-religious essence. Farmers' movements, led by dominant castes and dominant economic interests in the countryside but mobilizing literally millions of their poorer agrarian brethren, have emerged as the bridge between the right and the left.

Farmers' movements fit the classic mold of populism that mobilizes "the people" as a whole, while actively papering over the real conflicts of caste and class among them. These new agrarian movements explicitly reject the "old" social movements' (read Marxist) understanding of the peasantry as divided into antagonistic classes/castes. Farmers' movements downplay, and even deny, class differentiation in favor of the agrarian myth whose subject is "a homogeneous peasantry, the economic identity of which is linked to small-scale family farming in the village community, [and] its non-economic identity is mainly ethnic/cultural/national" (Brass 2000, 15). Farmers' movements have redefined exploitation as the exploitation of the entire traditional-agrarian sector as a whole, by the entire urban-modern sector as a whole. They claim that the modern Indian state and its Westernized, scientistic, and technocratic bureaucracy is biased in favor of cities and industries and is exploiting the entire countryside to benefit urban-industrial development (see Omvedt 1993, chapter 5). Arguing that the state is "looting" the innocent, ecologically sustainable, and communitarian traditional society in order to support the morally decadent urban and industrial sector, farmers' movements have tried to portray all of modern *India* as the enemy of all of the traditional *Bharat*. Not class or caste, but the urban bias of the "colonized," "Westernized" state which looks down upon the noble traditions and wisdom of Bharat's hardworking farmers becomes the common enemy of the entire village community.

The anti-statism and anti-modernist bias of the left-identified post-developmentalist movements neatly supplements this diagnosis. On the other side, the Bharat-India divide is one the central planks of the Hindu-right's critique of Nehru's modernist, socialist planning.

The Bharat-India divide admittedly predates the postcolonialist critiques. It is rather the critique of "urban bias" (associated with Michael Lipton's work), the Gandhian, and Congress-socialist advocacy of the "rural bias" that have been most widely cited justifications of farmers' agitations demanding more favorable terms of trade against the urban-industrial sector. The more recent critiques of science and modernity serve as ideological and rhetorical supplements to these preexisting tendencies. Moreover, as Akhil Gupta has recently pointed out, the intellectual justification of this discourse of the virtuous self and a malevolent and alien other has "not only been articulated by leaders within the movement, but also by crucial intermediaries working in economic institutes, regional colleges, and universities and journalists writing for the English-language and vernacular press. There are thus multiple avenues for the flow of development discourse from popular movements to scholarly circles" (1998, 327). Given the increasingly globalized connections between the activists and "scholarly circles" in the West and in the Third World, the critiques of science and development in Western academe become relevant to the activists and members of new social movements who appropriate these ideas through precisely the kind of intellectual intermediaries that Gupta mentions.

Who participates in farmers' movements and why? Organized movements mobilizing tens of thousands of big and small farmers in order to demand higher crop prices and cheaper seeds, agrochemicals, and other inputs began to make their presence felt in Indian politics at the national level in the late 1970s. The new agrarianism was the a result of a shift of political power from the vanishing breed of big landlords to the so-called "bullock capitalists" or the "small to medium-sized self-employed independent agricultural producers" (Rudolph and Rudolph 1987, 50), the beneficiaries of land reform and the Green Revolution. These bullock capitalists have radically altered the complexion of rural politics, replacing land redistribution with profitability of farming as the major agrarian demand. As the state is the major buyer of farmers' surplus and the major supplier of inputs (through subsidies to public-sector chemical and seed firms and through state-run electricity and water boards), the farmers' movements target the national and state-level governments, which they see as biased toward urban-industrial interests, despite any definitive evidence of such urban bias.[2] Major farmers' movements have arisen in the relatively prosperous regions of the nation where the Green Revolution has made the deepest inroads. These include the Bharatiya Kisan Union (BKU) including Punjab and Uttar Pradesh in the north, the Shetkari Sanghathana (SS) and Bharatiya Kisan Sangh (BKS) in Maharashtra and Gujarat in the west, the Karnataka Rajya Ratitha Sangha (KRRS) in Karnataka, and Tamilaga Vyavasavavigal Sangham (TVS) in Tamil Nadu in the south.

In terms of membership, these movements are not "rich peasant movements." Landowning farmers, including medium and small and even marginal landowners, *all* participate. Although multi-class in membership, these movements are led by sections of rural elite of big to medium surplus-producing farmers who stand to gain disproportionately from the price reforms and withdrawal of the state. In Maharashtra, the core support of Shetkari Sanghathana comes from farmers producing commercial crops on five to 15 acres of land (Banaji 1994, 230). In Uttar Pradesh, surplus farmers owning over eight acres of land participate most actively (Hasan 1994, 177). In Punjab likewise, the class of farmers who have been the biggest beneficiaries of the Green Revolution, and who find further avenues of growth blocked, have been the most active supporters (Gill 1994, 205). This class of farmers already dominates the institutional space of rural society—the local village councils, rural banks and credit corporations. And they use their dominance to augment their economic gains and their already considerable power over workers, often leading to caste-enforced contractual agreements for attached labor through debt peonage (Brass 1990), and through patriarchal control over the unpaid work of women (Chowdhry 1994).

If big and medium farmers set the agenda and reap the most benefits, why do small and marginal farmers, and sometimes even the landless laborers, participate in these agitations? The answer lies in the dynamic of the state-subsidized Green Revolution. The widespread adoption of Green Revolution technology by small to marginal farmers owning less than 2.5 ha of land made it necessary for even the small farmers to participate, as much as the big ones, in the market for inputs, labor, and for the sale of their surplus (Varshney 1995; Omvedt 1993). But this is only a part of a very complex pattern of what can be best described as "partial proletarianization" of small peasants. The owners-cultivators of marginal (less than 2.5 ha) plots have to sell their labor (or their wives' and children's labor) to bigger farmers in order to buy seeds, fertilizers, and other purchased inputs that Green Revolution farming requires. Thus, marginal farmers are both farmers and wage workers at the same time. At one level, objective conditions do exist which make it rational for the small to marginal peasants to align with the big farmers to demand higher profitability for their crops through an increase in procurement prices and lower input costs. But because the small and marginal peasants simultaneously sell their labor to larger farmers, depend on cooperatives, banks, village councils dominated by the big farmers, and buy their food from the market (which will become more expensive if state procurement prices go up) it is *also* in their interest to align themselves with the landless poor against the surplus farmers to demand fair wages, laws protecting the workers, and other development-oriented state interventions which create alternate job opportunities and target the poor through welfare services (Banaji 1994; and Balgopal 1987). In other words, just because more and more farmers participate in the state-controlled market and industrial economy does not make the class divisions in the rural economy "non-antagonistic," or make the state the main exploiter of the entire rural sector, as the supporters of farmers' movements claim (Omvedt 1993).

Regardless of whether they are big, medium, or small farmers, members of these movements are openly hostile to landless farm laborers, most of them come from lower and untouchable castes. To these landless and the near-landless marginal farmers who also sell their labor, the farmers' movements offer nothing more than a promise of a trickle-down from increased incomes; it is assumed that as farmers' profit margins improve, they will be willing to hire more workers and pay them higher wages. (After initial resistance, almost all major farmers movements have now included a higher minimum wage for workers in their charter of demands Omvedt 1994, Varshney 1995, 119.)[3] But this optimistic view contradicts most available evidence showing that farm wages do not keep up with farm incomes and profits. Indeed, as Hanumantha Rao (1994, 56–57) concludes in his recent review of Indian agriculture, "the share of wages in output has gone down steeply in areas undergoing rapid agricultural growth. The net domestic product in agriculture has risen much faster than real wages in such states." The stagnation of wages is partly because of mechanization and partly because of culturally sanctioned, caste and patriarchy-enforced modes of intimidation, segmentation (by gender, caste, and age), and relationships of bondage (Harriss 1992; Brass 1990). The two factors that have helped the landless poor the most—availability of cheaper food and state-run employment programs which have tightened the labor markets and helped increase the wages—are both threatened by the farmers' demand for higher crop prices which will raise the price of food and cut into the subsidies available for employment generation and other development plans. It is for this reason that the agricultural workers and their organizations have been lukewarm at best to farmers' demands (Banaji 1994; Brass 1994).

This brief detour into the multiple divisions among rural classes was necessary to show the crucial importance of a mobilizing ideology, a suitably contemporized agrarian myth, that can paper over the deep class and caste divisions. The relatively rich farmers with surplus to sell *need* the support of the majority of poorer farmers and landless workers in order to pressure the state for subsidies and procurement prices that will mostly help the richer farmers. In order to mobilize the poorer members of the agrarian economy in a movement which will serve the interests of the richer members, they *need* an ideology that can present the entire village "community" as a victim of the modernist state.

### Post-development, Ecofeminism, and the Farmers' Movements

What do post-development ecofeminist theories have to do with farmer's movements? The answer is simple: they have provided the much needed agrarian myth to the farmers' movements.

Simply put, post-developmentalist critique of modernity in the Third World as mimetic, culturally alien, and elitist helps to displace class/caste distinctions in favor of a shared cultural victim-hood of the entire village "community" at the hands of the Westernized urban-industrial interests allied with global capitalists. The "ecological analysis" of the linked oppression of nature and women by ecofeminists helps to reinstate women and their special relation with nature as natural allies of *Bharat*, the traditional India. This defense of the village community easily takes on nationalist and patriotic connotations, as the village is made to stand in as a symbol of Indian civilization threatened by Western science and technology. As even the advocates of new agrarianism recognize (Rudolph and Rudolph 1987, 358), such appeals to the village community as a bulwark of traditional virtues carries an enormous emotional appeal with the small peasants who have not yet become a part of the market for discretionary consumer goods (which they simultaneously desire and feel threatened by).

Farmers' movements, like Hindu nationalist movements, are Janus-faced: they simultaneously act as social movements *and* intervene in the electoral politics as political parties do (by either fielding their own candidates or more commonly, endorsing political parties). As social movements, farmers' movements take on a larger, populist social agenda which cuts across their own interests, while as political lobbies, these movements agitate for demands which serve the dominant class interests (Basu 2002). In their populist, social movement phase, farmers' movements have built alliances with other social movements, and even co-opted their agendas, in order to obtain the clout that comes with sheer numbers in a democracy and in order to shine in the reflected prestige of these other movements.

Ecofeminist movements in India are a case in point. The Third World ecofeminist discourse that combines all the elements of post-development—the critique of science as alien, masculine, and colonialist, the celebration of the feminine, the local, and the traditional—has become the ideology of choice for the social component of the farmers' movements. At least three major movements, with widely varying ideologies, have adopted ecofeminist ideas: the pro-GATT, neo-liberal Shetkari Sanghathana of Maharashtra led by Sharad Joshi, the staunchly anti-imperialistic and Gandhian Karnataka Rajya Ratitha Sangh in the state of Karnataka led by Nanjundswamy, and the anti-imperialist but non-Gandhian Bharatiya Kisan Union in the state of Uttar Pradesh led by M. S. Tikait. India's leading ecofeminist, Vandana Shiva, has served as an advisor to the Gandhian Karnataka Rajya Ratitha Sangh and the famously rustic and patriarchal Bharatiya Kisan Union. But as Omvedt describes in her survey of new social movements, "Shiva's articulation of the feminine principle . . . finds its echo in the themes of Stri Shakti [women's power] within the women's movement connected with the Shetkari Sanghathana and other rural organizations" (1993, 316).

Third World ecofeminism, which places nature-maintaining, subsistence-based rural women at the center of production of value and reproduction of life, is all things to all movements challenging the modern-industrial paradigm of modernization. It fits in with nearly all the populist juxtapositions: Bharat against India, small against big, folk wisdom against abstract knowledge, nurture against pollution, feminine against masculine. While ecofeminist ideas have been appropriated by the Karnataka Rajya Ratitha Sangh and Bharatiya Kisan Sangh for campaigns against multinational corporations, it is the Shetkari Sanghathana that has built a cohesive social program around the idea of the feminine principle and tried to integrate it with neo-liberal economic reforms. Correspondingly, the Shetkari Sanghathana seems to have won the support of many prominent feminists as a harbinger of a "genuine historical materialism for women" (Omvedt 1993, 316).

Shetkari Sanghathana's social program subsumes the environmental and women's movements' demands in a way that ties them to the interests of the landed peasants described here. Following the feminine principle of ecofeminism, Sharad Joshi declared women's unpaid work in families as the true creator of agriculture, thus displacing the contradictions of class with gender: women—all women—came to occupy the position of the proletariat. Furthermore, the feminine principle was seen as more in tune with Indian culture for it was not anti-men and incorporated the local knowledge of women. This analysis, with pressure from women activists in Shetkari Sanghathana's, led to the "Laxmi Mukti" ("Liberating the Goddess of Wealth") campaign in 1990. In this campaign, men are urged to voluntarily gift a portion of their land to their wives with the proviso that they (the wives) could only use traditional organic farming techniques on their portion of the land. In 1990, 127 women received shares from their husband's property ranging from half-an-acre to seven acres (Guru 1992).

The Laxmi Mukti program has been hailed by feminists and environmentalists as an exemplar of a new kind of activism. There is no denying the force of the arguments supporting a woman's right to own land in her own name (Bina Agarwal 1994). Programs like Laxmi Mukti do strengthen the position of women in their families and give them a fair amount of income security (although it is not clear if the beneficiaries can sell the land or mortgage it without their husband's permission).

Even though its limited usefulness can be readily granted, these kinds of social programs serve the tactical needs of the farmers' movements more than the real needs of all women. To begin with, if land rights are so important for women's independence and security, they are all the more important for landless women. The "gift" of land in Laxmi Mukti is only within the family of landowning farmers; the landless women are by definition excluded. In fact, this program performs an extremely potent ideological function by making the intra-family gift serve as a substitute for land redistribution to the landless.[4] Second, there are reports of some relatively large farmers using this program to evade the legal limits on land holdings (Guru 1992). Third, the sexual division of labor continues in the form of subsistence agriculture for women on their patch of land, while commercial crops using modern inputs continue on the men's land. This division of labor may not be in the economic interest of women. Finally, this and all other social programs (which include appropriate technology projects, and food processing) have to be judged in terms of justice for all the classes of the rural community. Given the continued atrocities against landless and migrant workers, sometimes even by members of some of the farmers' movements (Assadi 1994; Gupta 1997), it is hard to cheer the rhetoric of the feminine and the local as a genuine alternative to development.[5]

Clearly, those activists who are enthused by the ecofeminist and/or spiritual vision of new politics centered around the "production of life" by women, peasants, and defenders of organic life processes, are by no means backward looking. As the participant-chronicler of India's new social movements, Gail Omvedt (1993, 300) insists, farmers' and environmental movements' "radical use of traditional symbolism looks not backward but forward to a less exploitative as well as sustainable community . . . it is only from a centralist, statist and secularist position that any reference of community identity is seen as traditionalism." Yet, the fact remains that the "production of life" perspective is perfectly compatible with a conservative ruralism which accepts the hard labor of women and the lower orders as a part of the natural, organic web of relations, sanctified by the presence of God in all.

Another more recent and more interesting example of the coming-together of the left and the right is the famous *bija satyagrah* or "seed non-cooperation/resistance" started by Vandana Shiva in 1999. This example is full of ironies, for even while the political supporters of an imagined united and rural Bharat were condemning transgenic seeds, some of the inhabitants of Bharat, that is, commercial cotton farmers, were demanding the right to use these seeds! So much for the populist myth of shared interests of all those who live in Bharat. (For details, see Herring 2001.) Genetically modified seeds have come to stand for the entire paradigm of corporate, reductionist science, while the traditional seeds are made to represent all the virtues of a self-provisioning, nature-close peasantry. Predictably, Shiva has reached out to a variety of groups openly affiliated with the Rashtriya Swayamsewak Sangh. The connecting thread is the defense of the traditional way of life.

### Hindu Nationalism and Farmers' Movements

For a while, there were reasons to hope that agrarian populism might be able to stem the tide of Hindu nationalism. The attempt to unify Bharat against India might be able to unify all those who lived in Bharat, including Muslims, Christians, and other non-Hindus. Initially, major farmers' movements, especially the Shetkari Sanghathana's in Maharashtra and even the Bharatiya Kisan Union in Uttar Pradesh did oppose those parties that tried to create divisions on the basis of religion. The entire peasantry was to be included in the community of Bharat.

But it did not take long before farmers' agitations for concessions and Hindutva agitations for the Rāma temple discovered each other's usefulness. They discovered that they spoke the same language of Hindu traditionalism and could work together.

Like the farmers' movement, the Bharatiya Janata Party, the political party of the Hindu right is also Janus-faced: it has a moderate populist face, and it has an extremist rabble-rousing face. It switches between the two depending upon its own survival needs and the mood of the electorate (Jaffrelot 1996). In its moderate populist phase, the Bharatiya Janata Party is well-known to downplay its Hindu nationalism and take on "rural development," complete with all the demands of farmers' movements for higher prices, more subsidies. The Bharatiya Janata Party, with its sister Hindutva groups, shares the agrarian myth of "village republics" as nurseries of traditional virtues, which need to be defended from the invasion of modern vices. The problem for these groups is how to balance their ruralism, which inclines them toward protectionism, with the interests of the urban small and middle classes, who are clamoring for more Western goods. These conflicts are creating considerable tension in the Hindutva family (see the following paragraphs). But Hindutva ideology is a potent brew which can unite the populous and fractious countryside around religious identity *and* speak their language of rural virtues. What is worse, its anti-foreign, protectionist rhetoric can even co-opt the left opponents of globalization and modern technology. By making the alien-ness of the modern West as the enemy, the left has played into the hands of the religious right.

There is ample evidence that the Bharatiya Janata Party rode to power in Uttar Pradesh, India's most populous state, in 1991 with full support from that state's farmers' movement (Bharatiya Kisan Union). According to Zoya Hasan (1998, 107) who has studied the farmers' movements in this region in great depth, "a combination of ideological activity by the BJP-VHP-RSS combine followed by violence and rioting played a decisive role in BJP victory in 1991. In western UP, the BJP won 11 out of 17 seats, all in riot-hit towns . . . the BKU's endorsement helped the party establish its credentials among farmers who might not otherwise have been sufficiently inspired by the BJP." (But later elections show that the farmers do not vote entirely on the basis of religious identity. The Indian electorate has a strong instinct for judging the candidates on performance. The Bharatiya Janata Party has suffered setbacks due to anti-incumbency and caste divisions. It now rules Uttar Pradesh with the help of a dalit party which has sought this strange alliance in order to oppose backward castes.) Even in those regions where farmers' movements were considered most progressive on issues of inclusiveness of women, as in Maharashtra where the Shetkari Sanghathana's won warm praise from ecofeminists, communal frenzy overtook the countryside. While the leadership of the Shetkari Sanghathana's remained opposed to the Hindu right, the rank and file got busy electioneering for Hindu parties (Banaji 1994, 237).

India is too complex a society, with too many internal divisions to assure *any* political party a permanent majority. That is indeed its saving grace. Moreover, the country still has an active and vocal opposition, a relatively free press and an independent judiciary. Hindutva's sway over the electorate cannot be taken for granted. But Hindutva ideology has the right mix of traditions and modernity to mobilize masses across a wide spectrum of regions, classes, and castes for its own chauvinistic and anti-minority agenda. Fundamentally, the Hindu right deploys the grammar of India against Bharat, traditions-under-threat from alien ideas, which finds a deep and wide resonance among all sectors of a fast-changing society. Coupled with the ideology, the Hindu right has learned that provoking communal passions pays electoral dividends.

This grammar, however, is not without its own problems. For the fact is that the agrarian populism of the capitalist farmers is an ideological ploy to extract more concessions from the state—which include modern agricultural technology, modern conveniences, and services. The cry of Bharat being exploited by India invokes the virtues of Bharat to demand the technologies of India. Similarly, the urban supporters of Hindu parties are simultaneously insatiable consumers of goods and services that globalization is bringing into India.

How to talk the talk of traditions while opening up to the reviled modern Western evils is creating tensions and divisions in the Hindutva family. Ironically, the left-inclined anti-modernist movements that we have examined in this book are lining up with the more fanatical, more dangerous, true-believers of the Rashtriya Swayamsewak Sangh, against the more pragmatic (though no less dangerous) Bharatiya Janata Party.

Now that the Bharatiya Janata Party is running the government, it has diluted its commitment to "Swadeshi" which stands for the indigenous and for self-reliance. It has been active in selling off loss-making public sector industries and opening the economy to foreign investment and imports of goods and services. In the meantime, the Rashtriya Swayamsewak Sangh, the moral leader of the Bharatiya Janata Party, has become aggressively anti-globalization. Rashtriya Swayamsewak Sangh-affiliated groups are threatening agitation against the Bharatiya Janata Party's neo-liberal economic policies. Interestingly, the Rashtriya Swayamsewak Sangh's arguments against globalization are indistinguishable from those of the Marxist and Gandhian left. As one Bharatiya Janata Party sympathizer put it, "the RSS has already begun questioning the erosion of national sovereignty under the WTO regime. RSS chief Sudarshan has imbibed the mass of literature circulated by the anti-WTO protesters in Seattle and is busy disseminating the message in the shakhas [RSS training schools]. The RSS has also linked nationalism with uncompromising opposition to foreign investment in consumer goods" (Dasgupta 2001).

If the new social movements of the Marxist and non-Marxist left find themselves pushed into the arms of the Hindu communalists, it is because they themselves have for so long played the anti-imperialist card so indiscriminately. Non-Marxist movements, or rather "post-Marxists," as they call themselves, have imputed imperialism to ideas themselves, and not just to economic relations (however exaggerated even this might be). Authenticity and indigenousness, and not objective truth, have been used as criteria of acceptance. Any critique of the indigenous has been scoffed at. "Tradition under attack" has been the slogan of Gandhian and postmodernist critics of scientific temper and modernity. The Marxist left has not attacked modern science and modernity as such, but it has not been able to resist the postmodernist onslaught. Indeed, the Marxist analysis of globalization as "imperialism" has put Marxists in a compromising position vis-à-vis those rejecting modern science as "mental imperialism;" they could not simultaneously oppose economic imperialism and oppose those movements that were agitating against cultural imperialism.

## Conclusion

At first sight, this chapter about farmers' movements—and the alphabet soup of their acronyms—may appear to be a far cry from the sophisticated and highly theoretical discourse of social constructivist feminist or postcolonial epistemologists. We are talking here of such down-to-earth issues like remunerative prices, and prices of pesticides and fertilizers, a far cry indeed from the refined academic debates over the nature of modernity and science.

But this "far cry" is precisely the point, that is, that ideas have consequences, and more often than not, these consequences are unintended.

The cultural despair of Indian intellectuals, I have tried to show in this chapter, has given birth to a set of ideas for cultural rejuvenation that are finding a great resonance in the agrarian populism of farmers' movements. These populist movements involving tens of millions of farmers and farmworkers are increasingly serving as vote banks for the Hindu right wing. Hindutva parties are able to combine the economistic demands of the farmers (for subsidies and higher prices), while at the same time papering over the caste and class distinctions under the cover of a unified, Rāma-oriented Hinduism. Recent trends show that Hindu nationalism is no longer limited to the small property owners, government employees, and other lower middle classes in the cities, but is also spreading to the countryside and the homelands of the tribal peoples.

By impeaching modernity itself as the enemy of the people, the secular alternative science movements lent a new respectability to the old arguments for rural populism. The old "urban bias" which favored a family-farm based strategy for modernization over industrialization and urbanization, found new adherents among urban intellectuals influenced by new ideas. The notion that urban, modern, and mentally-colonized "India" as a whole is exploiting the rural, non-modern, and authentic "Bharat" as a whole, gained many adherents among social movements. When ideas themselves are held responsible for "epistemic violence," it is easy to overlook distinctions of caste, class, and gender which run deep in both the urban and rural sectors.

Under the influence of anti-modernist ideas, many social movements sought the vanguard of a new kind of modernity from within the presumably "non-modern" rural traditions and institutions. Farmers' movements continue to make strategic use of green and feminist ideas, as, for example, in the case of setting up organic farms for the wives of the farmers, or alliance with anti-globalization movements to oppose foreign imports of farm products which Indian farmers cannot compete against.

But culturally, it is the message of the religious right that resonates with the rural constituency. The cultural symbols of the religious right have an emotional resonance with the rural populations that cut across class and even caste. None of the agrarian movements have been able to withstand or oppose the Hindutva message of defense of traditions, more so because like all other parties, the Bharatiya Janata Party and its allies have promised to meet the economic demands of the agrarian sector, at the cost of raising the food prices for the really poor.

In short, the protest against modernity itself has encouraged a religious turn in politics in the agrarian mobilizations we have examined in this chapter.

## Eleven

# Conclusion

### Prophets Facing Forward

*A culture which permits science to destroy traditional
values, but which distrusts its power to create new
ones, is a culture that is destroying itself.*
—*John Dewey,* Freedom and Culture

*We are going to enter a life of contradictions. In politics
we shall have equality and in social and economic life
we will have inequality. We must remove this
contradiction at the earliest possible moment . . .*
—*Bhimrao Ambedkar,* Constituent Assembly Debates

We have examined the dreary record of the prophets of postmodernity
facing backward toward premodern traditions. As we come to the close of our inquiry, it is time we turn our faces to the future, toward a renewed hope for the
possibility of reasoned change, not just in the technological hardware, but also in
the cultural software of our societies. The latter includes such things as our ways
of adjudicating beliefs, our conception of nature, human beings, and God. This re-
orientation to the future will require the rediscovery of the unfulfilled potential of
modern science as a weapon against the forces of reactionary modernism that
stare us in the face, not just in India but around the world.

The secular-humanist Dewey and the Buddhist-liberal Ambedkar, the teacher-
student pair we have met earlier in the book, are ideal guides to help us readjust
to a forward-facing vision that neither despairs of modern science, nor celebrates
it mindlessly. Together they show the dangers of reactionary modernism and
suggest a way to overcome them.

Modernity turns reactionary in those cultures which, as Dewey points out,
"permit science to destroy traditional values, but . . . distrust its power to create
new ones" (see epigraph at beginning of chapter). These cultures allow modern
technology, industry, and capitalism to destabilize the traditional moral economies
and social structures. But these cultures also actively insulate their "traditional
values" from the new view of nature and new processes of knowing which
emerged with modern science, and which have made the material infrastructure
of modern age possible. While everything else is changing, there is a tendency to
try to hold on to erosion-resistant values of traditions and most importantly, reli-

gion. Unless this cultural inertia is actively challenged, either by a developmentalist state, in cooperation with or against religious institutions themselves (depending upon the nature of the religion in question and its historic relationship with state and civil society) and by secular intellectuals in the civil society, a lag between the technological and cultural realms is fairly common in the process of modernization.

If this gap between the realm of production, economics, and state and that of culture, meanings, and values is left to grow unchecked, a very fractured, distorted modernity—what I (following Jeffrey Herf) have called reactionary modernity—results. As those millions rendered homeless (both physically and psychically) by the gale of modernization begin to wake up to the new world, they do not have the cultural resources to make themselves at home (physically, but mostly psychically) in the world they find themselves in. Suitably engineered religious ideologies fill the void, offering at least some connection with the world that is in the process of being destroyed. *This insulation of traditional values from rationalization and secularization, while traditional economies are falling apart, is the root of all reactionary modernisms.* What is happening in India is not unique at all. Such reactionary modernism lies at the heart of radical Islam in most of the Middle East as well.

Modernity in India has, from the very start, been full of deep contradictions. As Ambedkar observed, most presciently on the eve of India giving herself a new Constitution that he himself helped craft, "we are going to enter a life of contradictions. In politics we shall have equality and in social and economic life we will have inequality." These contradictions run through the entire fabric of social life in India. India wants to, and professes to, live by modern values of democracy, tolerance, and scientific reason. The trouble is that *in reality*, these values have hardly any purchase in the lived life of ordinary people. The words "democracy," "tolerance," "reason" carry meanings which imply a non-individualist community, hierarchy, and mystical authoritarianism, respectively. The *modern* meanings of democracy, tolerance, and reason, as they are generally understood in the liberal-secular sense, in fact, go *against* the dominant strain of Indian traditions and cultural heritage.

The question that Indian social movements and intellectuals have had to grapple with from the very start has been this: How can the gap between the ideals of a liberal democracy and the reality of an inegalitarian caste society be closed? The Indian Constitution enshrines the three ideals of modernity—equal citizenship, a secular state that respects and cultivates scientific reason in affairs of the state and society. The reality that modern India inherited, however, was that of a caste society, overlaid with distortions introduced by over two centuries of British colonialism. The question that occupied Ambedkar and others was how best to bridge the gap? How could modern ideals be made a reality in the social life of the people?

In theory, there are only two ways to bridge the gap, although in practice there will always be a little bit of both. The first way is what de Tocqueville called "educating democracy," which involves "purifying the mores of a society . . . changing the laws, customs and mores needed to make democratic revolution profitable" (see chapter 1). This involves a critical engagement with, and modification of, the existing beliefs, even at the cost of discarding some of them altogether, in favor of new mores that are more attuned to a democratic style of thinking and living. Educating democracy requires, at a minimum, accepting the legitimacy of modern ideals and using them as critical vantage points to reform tradition. The second way is that of "reinterpreting modernity" which is what David Kopf refers to as "pouring the new wine of modern functions into the old bottles of Indian culture" (see chapter 3). This amounts to keeping the modern words, but using them to mean what tradition says they mean. Only the rhetoric becomes modern, while the ideas continue, largely, to resonate with the old meanings. In contrast to the first option in which modern ideas are the standards against which tradition is judged, the second alternative turns the tables and judges modern ideals from the vantage point of traditions.

In India, the postmodernist stress on "alternative modernity," or even on the less essentialist, pastiche-like state of "hybridity," has served the function of reinterpreting modernity, of pouring new wine in old bottles. The only hope India has for recovering the ground lost to reactionary modernity is for Indian intellectuals to heed the call of Ambedkar to reduce the contradictions between the sphere of production and the sphere of lived life, ideas, and meanings. For that, I want to suggest in closing, Indian intellectuals have no choice but to confront their religious heritage with an uncompromising honesty, something they have avoided doing for a very long time.

### Postmodernism and Hindutva: A Brief Recapitulation

But before I move on to the necessity of religious reformation and secularization in Hinduism for the development of the spirit of liberalism in Indian culture, I want to close the book on the cult of postmodernity that spans the boundaries between the Indian and Western intellectuals. I have already said most of what needs to be said in the book at great length. Here I will only recapitulate the highlights.

The essential insight of postmodernism and social constructivism is perfectly valid, namely, entities, ideas, institutions do not come with fixed timeless essences, but their meanings change with the context and culturally condoned usage. But when made into a dogma, this anti-essentialism runs the risk of denying legitimate distinctions, and giving the extant cultures the right to give any meaning to any idea—"turning all cows equally grey," as the late Ernest Gellner used to quip. Thus, the critics of modernity have denied that there is any essence to the modern experience, suggesting, therefore, that other cultures can become modern in "their own ways," without becoming secular. Likewise, the critics of modern science have denied that modern science is at the same time a continuation of the fundamental human capacity to reason, and *also* a revolutionary break from all previous forms of formulating and testing hypotheses about the natural world. This leads them to suggest that other cultures can continue to do science "in their own way," without the necessity of accepting the disenchantment of nature brought about by the cumulative result of physical and biological sciences over the last four centuries.

This radical anti-essentialism has brought in the vogue of "alternative sciences." This stance of "epistemic charity," as we have seen (chapter 5, and passim), insists upon a radical symmetry between all ways of knowing, from their overall reasonableness to their rules for evaluating sensory experience, and adjudicating between rival hypotheses. All the relevant cognitive rules and values are explained as obtaining their ability to rationally convince from the wider culture and social structures in which scientific institutions are embedded. The alternative (which I have tried to argue in chapter 7), that modern science itself, by learning by doing, has evolved cognitive values which are justified by success in scientific practice in the past, and that modern science has gradually become relatively autonomous from the rest of the culture, is not fully appreciated by the critics. There is an almost desperate urgency, especially among the postcolonial critics of science, to accept the most radical—and also the most relativist—views of modern science, in order to make room for locally grounded alternatives.

Let me mention in passing that I have not just criticized, but have also done my best to understand this desperate urgency to defend "local knowledge" among Indian intellectuals. Following Edward Shils's dated but still valid observations, I put it down to the legacy of colonialism and caste society which has made Indian intellectuals feel inordinately guilty for the advantages they—as all intellectuals in all societies—do enjoy (chapters, 1, 3, and 8, and passim). The high degree of populism in India is partly an over-compensation for feeling alienated by the advantages of a modern, cosmopolitan education Indian intellectuals have enjoyed, and partly an over-compensation for historical advantages of caste.

Even though I was not convinced of the arguments of the critics of science and modernity in the first place, the rising right-wing religious politics in India turned my intellectual discomfort into a grave political concern. I found the romantic Gandhian and postmodernist nostrums being repeated, almost verbatim, in the Hindu nationalist discourse. Through my fairly detailed analysis of these tendencies in social constructivist theories and in new social movements, I arrive at this conclusion: *The tendency to remove distinctions between modern science and premodern and other folk sciences has led the postmodernist left to, in effect, reiterate the First Principle of neo-Hindu nationalists, namely, the "equality of all truths."*

The Hindutva proponents of Vedic science, as we have seen (chapters 3 and 4), assume that all sciences are equally true, for they all converge at the "same truth" by different routes. Needless to say, they see the "same truth" of all sciences contained in the Vedic and Upanishadic view of the material world as a construct of, and indistinguishable from, the Absolute Soul, or Brahman. This doctrine of equality, of "no real differences" of all sciences, enables the Hindu ideologues to erase distinctions between myths and science. Consequently, Hindutva ideology presents myths of sacred Hindu texts as containing propositional knowledge about the natural world, and presents the method of introspection (yoga) as "scientific" within the metaphysical assumptions of Hinduism.

Under the spell of postmodernism, there has been very little, almost insignificant, organized resistance in India to this ideology of Hindu supremacy in science, which has now become the state-supported educational policy. That the repudiation of modern science and secularism in favor of local sciences has led the left-inclined new social movements to a dead-end is obvious. Nothing more needs to be said, except that the Indian experience should serve as food for thought for science studies and other proponents of multiculturalism and theories of "difference." The internal critics of the West have for long picked and chosen those ideas from the non-Western world which affirm their own discomfort with their own hyper-rationalized societies where instrumental reason (in the shape of profit motive and efficiency) very often elbows out consideration of collective good and social ties. In this process, they have not paid sufficient attention to the role these ideas might have played in the non-Western cultures themselves. Many of the "local knowledges" which look like "epistemologies of the oppressed" from the vantage point of the West, are actually part of the ruling ideologies in many non-Western societies. They might be useful as "resistance" against the West, but often at the price of lending them even more legitimacy in the societies of their origin. The Western friends of the Third World have an obligation to understand the complete social history of ideas in situ in other cultures.

### Resuming the Enlightenment Quest

Reinterpreting modernity to conform to the traditional meanings has not worked. It has led to religious revival, espoused by the family of Hindu nationalist parties. It is time to return to the task of "educating democracy," which involves the creation of new norms and new mores.

The way I see it, Indian democracy has come as far as it could in the absence of an Enlightenment-style cultural revolution. In the absence of a fundamental religious and cultural reformation, it can only go the way of Hindutva.

For all its limitations, democracy in India has chalked up considerable successes. The right to vote has been the single most important tool for social change in India. Without doubt, social change has been quite remarkable. Dalits, women, religious minorities, and others on the bottom of the heap have learned to make use of their right to change governments and hold them accountable. Even the most extreme of the Hindu right wing does not dare to (and does not want to) go back to the days of untouchablity, even though the Hindu right continues to espouse the virtues of the caste society, sans its excesses. For all my criticism of what is happening, I do not deny the progress India has made.

But the mobilization of the popular masses, *without a corresponding attention to the cultivation of democratic and secular cultural mores*, is now turning democracy into a mob rule. It is not that checks and balances in the legislature and the judiciary, such as they were, are being dismantled. They are "only" being reinterpreted into a Hindu idiom to allow the mixing up of religion with the most sectarian, Hindu-supremacist politics targeted at religious minorities. People are being mobilized to legitimize political power, but the mobilization is taking place by addressing them not as citizens, but as members of religious communities. The trouble is that given the large and continuing presence of religion in *all* aspects of people's lives, the invocation of religious symbols resonates strongly with the electorate. Even though all those who mobilize to build the Rāma temple, for example, may not vote for Hindu nationalist parties in each and every election, yet, they do heed the call to come to Ayodhya, or send in money for the temple, or do a ritual prayer—all lending legitimacy to the Hindu nationalist cause.

I have ascribed the rise of Hindu nationalism and its reactionary brand of modernity to the weakness of the Enlightenment and secularization in India (chapter 2). Indian intellectuals, I have argued, have failed to challenge the worldview sanctioned by the dominant Brahminical traditions of Hinduism. By and large, even the supposedly secular, left-wing intellectuals have not shown any urgency to subject the inherited cosmology or epistemology to a rational critique in the light of science. Ever so awake to the "epistemic violence" of modern science, and colonialism, Indian intellectuals have shown a strange reluctance to even see the real deprivations and material violence against the discriminated castes and women justified by elements of the dominant religious traditions in India. Indian intellectuals are still fighting a battle against the British Raj rather than confronting the problems internal to India. Anti-colonial nationalism, the Marxist faith in the primacy of changing material conditions, caste belongingness, populism, and the excessive dependence on Western academic fashions (e.g., postmodernism) have all played a role in creating and nurturing this blind spot regarding the role of religion in sustaining internal oppressions.

But now, faced with the horror of a virulent Hindu nationalism, Indian intellectuals have no choice but to awaken to Ambedkar's message (chapter 7). When Ambedkar called for the "annihilation of caste" he was calling for a rationalization and secularization of the cultural common sense. When he, following Dewey, called for scientific temper, he was calling for creating a new democratic ethos that asked questions, and did not take the mystagogues at their word. Rather than declaring a spurious "equality of all truths," he was pointing to the need for updating our conception of truth—from metaphysical truths, to truths that are accessible to all human beings, endowed with the limited abilities we as a species have. It is in this regard his recourse to the Buddha was so revolutionary: the Buddha and his teachings pointed to the suppressed traditions within the Indian culture that value naturalism and empiricism over the metaphysical flights of the Vedic and Vedāntic Hinduism.

This kind of critical engagement with the fundamental ontology and epistemology of Hinduism is what is required for a thoroughgoing secularization of the imagination. This is what the Enlightenment is all about.

Everyone in India today is for secularism, even the Hindu right wing. But not many are willing to even admit the need for a serious rethinking of the religiously sanctioned cognitive, emotional, and ethical values. A questioning of religious cosmology and epistemology is put down as preaching atheism, or as insulting the religious sensibilities of the people. But secularization has never meant the death of God. It is a demand for God to vacate those areas of life—nature and politics, for starters—where He/She does not belong. Secularization is a matter of delimiting, but not eliminating God's province.

For Indian intellectuals to take on the project of the Enlightenment, they will have to get over at least two postmodernist habits: First, they will have to admit that modernity *does* have an essence and that is secularization of society and a disenchantment of nature. Not every way of accommodating the modern world constitutes "alternative modernity." For a society to become modern, it has to remove God from nature, and from other profane aspects of life. Without this secularization of nature and science, priests, charlatans, and magicians will always claim extraordinary powers and extra-parliamentary authority over people. Without secularization at *this* level of ontology and epistemology, Indian people will always remain at the mercy of false prophets.

Second, they will have to rediscover the vocation of all intellectuals: to agitate on behalf of universal values. Their talk of "alternative sciences" has only emboldened those who speak the language of blood and soil.

Modern science combines in it the power of disenchantment and universalism. It is time it was recognized, once again, as an ally of social justice, peace, and advancement all around the world.

# Acknowledgments

I have incurred many debts as I have moved along various stages of the dissertation and then the book.

I thank the faculty, students, and staff at the Science and Technology Studies department at the Rensselaer Polytechnic Institute for giving space to my dissenting views. For the most part, I was treated fairly and cordially.

The post-dissertation work on the book, which was extensive, was made possible by a generous fellowship (2000–2001) from the American Council of Learned Societies. I remain grateful for the timely help, both financial and moral, the fellowship provided.

My special thanks go to my dissertation adviser and good friend, Langdon Winner. He saw right from the start where I was coming from, and encouraged me to persevere. Deborah Johnson, likewise, showed enormous sympathy for my position and was always there to lend an ear. Andrea Rusnock's insights and her cheerful, supportive presence were a great help.

As I was beginning to develop my own critique of social constructivism, Paul Gross and Norman Levitt's book, *Higher Superstition* came along in 1994. Paul and Norm became good friends and helped me find other like-minded scientists and philosophers of science. I will never forget the phone call from Professor Robert Merton and his kind and supportive words. Noretta Koertge, Philip Kitcher, Susan Haack, Gerald Holton, Margaret Jacob, and Daniel Dennett have been most supportive and generous with their time. Noretta Koertge, Cassandra Pinnick, and Jim Maffie were kind enough to include my work in their edited volumes.

My special thanks go to Alan Sokal and his wonderful wife, Marina Papa. Both have been supportive friends and great hosts on my numerous visits to New York City. They, along with Norm Levitt and Rene Greene, have provided me with a sense of community that I so badly missed.

Outside the circle of scientists and philosophers of science, I am deeply indebted to Stephen Bronner who gave his time and attention to my dissertation. I appreciate the moral and intellectual support of Russell Jacoby, Irene Gendzier, Jay Mandle, and Chandler Davis. Encouraging remarks from Noam Chomsky and Martha Nussbaum helped me to keep on going. Ellen Meiksins Wood, Leo

Panitch, and Rosemary Hennessey were kind enough to invite me to contribute to their edited volumes.

I cannot thank enough the steadfast friendship of Daphne Patai. She was always there in my time of need. Martin Lewis has been a source of encouragement and support right through this project.

Many among my South Asian friends, here and in India, extended a hand of friendship. I cannot thank Sanjib Baruah enough for his consistent and constant help and advice. Old and new friends in India read my work and gave me useful feedback. I found my communications with Achin Vanaik, Bipan Chandra, K. N. Panikkar, Pervez Hoodbhoy, T. Jayaraman, Maithreyi Krishnaraj, Sharmila Rege, and Gopal Guru and Eleanor Zelliot most enlightening. My critics Gita Chadha and Sundar Sarukkai deserve a special word of thanks for making me think harder. My fellow rationalist, Dr. Visvanathan shared his work on Indian medicine with me.

I want to thank Zainal Abidin for inviting me to Yogyakarta, Indonesia. I will forever remember the wonderful experience I had with the students of Gadjah Mada University and the beautiful and generous people of Yogyakarta.

Thanks are also due to Krishna Raj, the editor of the *Economic and Political Weekly*. He has always been most generous in giving me space in the venerable periodical. Asad Zaidi and Nalini Taneja were kind enough to include my work in the first batch of books for their new publishing house, Three Essays Collective. *Breaking the Spell of Dharma and Other Essays* owes its existence to them.

Working with Audra Wolfe at Rutgers University Press has been a pleasure. This manuscript has gained tremendously from her incisive comments and very practical suggestions. My thanks go to Adi Hovav of Rutgers for her always prompt and cheerful help. Jaya Dalal's careful copyediting has saved me from many embarrassing errors.

# Notes

## One Prophets Facing Backward: Betrayal of the Clerks

1. "Fashionable nonsense" is the title of the 1998 book by Alan Sokal and Jean Bricmont.
2. Detailed accounts of the historical evolution and organizational structures of Hindu nationalist organizations can be found in Basu et al. (1993) and Banerjee (1998).
3. "Hindutva ran like a vital spinal cord through our whole body-politic and made the Nayars of Malabar weep over the sufferings of the Brahmins in Kashmir" (Savarkar 1989, 46).
4. See a recent essay by Eisenstadt (2000) for a clear statement of the common core of modernity.
5. The violence is Gujarat began on February 27, 2002 when a Muslim mob in the town of Godhara attacked and set fire to two train carriages carrying Hindu activists. Fifty-eight people were killed, many of them women and children. The activists were returning from the city of Ayodhya in Uttar Pradesh, where they had gone to support a campaign led by the Vishva Hindu Parishad to construct a temple to the Hindu god Rāma on the site of a sixteenth-century mosque destroyed by Hindu militants in 1992.

   A murderous retaliation followed. Organized Hindu mobs attacked Muslim homes, properties, and places of worship, leaving hundreds dead and tens of thousands homeless and dispossessed. The Gujarat government chose to characterize the violence as "spontaneous reaction" to the incidents in Godhara. Investigations by Human Rights Watch, numerous Indian human rights and civil liberties groups, and most of the Indian press clearly showed that the attacks on Muslims were planned and organized with extensive police participation and in close cooperation with officials of the Bharatiya Janata Party state government. See Human Rights Watch (2002) for more details.
6. Almost all major critical writings on the issue of the fascist tendencies of Hindutva have been compiled on the internet. Full texts are available on the website *http://www. geocities.com/indianfascism/*.
7. Marty and Appleby use the term "fundamentalist" in the same spirit as I use the term "nationalist" here, that is, they, too, see political activism and not a literal belief in the doctrine as the hallmark of fundamentalism.
8. National rebirth is now accepted as the mythic core of fascism. Apart from Griffin, see also Eatwell (1996) and Paxton (1998).
9. This is a vast and still emerging literature. For an extensive bibliography, see the website of the Clero-Fascist Studies Project (*http://www.home.earthlink.net/~velid/cf*).
10. "Fascism is inconceivable without democracy or without . . . the entrance of the masses onto the historical stage. Fascism is an attempt to make reaction and conservatism popular and plebian . . ." (Moore 1966, 447).

11. India as the birthplace of the "Aryans" and the cradle of all other civilizations is argued recently in a book bearing the title *In Search of the Cradle of Civilization* by George Feuerstein, David Frawley, and Subhash Kak (1995).
12. The role of intellectuals in naming and creating new ideals for popular aspirations is derived from Yack (1992).
13. See Rosenau (1992, chapter 1) and also, Best and Keller (1991) for succinct accounts of postmodernism.
14. Quoted from Lambert (1999).
15. See Jaffrelot (1996) for an account of Gandhian populism in the "JP movement" during the Indira Gandhi years.
16. Arif Dirlik (1994), in a highly prescient critique of postcolonial intellectuals, argues that it is the shared postmodern discourse, rather than any geographical location, that defines who is a postcolonial intellectual. While the self-identification as "Third World intellectuals" still tied the diasporic intellectuals to their native lands, the identity of "postcolonial intellectuals" is no longer structural but discursive.
17. See Pierre Bourdieu (1991) for a succinct history of the modern intellectual.
18. The number of poor in 1999–2000 varies from 26.1 to 23.3 percent of the total population, depending upon whether the food intake is measured on the basis of a 30-day recall or a 7-day recall, respectively.
19. Here are some figures from the UN's Human Development Report (United Nations Development Program 1998) for 1998:

> Average life expectancy went up from 44 years in 1960 to 61.6 years in 1995
> Infant mortality declined from 165 to 73 per 1000 births from 1960 to 1995
> Percentage of underweight children declined from 71 in 1960 to 53 in 1995
> Adult literacy went up from 34 percent in 1960 to 52 percent in 1995.

## *Two*   Dharma and the Bomb: Reactionary Modernism in India

1. The phrase "tryst with destiny" is from Jawaharlal Nehru's famous speech on the eve of India's independence on August 15, 1947.
2. The "Shah Bano affair" that rocked Indian politics had to with a pittance of an alimony that the Indian Supreme Court ordered be paid to Shah Bano, an impoverished old Muslim woman, by her ex-husband. The conservative Muslim factions opposed the ruling as it contravened the *sharia*, or the Islamic law. Rajiv Gandhi gave in to the conservative demands and got a new bill passed by Parliament that annulled the Supreme Court decision. See Jaffrelot (1996, 334–336) for more details.
3. For a historical account of the rise of the Rāma temple movement, see Jaffrelot (1996), especially part IV.
4. The letter can be found at *http://www.acusd.edu/theo/risa-l/archive/msg00782.html*.
5. Jaffrelot at least shows some ambiguity on this issue. He acknowledges that religious faith itself is part of the motivation. But he hedges it with many materialist explanations which make the religious motive look like an excuse.
6. I have been influenced in my understanding of neo-Hinduism by Paul Hacker (1995), Wilhelm Halbfass (1988), Gerald Larson (1995), Richard Fox (1989), and Ainslee Embree (1990).
7. The separation of the process of secularization from the consequences is forcefully argued by Jose Casanova (1994). The factors influencing the difference in the scope and depth of secularization in different societies are from Casanova as well.

8. For a popular account of liberal Protestantism in America, see Karen Armstrong (2000). On liberalism in Protestant churches, see Keith Yandell (1986).

9. This is a manifesto of a holist society written by Deendayal Upadhyaya in 1965.

10. The relevant ARTICLES are ARTICLES 25 and 26 for freedom of religion, ARTICLES 15 and 16 for no state discrimination on the grounds of religion, and ARTICLES 25(2), 27, 290A on the separation of state and religion. See Smith (1998) for the full text of these articles.

11. See Rajeev Dhavan (2001) for a succinct description of the strengths and weaknesses of India's Constitution.

12. This example is from Louis Dumont (1980, 191).

13. The term "religionization" is from Larson (1995, 199) who uses it to describe " the transformation of issues regarding governance and distribution of power into issues directly related to the ultimate meaning and significance of human existence."

14. *Vastu* is ancient Hindu architecture which is supposed to align buildings with "cosmic energy." Buildings constructed according to the principles of *vastu* are supposed to bring good fortune to those who live in them.

## *Three*    Vedic Science, Part One: Legitimation of the Hindu Nationalist Worldview

1. As far as I know, the term was first used by David Frawley, an American Hindu, who goes by the name of Vamadeva Shastri and runs a Vedic astrology ashram in California. His book, *Awaken Bharat* (1998) is where the term first made its appearance.

2. As were the Nazis. Hindu nationalists share a penchant for combining contraries without worrying about contradictions with German fascism, which was notorious for being, in the words of Ernst Troeltsch, "a queer mixture of mysticism and brutality." [Quoted here from Viereck (1965, 7)]. See the writings of George Mosse (1999) and Roger Eatwell (1996) for a good treatment of syncretic aspects of fascism. The phrase "scavenger style" is borrowed from Mosse.

3. The expression "letting the independent causal structures of the world push back against his [scientist's] own hypothesizing" is borrowed from Robert Klee (1997, 5). My own understanding of a moderate skeptical, pragmatic realism is shaped by the work of Philip Kitcher (1993), Nicholas Rescher (2000), and Susan Haack (1993).

4. Bharati has described his experiences as a Hindu ascetic, first with the Ramakrishna Mission and later as a wandering monk in his 1980 autobiography, *The Ochre Robe*.

5. See "UGC Goes Saffron: Courses in Vedic Astrology" at *http://www.Tehelka.com*, November 14, 2000; "Future is an MA in Rituals and Astrology—UGC Chief," *The Indian Express*, July 26, 2000. This issue was covered extensively in the Indian media. For further details, see May through July issues of *Frontline*, one of India's leading newsmagazines.

6. For a defense of the mass hysteria that took over Hindus around the world over the miracle of Ganesh idols drinking milk, see "'It's a Miracle!' Rejoice Millions as Lord Ganesha Receives Milk," *Hinduism Today*, volume 17, November (*http://www.hinduismtoday.com*). For a case of political abuse of *vastu shastra*s by a politician, see Nanda (1997).

7. Dharma Hinduja International Center for Indic Research in Delhi, India has recently published a series of seven "exploratory papers" on topics related to science and the Vedas. These examples are taken from these exploratory papers. Prajna Bharati organized a conference on "Indianization of Science and Technology Education" in October,

1994. Some examples come from the proceedings of this conference which are available on the web at *http://www.pragna.org*.

8. The National Curriculum Framework for School Education is available at *http://www.nic.in/vseducation/ncert/cfcontents.html/*.

9. For a stimulating discussion of the nature of religion and its interaction with science, see Willem Drees (1996) and Steven Bruce (1996).

10. For powerful statements regarding the importance of the change of doxa over the change in doctrines, see Charles Taylor (1995), John Hedley Brooke (1991), and Steve Bruce (1996).

11. The history of the Scientific Revolution and its cultural impact hardly needs rehearsing. My understanding of the change in the conception of nature and God is influenced by the following sources: Deason (1986), Westfall (1958), Brooke (1991), Jacob (1988).

12. My understanding of Vedanta is derived from my own experiential understanding of Hinduism, the religion I was born into. I have also consulted secondary sources, including such well-known exponents as Hiriyanna (1996), Dasgupta (1969), Zaehner (1962), Zimmer (1951), Radhakrishnan (1993), and also Marxist philosophers, notably, Chattopadhyaya (1976) and Damodaran (1967).

13. The term "affirmative Orientalism" is borrowed from Richard Fox (1989).

14. This summary of British Orientalists is derived from David Kopf (1969), Wilhelm Halfbass (1988), and Richard King (1999). On the question of the origin of Aryan culture, see Edwin Bryant (2001).

15. See Panikkar (1995) and Sarkar (1975) for Roy's retreat from the modern conception of reason.

### *Four*    Vedic Science, Part Two: Philosophical Justification of the Vedic Science

1. For an Evangelical Christian defense of postmodernism see Stanley Grenz (1996), for an Islamic perspective, see Akbar Ahmed (1992) and for a postmodern theology that can accommodate an enchanted view of nature, see David Ray Griffin (1988).

2. His two other criteria are: explanation of particulars using general laws, and conservation of energy and matter. These criteria which he lays out in his "Reason and Religion" lecture seem to be of great importance to Vivekananda, for he repeats them many times through his numerous lectures.

3. Despite his controversial thesis of the cultural origins of quantum mechanics, Paul Forman (1971) remains the best source for understanding Spengler's influence on the scientific community.

4. Even though Nandy is a proponent of the Spenglerian notion that "every culture produces its own science," he does not approve of Bose's attempt to purposefully turn to Vedanta as a working assumption in his experimental work. He prefers the quiet unself-conscious orthodoxy of Ramanujam. Nandy sees Bose as acting out of an aggressive inferiority complex created by colonialism.

5. I have found works by Victor Stenger, especially his 1995 book, *The Unconscious Quantum*, very instructive. Stenger is a professor of physics and astronomy at the University of Hawaii. Also useful is the Nobel physicist, Steven Weinberg's 1992 book, *Dreams of a Final Theory*. Richard Jones (1986) has written a very knowledgeable and level-headed book comparing Hindu and Buddhist mystical traditions with science.

6. Although there is a significant presence of those with degrees in science and engineering among Hindu nationalists. Indeed, the Bharatiya Janata Partys's minister for edu-

cation and human resources development, Murli Manohar Joshi, holds a doctorate in physics.

7. *In Search of the Cradle of Civilization* (Feuerstein et al. 1995) is the title of the book in which one of the Vedic physicists, Subhash Kak, argues his case for finding astronomical findings coded in the Vedas.

8. Witzel (2001) and Chattopadhyaya (1986) cite evidence showing that the Aryan tribes did make use of the native population for non-ritual jobs.

9. There is also an Islamic brand of creationism that uses the work of Henry Morris's Institute for Creation Research for its own purpose. See Mackenzie-Brown (2002) for more references on Vedic and Islamic creationists.

10. Mackenzie-Brown (2002) provides a good summary of the theology behind Vedic creationism. These ideas are not fully explained in *Forbidden Archeology*, but can be found in Thompson's 1981 book, *Mechanistic and Non-mechanistic Science*.

11. For a review of how Cremo and Thompson use social constructivism, see Wodak and Oldroyd (1996).

## *Five*    Epistemic Charity: Equality of All "Ethnosciences"

1. Ian Jarvie (1984, 48) refers to "minirat" as weak rationality, or Rationality$_0$. It is the kind of rationality that learns the least from experience. It is action directed toward some end and predicated on some body of ideas and information, regardless of their validity. Cargo cults and rain dances are Rationality$_0$. Rationality$_1$ is goal-directed action based on the best ideas and information according to some explicit standard. Rationality$_2$ is goal-directed action employing the best ideas and information assessed by the best standard.

2. The imagery of citadels and cyborgs is not mine. Anthropologists of science describe their project as "cyborg anthropology" aimed at opening the "citadel of science" to the rest of the society. See the introduction by Gary Downey and Joseph Dumit (1997) to the report of the landmark seminar in Santa Fe.

3. Michel Callon (1995, 51) defines inscription as: "all written marks [including] . . . graphic displays, laboratory notebooks, tables of data, brief reports, lengthier and more public articles and books. Inscriptions range from the crudest marks to the most explicit and carefully crafted statements."

4. Other critics have also noticed the connections between cyborg anthropology and a religious view of the world. Bloor (1999) points out the elements of Alfred North Whitehead's metaphysics in Latour's recent works. William Grassie (1996) finds Haraway eroding all boundaries between religion and science and moving toward the process metaphysics of Whitehead, while Michael Zimmerman (1994, 365) finds in Haraway elements of Mahayana Buddhism that holds that "entities lack fixed essences and are instead complex gestalts constituted by interactive perceptual and discursive processes."

5. This section was originally written for another essay, Nanda (n.d.), forthcoming in Pinnick et al. Reproduced here with permission.

6. Carolyn Merchant's 1980 book, *The Death of Nature* has acquired the status of a feminist classic. Important also are Genevive Lloyd's *The Man of Reason* (1984) and Evelyn Keller's *Reflections on Gender and Science* (1985).

7. For all her apparent differences with Harding, Longino agrees with Harding's argument for strong objectivity. The transformative interrogation by the scientific community that Longino uses as her shield against subjectivism will not prevent "community values to remain embedded in scientific reasoning" (Longino 1990, 216).

8. Harding believes that a "woman scientist is a contradiction in terms" (1986, 59). "Women of color" is a "cyborg" identity made of all those "refused stable membership in the social categories of race, sex and class" and joined only by its "oppositional consciousness" to Enlightenment humanism (Haraway 1991, 156).

9. The recent crop of overviews of the state of feminist science studies holds the bridging of subjectivity and objectivity as the distinguishing feature of feminist thought. See Code (1998), Hircshmann (1997), and Harding (2000).

10. This is a composite argument, made of many strands. The *locus classicus* for the Marxian strand is the work of Nancy Harstock and Sandra Harding, cited in the text. The psychoanalytical strand derives largely from Nancy Chodrow's 1978 classic, *Reproduction of Mothering.* Chodrow's object-relations theory was introduced into feminist epistemology by Evelyn Fox Keller (1982). The central text for the epistemic privilege of non-Western women is Vandana Shiva's *Staying Alive.* Also important are Harding (1998) and Marglin (1996).

11. Kuhn (1977) singled out accuracy, simplicity, internal and external consistence, breadth of scope, and fruitfulness as values that scientists use to guide their judgment between competing theories.

12. Longino has progressively loosened the constraints of the existing community standards. In a 1993 publication, she argued for "detaching scientific knowledge from consensus" (1993, 114) in order to make room for critical oppositional positions in mainstream science. In that article she proposed that different "sub-communities" in a given domain of inquiry be allowed to bring in their own models and metaphors that will allow them to "map" the relations posited in these models onto some portion of the experienced world, without seeking consensus regarding one true story (1993, 116–117).

In a bow to pluralism, Longino does not demand that the background assumptions/heuristic models of various "sub-communities" should have passed through the social process of "transformative criticism" that she admits is the hallmark of science. It can be readily shown that many of the models and metaphors of sub-communities she allows in her later work (e.g., ecofeminists) have never faced the tribunal of transformative criticism.

13. See Mosse (1961) for the influence of theosophy and the occult on Nazism.

## *Six*   We Are All Hybrids Now! Paths to Reactionary Modernism

1. For a fine treatment of Orientalist influences on Gandhi, see Richard Fox (1989).

2. My understanding of the postcolonial theory of hybridity is derived largely from Akhil Gupta (1998), Robert Young (1990) and Moore-Gilbert (1997).

3. This phrase is from Vasavi (1999).

4. This episode is taken from Partha Chatterjee (1986, 96–97).

## *Seven*   A Dalit Defense of the Deweyan-Buddhist View of Science

This chapter incorporates parts of two of my previously published essays, Nanda (2000) and Nanda (forthcoming). Reprinted here with permission.

1. One hundred and sixty million men, women, and children—nearly one-sixth of India's one billion people—belong to castes (jatis) which are considered occupationally, ritually, and inherently polluted. India's Constitution makes untouchablity illegal, but the practice still prevails. For the state of dalits in contemporary India, see *Human Rights*

*Watch* (1999), Mendelsohn and Vicziany (1998), Fuller (1996), and Srinivas (1996). The website *http://www.ambedkar.org* is another valuable resource.

"Dalit" is the self-chosen name of ex-untouchables. The word carries connotations of proud militancy and self-respect in contrast to the more paternalistic name of Harijan (meaning "god's children") given to them by Mahatma Gandhi.

2. Eleanor Zelliot, the American scholar of dalit movements, goes the farthest in exploring Ambedkar's American experience. But even she gives it very little importance: "American influence on Ambedkar really counted for very little. It is more likely that in those early years in America, his own natural proclivities and interests found a healthy soil for growth . . . and strengthened him in his lifelong battle for dignity and equality of his people" (1992, 85).

3. For an overview of dalit intellectual activity, with a strong expression of concern with postmodernist tendencies, see Guru and Geetha (2000).

4. By now there is a large volume of critical literature that argues that the Nietzschean, postmodernist reading of Dewey made popular by Richard Rorty in his Mirror of Nature (1979) does not capture the spirit of Dewey or the rest of the pragmatists. Indeed, Rorty himself has acknowledged that his postmodernist Dewey is a "hypothetical Dewey", and an "imaginary playmate," who says the sort of things Dewey would have said had he made the linguistic turn (Westbrook 1998, 128). See also, Haack (1998) for a critique of Rorty.

5. Dewey also defines scientific temper negatively "as freedom from control by routine, prejudice, dogma, unexamined traditions, sheer self-interest" (1955b, 31).

6. Even though Laudan only very rarely mentions Dewey, I read his normative naturalism as an operationalization of Dewey's ideas.

7. Mass conversions continued after his death. The 1951 census lists merely 2,500 Buddhists in India. The number jumped to 3 million in the 1961 census, a 1671 percent increase (Deliege 1999). For a moving eyewitness account of the conversion ceremony, see Moon (2001).

8. Mahars were general-purpose village servants whose caste-duties included cutting wood for cremation and removing dead cattle, among other things. The touch of a Mahar was considered polluting. The book-length political biography by Gail Omvedt (1994), the writings of Eleanor Zelliot (1992, 1994) and Deliege (1999) are good resources for Ambedkar's biography.

9. As the British did not observe caste, they had no hesitation in making use of caste divisions to serve their own interests. In the process, they unintentionally opened up avenues of education and employment (as soldiers, cooks, waiters, etc.) which used to be closed to the lower castes. But, even though they did not give ideological support to caste, the British did not actively oppose it either. Under the pretext of respect for local religion and customs, the British allowed the Brahmin elite to define social norms.

10. Manu, the mythical author of Manusmriti, to take one leading but hardly the only example of Indian sacred texts that make inequality appear sacred and venerable, reserves his severest condemnation for a Brahmin woman marrying a Sudra man: the children born of such hypogamous or pratiloma (literally, "against the hair," or contrary to the natural order) unions are consigned to the lowest of the low caste of Chandala. To maintain the purity of the caste lineage, Manu only allows marriage within the caste group, or at most permits women from inferior castes to marry into slightly higher castes. Hypergamous, or anuloma (literally, "following the hair," or in accord with natural order) are acceptable because women, being naturally inferior to men, can be "uplifted" (Chakravarti 1993).

11. Sānkhya originated in the same northeastern region of the Indian subcontinent, now in Nepal, where the Buddha came from. Buddha's birthplace, a village called Lumbini, was located near the ancient town of Kapilvastu, the abode of Kapila, the original Sānkhya philosopher. This same region is also home to other proto-materialist religions of mother right and tantra. Chattopadhyaya (1959) hypothesizes a pre-Aryan origin of these ancient naturalist philosophies.
12. For an illuminating exposition of all varieties of Indian idealism, see the classic work of Surendranath Dasgupta (1969).
13. Complete references to the Amnesty and Human Rights Watch reports can be found on *http://www.Ambedkar.org.*

## *Eight*    The Battle for Scientific Temper in India's New Social Movements

1. Nandy's "Counter-statement" should be read together with the M. N. Roy Memorial Lecture he delivered at the Gandhi Peace Foundation in March 1980. An updated version of that lecture, titled " Science, Authoritarianism and Culture" is available in Nandy (1987).
2. Michael Friedman (1999) argues that in the context of German idealism that pervaded Europe in the nineteenth and early twentieth centuries, the positivists offered a truly revolutionary agenda for socio-cultural movements. It is only when Nazism forced the dispersion of the Vienna Circle intellectuals around the world that positivism got domesticated, and lost its anti-metaphysical component.
3. Radhika Desai (2002) provides a good summary of the origins and ideology of CSDS.
4. I have used the writings of Ayesha Jalal (1995) and Christophe Jaffrelot (1996) for my understanding of Indira Gandhi and the JP movement.
5. I take the paper by Issac and his co-authors as a definitive account. This paper incorporates parts of a book Thomas Issac wrote with B. Ekbal titled, *Science for Social Revolution: The Experience of KSSP.* This book is considered the most authoritative account of the Kerela Sastra Sahitya Parishad's history. M. P. Parameswaran, the third author of the 1997 paper, is one of the most respected senior members of the Kerela Sastra Sahitya Parishad.
6. I have had many conversations with my Indian friends in which I have observed this contradictory posture. Even though those sympathetic to the people's science movements agenda disagree with the irrationalism of alternative science movements, they still defend them in public, for they are seen as fighting the good fight against the West.
7. Issac et al. (1997) arrive at a similar conclusion, although they commend the Kerela Sastra Sahitya Parishad for this "evolution."

## *Nine*    The Ecofeminist Critique of the Green Revolution

Parts of this chapter first appeared in an anthology of essays on material feminism edited by Hennessy and Ingraham (1997). Reprinted here with permission.
1. As an example of what such a "subsistence perspective" would involve, Mies finds reasons to hope in the current economic and energy crisis in Cuba which has forced the Castro government to replace tractors with oxen and buses and cars with bikes. She believes that this "compulsory retreat to subsistence production [should be] seen as chance rather than a defeat" and it may offer a new model for the new nation states born out of the collapse of the Soviet Union and the struggling nations of Africa. Thus,

while Mies has ruled out material equality between the Third and the First World as a "logical impossibility," which is "not even desirable" (Mies and Shiva 1993, 300), she suggests that the rest of the world lower its expectations to the bare survival that subsistence economy delivers.

2. "The woman peasant works invisibly with the earthworm in building soil fertility . . . women have a major productive role in maintaining the food cycle. In feeding animals from trees or crop by-products, in nurturing cows and animals, in composting and fertilizing fields with organic manure, in managing mixed and rotation cropping, this critical work of maintaining ecological cycles was done by women, in partnership with the land, with trees, with animals and with men . . ." (Shiva 1988, 109).

3. See William Newman (1998) for a critique of Merchant and Keller for ignoring the patriarchal and misogynist aspects of the hermetic traditions.

4. Capabilities like "living one's own life" often serve as the reminders of the "limits against which we press" instead of being included in the "capabilities through which we aspire" (Nussbaum 1995, 75). Whether a capability has become a part of the already available repertoire of capabilities of a group of people, or it is still a dimly perceived desire beyond the limits of the possible, depends upon the historical context. But the point the critical universalists would hold on to is that some capabilities, even those which may not be attainable under certain contexts, are part of any life we will count as a human life. Nussbaum lists 11 capabilities in the first level and 10 in the second level. See Nussbaum (1995) for a complete description.

5. On translating these figures of the total women who are missing into ratios of the number of missing women to the number of actual women in a country, for Pakistan we get 12.9 percent, for India 9.5 percent, Bangladesh 8.7 percent, China 8.6 percent, Iran 8.5 percent, North Africa 3.9 percent, Latin America 2.2 percent and Southeast Asia, 1.2 percent (Nussbaum 1995, 3).

6. Cecile Jackson (1993) cites many instances from her experience in Africa of men, because of ownership interest, making an investment in conservation. In certain parts of India (Jharkhand), men are as involved in gathering activities as women and contribute up to 40 percent of labor in gathering (DN 1990). DN also points out that ecological knowledge is not a sole province of women in traditional cultures: men actually monopolize the useful and specialized knowledge of medicinal plants in Jharkhand.

7. Though not entirely. As Kalpana Bardhan (1989) has shown, the rural labor markets continue to use traditional gender, caste, and kinship norms to decide the level of wages for different categories of workers.

## Ten   The "Hindu Left," Agrarian Populism, and the Hindu Right

1. Ashis Nandy, however, is a Christian, even though he writes from a Hindu perspective.

2. Debates about urban bias center around the claim that the state extracts surplus from the villages to make it available to industries and urban centers at artificially low prices. While the debate can be highly technical, the evidence for urban bias is not as clear-cut as farmers' movements make it out to be. There is evidence that since independence, terms of trade have actually moved in favor of agriculture. Going by one index of state agricultural policy—namely, subsidies—from 1970 to 1991, subsidies to agriculture have risen consistently (Varshney1995, 169). Between 1981/1982 and 1985/1986, direct subsidies for items such as food, fertilizers, and seeds paid by the central government doubled to Rs. 41, 880 million, an increase over and above subsidies paid for

infrastructural investments (Brass 1994, 55). Moreover, nearly 90 percent of subsidies earmarked for agriculture were spent on input subsidies by the late 1980s, leaving very little for other services and asset creation (Rao 1994, 233–234).

3. The inclusion of minimum wage in the demand for remunerative prices is motivated by this logic: the farmers want the state to base the producer price on the legal minimum wage set at 25 rupees, or a little more than half a United States dollar. The catch is that in most cases, the *actual* wages are far below the minimum wage. This enables the farmers to demand higher procurement prices, without paying adequate wages. There being hardly any enforcing mechanism, the landless have nothing more than a promise of future rise in wages.

4. When challenged on Shetkari Sanghathana's lack of concern for landless labor, Gail Omvedt (1987) responds by pointing to the progressive policies of the Shetkari Sanghathana for "the largest section of landless in the rural areas, women."

5. Assadi (1994) describes physical attacks on dalit laborers by members of the Karnataka Rajya Ratitha Sangh for not accepting lower wages and not agreeing to live in segregated quarters. Dipankar Gupta (1997) describes most extreme patriarchal views held by the leaders of the Bharatiya Kisan Union.

# Bibliography

Agarwal, Anil. 1994. An Indian environmentalist's credo. In *Social ecology*, edited by R. Guha, 346–384. Delhi: Oxford University Press.

Agarwal, Bina. 1988. Neither sustenance nor sustainability: Agricultural strategies, ecological degradation and Indian women in poverty. In *Structures of patriarchy: The state, community and household in modernizing Asia*, edited by Bina Agarwal, 1–28. London: Zed.

———. 1992. Rural women, poverty and natural resources: Sustenance, sustainability and struggle for change. In *Poverty in India: Research and policy,* edited by B. Harriss, S. Guhan, and R. H. Cassen, 390–432. Bombay: Oxford University Press.

———. 1994. *A field of her own: Gender and land rights in South Asia.* Cambridge: Cambridge University Press.

Agarwal, Dinesh. 1999. RSS is the torch-bearer of Mahatma Gandhi's legacy. *http://rss.rg/rss/library/gandhi-torch.html.*

Ahir, D. C., ed. 1997. *Selected speeches of Dr. B. R. Ambedkar.* New Delhi: Blumoon Books.

Ahmad, Aijaz. 1992. *In theory.* London: Verso.

———. 2002. *On communalism and globalization: Offensives of the far right.* New Delhi: Three Essays Press.

Ahmed, Akbar. 1992. *Postmodernism and Islam: Predicament and promise.* New York: Routledge.

Al-'Azm. 1992. A criticism of religious thought. In *Islam in transition: Muslim perspectives,* edited by J. L. Donohue and J. L. Espostio, 113–119. New York: Oxford University Press.

Alberuni. 1971. *Alberuni's India,* translated by Edward Sachau. New York: W. W. Norton & Co.

Almond, Gabriel, Emmanuel Sivan, and R. Scott Appleby. 1995. Fundamentalism: Genus and species. In *Fundamentalisms comprehended,* edited by M. Marty and R. Scott Appleby, 399–424. Chicago: University of Chicago Press.

Alvares, Claude. 1992. *Science, development and violence: The twilight of modernity.* Delhi: Oxford University Press.

Ambedkar, Bhimrao Ramji. 1936. *Annihilation of caste.* Jalandhar: Bhim Patrika Publications.

———. 1992. *The Buddha and his Dhamma: Writings and speeches.* Vol. 11. Bombay: Government of Maharashtra. First published, 1957.

Anandhi, S. 1995. *Contending identities: Dalits and secular politics in Madras slums.* New Delhi: Indian Social Institute.

Antonio, Roberto. 2000. After postmodernism: Reactionary tribalism. *American Journal of Sociology,* 106(2): 40–87.

Armstrong, Karen. 2000. *The battle for God.* New York: Ballantine Books.

Assadi, Muzaffar. 1994. Khadi curtain. Weak capitalism and 'Operation Ryot': Some ambiguities in farmers' discourse, Karnataka and Maharashtra. *Journal of Peasant Studies*, 21(3/4): 212–227.

Babb, Lawrence. 1983. Destiny and responsibility: Karma in popular Hinduism. In *Karma: An anthropological inquiry*, edited by C. Keyes and V. Daniel, 163–181. Berkeley: University of California Press.

Balgopal, K. 1987. An ideology for the provincial propertied class. *Economic and Political Weekly*, September 5, 1544–1546.

Banaji, Jairus. 1994. The farmers' movements: A critique of conservative rural coalitions. *Journal of Peasant Studies*, 21(3/4): 227–245.

Banerjee, Partha. 1998. *In the belly of the beast: An insider's story*. Delhi: Ajanta Books.

Barber, Benjamin. 1996. *Jihad and McWorld*. New York: Basic Books.

Barbour, Ian. 1997. *Science and religion: Historical and contemporary issues*. San Francisco: Harpers.

Bardhan, Kalpana. 1989. Poverty, growth and rural labor markets in India. *Economic and Political Weekly*, March 25, A21–A38.

Barnes, Barry. 1991. How not to do the sociology of knowledge. *Annals of Scholarship*, 8(3–4): 321–336.

Barnes, Barry and David Bloor. 1982. Relativism, rationalism and sociology of knowledge. In *Rationality and relativism*, edited by M. Hollis and S. Lukes, 21–47. Cambridge: MIT Press.

Basu, Amrita. 2002. The dialectics of Hindu nationalism. In *The success of India's democracy*, edited by A. Kohli, 163–190. Cambridge, UK: Cambridge University Press.

Basu, Tapan, Pradip Datta, Sumit Sarkar, Tanika Sarkar, and Sambuddha Sen. 1993. *Khaki shorts, saffron flags*. New Delhi: Orient Longmans.

Bauman, Zygmunt. 1987. Legislators and interpreters. Ithaca: Cornell University Press.

Baviskar, Amita. 1997. *In the belly of the river*. New Delhi: Oxford University Press.

Bell, Daniel. 1996. *The cultural contradictions of capitalism*. New York: Basic Books.

Benda, Julien. 1928. *The betrayal of the intellectuals*. Boston: Beacon Press.

Berger, Peter. 1967. *The sacred canopy*. New York: Doubleday.

———. 1999. *Desecularization of the world*. Grand Rapids, Michigan: William Eerdmans Publishing Co.

Bernstein, Richard J. 1971. *Praxis and action*. Philadelphia: University of Pennsylvania Press.

Best, Steven and Douglas Keller. 1991. *Postmodern theory: Critical interrogations*. New York: Guilford Press.

Bharati, Agehananda. 1970. The Hindu renaissance and its apologetic patterns. *Journal of Asian Studies*, 29(2): 267–287.

———. 1980. *The ochre robe*. Santa Barbara: Ross Erikson.

Bharatiya Janata Party (BJP), 1993. *BJP's white paper on Ayodhya and the Rama temple movement*. New Delhi (available at *http://www.hvk.org*).

Bharucha, Rustom. 1998. *In the name of the secular: Contemporary cultural activism in India*. New Delhi: Oxford University Press.

Bidwai, Praful. 2002. Dalits and adivasis: Canon fodder for Hindutva? *Himal Magazine*, May.

Bidwai, Praful and Achin Vanaik. 1999. *New nukes: India, Pakistan and global nuclear disarmament*. New York: Interlink Publishing House.

Biehl, Janet. 1995. Ecology and the modernization of fascism in the German ultra-right. In *Ecofascism: Lessons from the German experience*, edited by J. Biehl and P. Staudenmaier. San Francisco: A K Press.

Bloor, David. 1991. *Knowledge and social imagery*. 2d edition. Chicago: University of Chicago Press.

———. 1999. Anti-Latour. *Studies in History and Philosophy of Science*, 30(1): 81–112.

Bourdieu, Pierre. 1991. Universal corporatism: The role of intellectuals in the modern world. *Poetics Today*, 12(4): 655–669.

Bowes, Pratima. 1977. *Hindu intellectual tradition*. Columbia, Missouri: South Asia Books.

Braidotti, Rosi, Ewa Charkiewicz, Sabine Hausler, and Saskia Wieringa. 1994. *Women, the environment and sustainable development: Towards a theoretical synthesis*. London: Zed.

Brass, Tom. 1990. Class struggle and the deproletarianisation of agricultural labor in Haryana (India). *Journal of Peasant Studies*, 18(1): 36–87.

———. 1994. The politics of gender, nature and nation in the discourse of the new farmers' movements. *Journal of Peasant Studies*, 21(3/4): 27–71.

———. 2000. *Peasants, populism and postmodernism: The return of the agrarian myth*. London: Frank Cass.

Brooke, John H. 1991. *Science and religion: Some historical perspectives*. Cambridge: Cambridge University Press.

Brown, James R. 2001. *Who rules in science: An opinionated guide to science wars*. Cambridge, MA: Harvard University Press.

Bruce, Steve. 1996. *Religion in the modern world: From cathedrals to cults*. New York: Oxford University Press.

Bryant, Edwin. 2001. *The quest for the origins of Vedic culture*. New York: Oxford University Press.

Burra, Neera. 1996. Buddhism, conversion and identity: A case study of village Mahars. In *Caste: Its twentieth century avatar*, edited by M. N. Srinivas, 152–173. New Delhi: Penguin Books.

Callon, Michel. 1995. Four models for the dynamics of science. In *Handbook of science and technology studies*, edited by S. Jasonoff, G. E. Markle, J. Petersen, T. Pinch, 29–63. Thousand Oaks, CA: Sage Publications.

Casanova, Jose. 1994. *Public religions in the modern world*. Chicago: University of Chicago Press.

Casolari, M. 2000. Hindutva's foreign tie-up in the 1930s: Archival evidence. *Economic and Political Weekly*, January 22, 218–228.

Chakrabarty, Dipesh. 1995. Radical histories and questions of enlightenment rationalism: Recent critiques of subaltern studies. *Economic and Political Weekly*, 30: 751–59.

———. 2000. *Provincializing Europe: Postcolonial thought and historical difference*. Princeton, NJ: Princeton University Press.

Chakravarti, Uma. 1993. Conceptualizing Brahminical patriarchy in early India: Gender, caste, class and state. *Economic and Political Weekly*. April 3, 579–585.

———. 1998. *Rewriting history: The life and times of Pandita Ramabai*. New Delhi: Kali for Women.

Chandler, Kenneth. 1989. Modern science and Vedic science: An introduction. *Modern Science and Vedic Science*, 1(2): v–xxvi.

Chandra, Vikram. 2000. The cult of authenticity. *Boston Review*, Febuary–March, 42–49.

Chatterjee, Partha. 1986. *Nationalist thought and the colonial world: A derivative discourse?* London: Zed Books.

Chattopadhyaya, Debiprasad. 1959. *Lokāyata: A study in ancient Indian materialism*. New Delhi: People's Publishing House.

———. 1976. *What is living and what is dead in Indian philosophy*. New Delhi: People's Publishing House.

————. 1978. *Science and society in ancient India*. Amsterdam: B.R. Gruner B.V.

————. 1986. *History of science and technology in ancient India: The beginnings*. Calcutta: Firma KLM Pvt. Ltd.

Chen, Martha. 1995. A matter of survival: Women's right to employment in India and Bangladesh. In *Women, culture and development*, edited by M. Nussbaum and J. Glover. Oxford: Clarendon Press.

Chodrow, Nancy. 1978. *The reproduction of mothering*. Berkeley: University of California Press.

Chowdhry, Prem. 1993. High participation, low evaluation: Women and work in rural Haryana. *Economic and Political Weekly*, December 25, A135–A148.

————. 1994. *The veiled women: Shifting gender equations in rural Haryana, 1880–1990*. New Delhi: Oxford University Press.

Clayton, Philip. 1997. *God and contemporary science*. Grand Rapids, MI: William B. Eerdman Publishing Co.

Code, Lorraine. 1981. Is the sex of the knower epistemologically significant? *Metaphilosophy*, 12: 267–276.

————.1991. *What can she know? Feminist theory and the construction of knowledge*. Ithaca: Cornell University Press.

————. 1998. Epistemology. In *A companion to feminist philosophy*, edited by A. M. Jagger and I. M. Young. Malden, MA: Blackwell.

Cremo, Michael and Richard Thompson. 1993. *Forbidden archeology: The hidden history of the human race*. San Diego: Bhaktivedanta Institute.

D.N. 1990. Women and forests. *Economic and Political Weekly*, April 14, 795–797.

Damodaran, K. 1967. *Indian thought: A critical survey*. New York: Asia Publishing House.

Dasgupta, Surendranath. 1969. *Indian idealism*. Cambridge, UK: Cambridge University Press. First Edition, 1933.

Dasgupta, Swapan. 2001. BJP: Up for grabs. *Seminar*, 497.

Datar, Chhaya. 1999. Non-Brahmin renderings of feminism in Maharashtra. *Economic and Political Weekly*, October 9, 2964–2968.

Davis, Richard. 1996. The iconography of Ram's chariot. In *Contesting the nation: Religion, community and the politics of democracy in India*, edited by D. Ludden, 27–54. Philadelphia: University of Pennsylvania Press.

Deason, Gary. 1986. Reformation theology and the mechanistic conception of nature. In *God and nature: Historical essays on the encounter between Christianity and science,* edited by D. Lindberg and R. Numbers, 167–191. Berkeley: University of California Press.

de Bary, Theodore., ed. 1958. *Sources of Indian tradition*. Vol. 1. New York: Columbia University Press.

de Tocqueville, Alexis. 1988. *Democracy in America*. Translated by George Lawrence, edited by J. P. Mayer. New York: HarperPerennial. First edition, 1966.

Deliege, Robert. 1999. *The untouchables of India*. Oxford: Berg.

Desai, Radhika. 2002. *Slouching toward Ayodhya*. New Delhi: Three Essays Press.

Dewey, John. 1925. *Experience and nature*. New York: Dover.

————. 1929. *Quest for certainty*. New York: Minton, Balch & Co.

————. 1934. *A common faith*. New Haven: Yale University Press.

————. 1955a. What I believe. In *Pragmatism and American culture*, edited by Gail Kennedy, 23–32. Boston: D. C. Heath and Co. First published, 1930.

————. 1955b. Unity of science as a social problem. In *Foundations of the unity of science: Toward an encyclopedia of unified science*, edited by O. Neurath, R. Carnap, and C. Morris, Vol. 1, Nos. 1–10, 29–38. Chicago: University of Chicago Press.

Dhavan, Rajeev. 2001. The road to Xanadu: India's quest for secularism. In *Religion and personal law in secular India: A call to judgment*, edited by G. Larson, 301–329. Bloomington: Indiana University Press.

Dirlik, Arif. 1994. The postcolonial aura: Third world criticism in the age of global capitalism. *Critical Inquiry*, 20: 328–356.

Doniger, Wendy. 1991. Hinduism by any other name. *Wilson Quarterly*, Summer, 35–41.

Doniger, Wendy and Brian Smith. 1991. *The laws of Manu*. London: Penguin Books.

Downey, Gary and Joseph Dumit, eds. 1997. *Cyborgs and citadels: Anthropological interventions in emerging sciences and technologies*. Santa Fe, New Mexico: School of American Research Press.

Drees, Willem. 1996. *Religion, science and naturalism*. Cambridge, UK: Cambridge University Press.

Duara, Prasanjit. 1991. The new politics of Hinduism. *Wilson Quarterly*, Summer, 42–50.

Dumont, Louis. 1980. *Homo hierarchicus: Caste system and its implications*. Chicago: University of Chicago Press.

Eatwell, Roger. 1996. On defining the 'fascist minimum': The centrality of ideology. *Journal of Political Ideologies*, 1(3): 303–319.

Eisenstadt, Samuel N. 1999. *Fundamentalism, sectarianism and revolution: The Jacobin dimension of modernity*. Cambridge, UK: Cambridge University Press

———. 2000. Multiple modernities. *Daedalus*, 129(1): 1–29.

Elst, Koenraad. 2001. *Decolonizing the Hindu mind: Ideological development of Hindu revivalism*. New Delhi: Rupa & Co.

Embree, Ainslie. 1990. *Utopias in conflict*. Berkeley: University of California Press.

Ferry, Luc and Alain Renaut. 1985. *French philosophy of the sixties: An essay on anti-humanism*. Amherst: University of Massachusetts Press.

Feuerstein, Georg, Subhash Kak, and David Frawley. 1995. *In search of the cradle of civilization*. Wheaton, Ill.: Quest Books (The Theosophical Publishing House).

Feyerabend, Paul. 1978. *Science in a free society*. London: Verso.

Fitzgerald, Timothy. 1994. Buddhism in Maharashtra: A tri-partite analysis. In *Dr. Ambedkar, Buddhism and social change*, edited by A. K. Narain and D. C. Ahir, 17–34. Delhi: B. R. Publishing Co.

Forman, Paul. 1971. Weimar culture, causality and quantum theory, 1918–1927. *Historical Studies in the Physical Sciences*, 3: 1–115.

Fox, Richard. 1987. Gandhian socialism and Hindu Nationalism: Cultural domination in the world system. *Journal of Commonwealth and Comparative Politics*, 25(3): 233–247.

———. 1989. *Gandhian utopia: Experiments with culture*. Boston: Beacon Press.

Frawley, David (Vamadeva Shastri). 1998. Awaken Bharat. Available online at *http://www.vedanet.com*.

———. 2001. *Hinduism and the clash of civilizations*. New Delhi: The Voice of India.

Freidman, Michael. 1999. *Reconsidering logical positivism*. Cambridge, UK: Cambridge University Press.

Fuller, Christopher J. 1992. *The camphor flame: Popular Hinduism and society in India*. Princeton, NJ: Princeton University Press.

———. 1996. *Caste today*. New Delhi: Oxford University Press.

———. 2001. The 'Vinayaka Chaturthi' Festival and Hindutva in Tamil Nadu. *Economic and Political Weekly*, May 12.

Gadgil, Madhav and Ramachandra Guha. 1992. *This fissured land: An ecological history of India*. Berkeley: University of California Press.

Gay, Peter. 1966. *The enlightenment: An interpretation. The rise of modern paganism.* New York: W.W. Norton & Co.

Geertz, Clifford. 1973. *The interpretation of cultures.* New York: Basic Books.

Geetha, V. and S. V. Rajadurai. 1998. *Towards a non-Brahman millennium: From Iyothee Thass to Periyar.* Calcutta: Samya.

Gellner, Ernest. 1982. The paradox in paradigms. *Times Literary Supplement,* April 23, 451–452.

Giddens, Anthony. 1990. *The consequences of modernity.* Stanford: Stanford University Press.

Gill, Sucha S. 1994. The farmers movement and agrarian change in the green revolution belt of northwest India. *Journal of Peasant Studies,* 21(3/4): 187–212.

Gold, Ann G. 1998. Sin and rain: Moral ecology in rural north India. In *Purifying the earthly body of God: Religion and ecology in Hindu India,* edited by L. E. Nelson, 165–196. Albany: State University of New York Press.

Goldman, Alvin. 1999. *Knowledge in a social world.* Oxford: Clarendon Press.

Goodrick-Clarke, N. 1998. *Hitler's preistess: Savitri Devi, the Hindu Aryan myth and neo-Nazism.* New York: New York University Press.

Goonatilake, Susantha. 1998. *Toward a global science.* Bloomington: Indiana University Press.

Gould, Stephen J. 1999. *The rock of ages: Science and religion in the fullness of life.* New York: The Ballantine Publishing Group.

Gouldner, Alvin. 1979. *The future of intellectuals and the rise of the new class.* New York: Seabury Press.

———. 1980. *The two Marxisms.* New York: Oxford University Press.

Govindacharya, K. N. 1993. Future vistas. In *Ayodhya and the future of India,* edited by Jitendra Bajaj, 181–212. Madras: Centre for Policy Studies.

Grassie, William. 1996. Donna Haraway's metatheory of science and religion: Cyborgs, tricksters and Hermes. *Zygon,* 31(2): 285–304.

Grenz, Stanley. 1996. *A primer on postmodernism.* Grand Rapids, MI: William B. Eerdman Publishing Co.

Griffin, David, R. ed. 1988. *The reenchantment of science : Postmodern proposals.* Albany: State University of New York Press.

Griffin, Roger. 1996. *The nature of fascism.* London: Routledge.

Guha, Ramachandra. 1988. The alternative science movement: An interim assessment. *Lokayan Bulletin,* 6(3): 7–25.

———. 1989. New social movements: The problem. *Seminar,* 355, March 12–15.

Gupta, Akhil. 1998. *Postcolonial developments: Agriculture in the making of modern India.* Durham: Duke University Press.

Gupta, Dipankar. 1997. *Rivalry and brotherhood: Politics in the life of farmers in northern India.* New Delhi: Oxford University Press.

Guru, Gopal. 1992. Shetkari Sanghatana and the pursuit of Laxmi Mukti. *Economic and Political Weekly,* Vol. 27, 28.

Guru, Gopal and V. Geetha. 2000. New phase of dalit-bahujan intellectual activity. *Economic and Political Weekly,* January 15.

Gurumurthy, S. 1993. The inclusive and the exclusive. In *Ayodhya and the future of India,* edited by J. Bajaj, 151–180. Madras: Center for Policy Studies.

Haack, Susan. 1993. *Evidence and inquiry: Towards reconstruction in epistemology.* Malden, MA: Blackwell.

———. 1998. *Manifesto of a passionate moderate: Unfashionable essays.* Chicago: University of Chicago Press.

Habermas, Jürgen. 1995. *The philosophical discourse of modernity.* 2d edition. Cambridge, MA: MIT Press.

Hacker, Paul. 1995. *Philology and confrontation: Paul Hacker on traditional and modern Vedanta,* edited by W. Halbfass. Albany: State University of New York Press.

Halbfass, Wilhelm. 1988. *India and Europe: An essay in understanding.* Albany: State University of New York Press.

Hansen, Thomas B. 2001. *Wages of violence: Naming and identity in postcolonial Bombay.* Princeton, N J: Princeton University Press.

Haraway, Donna. 1991. *Simians, cyborgs and women.* New York: Routledge.

Harding, Sandra. 1986. *The science question in feminism.* Ithaca: Cornell University Press.

———. 1991. *Whose science? Whose knowledge?* Ithaca: Cornell University Press.

———. 1994. Is science multicultural? Challenges, resources, opportunities, uncertainties. *Configurations,* 2: 301–330.

———. 1996. Science is "good to think with." *Social Text,* 46–47: 15–26.

———. 1998. *Is science multicultural?: Postcolonialisms, feminisms and epistemologies.* Bloomington: Indiana University.

———. 2000. Gender and science. In *The Routledge Encyclopedia of Philosophy* (at *http://www.rep.routledge.com*).

Harman, William. 2000. Speaking about Hinduism and speaking against it. *The Journal of the American Academy of Religion,* 68(4): 733–740.

Harrington, Anne. 1999. *Reenchanted science: Hoilsm in German culture from Wilhelm II to Hitler.* Princeton, N. J.: Princeton University Press.

Harriss, Barbara. 1992. Rural poverty in India: Microlevel evidence. In *Poverty in India: Research and policy,* edited by B. Harriss, S. Guhan, and R. H. Cassen. Bombay: Oxford University Press.

Harriss, John. 1992. Does the depressor still work? Agrarian structure and development in India: A review of evidence and argument. *Journal of Peasant Studies,* 19(2): 189–227.

Harstock, Nancy. 1999. *The feminist standpoint revisited and other essays.* Boulder, CO: Westview Press.

Hasan, Zoya. 1994. Shifting ground: Hindutva politics and the farmers' movement in Uttar Pradesh. *Journal of Peasant Studies,* 21(3/4): 165–194.

———. 1998. *Quest for power: Oppositional movements and post-Congress politics in Uttar Pradesh.* New Delhi: Oxford University Press.

Hawley, Jack. 2000. Who speaks for Hinduism—and who against? *Journal of American Academy of Religion,* 68(4): 711–720.

Heimsath, Charles. 1964. *Indian nationalism and Hindu social reform.* Princeton, NJ: Princeton University Press.

Hekman, Susan. 1997. Truth and method: Feminist standpoint theory revisited. *Signs,* 22, 341–365.

Hennessy, Rosemary and Chrys Ingraham. 1997. *Materialist feminism: A reader in class, difference, and women's lives.* New York: Routledge.

Herf, Jeffrey. 1984. *Reactionary modernism: Technology, culture and politics in Weimar and the Third Reich.* Cambridge: Cambridge University Press.

Herring, Ronald. 2001. Promethean science, Pandora's jug: Conflicts around genetically engineered organisms in India. The 2001 Mary Keatinge Das Lecture, Columbia University (at *http://www.einaudi.cornell.edu*).

Hess, David. 1995. *Science and technology in a multicultural world.* New York: Columbia University Press.

———. 1997. If you're thinking of living in STS: A guide for the perplexed. In *Cyborgs and citadels: Anthropological interventions in emerging sciences and technologies,* edited by G. Downey and J. Dumit, 143–164. Santa Fe, New Mexico: School of American Research Press.

Hiriyanna, M. 1996. *Essentials of Indian philosophy.* London: Diamond Books. First edition, 1949.

Hirschmann, Nancy. 1997. Feminist standpoint as postmodern strategy. *Women and politics,* 18: 73–92.

Human Rights Watch. 1999. *Broken people: Caste violence against India's "untouchables."* New York: Human Rights Watch.

———. 2002. *"We have no orders to save you:" State participation and complicity in communal violence in Gujarat* (at *http://www. hrw.org*).

Ilaiah, Kancha. 1998. Toward the dalitization of the nation. In *Wages of freedom: Fifty years of the Indian nation-state,* edited by Partha Chatterjee, 267–291. New Delhi: Oxford University Press.

Inden, Ronald. 1986. Orientalist constructions of India. *Modern Asian Studies,* 20(3): 401–446.

Israel, Jonathan. 2001. *Radical enlightenment: Philosophy and the making of modernity, 1650–1750.* New York: Oxford University Press.

Issac, Thomas, Richard Franke, and M. P. Parameswaran. 1997. From anti-feudalism to sustainable development: The Kerela people's science movement. *Bulletin of Concerned Asian Scholars,* 29(3): 34–44.

Jackson, Cecile. 1993. Women/nature or gender/history? A critique of ecofeminist "development." *Journal of Peasant Studies,* 20(3): 389–419.

Jackson, Cecile and Molly Chattopadhyay. 2000. Identities and livelihoods: Gender, ethnicity and nature in a south Bihar village. In *Agrarian environments: Resources, representations and rule in India,* edited by A. Agarwal and K. Sivaramkrishna, 147–170. Durham: Duke University Press.

Jacob, Margaret. 1988. *The cultural meaning of scientific revolution.* New York: Alfred A. Knopf.

———. 2001. *The enlightenment: A brief history with documents.* Boston: St. Martin's Press.

Jaffrelot, Christophe 1995. The idea of the Hindu race in the writings of Hindu nationalist ideologues in the 1920s and 1930s: A concept between two cultures. In *The concept of race in South Asia,* edited by P. Robb, 327–354. New Delhi: Oxford University Press.

———. 1996. *The Hindu nationalist movement in India.* New York: Columbia University Press.

Jain, Girilal. 1994. *The Hindu phenomenon.* Delhi: UBSPD.

Jalal, Ayesha. 1995. *Democracy and authoritarianism in South Asia.* Cambridge, UK: Cambridge University Press.

Jarvie, Ian. 1984. *Rationality and relativism.* Boston: Routledge & Kegan Paul.

Jitatmanada, Swami. 1993. *Holistic science and Vedanta.* Bombay: Bharatiya Vidya Bhavan.

Jodha, N. S. 1988. Poverty debate in India: A minority view. *Economic and Political Weekly,* Special Number, November, 2421–2428.

Jones, Richard. 1986. *Science and mysticism: A comparative study of western natural science, Theravada Buddhism, and Advaita Vedanta.* Lewisburg: Bucknell University Press.

Joshi, Murli Manohar. 1994. India has a hoary scientific tradition. In Roundtable on India: Synthesis of science and spirituality. *Pragna Journal* (renamed *Bharatiya Pragna*) (at *http://www.pragna.org*).

Jurgensmeyer, Mark. 1993. *The new cold war? Religious nationalism confronts the secular state*. Berkeley: University of California Press.

Kak, Subhash. 1994a. *India at century's end: Essays on history and politics*. New Delhi: Voice of India.

———. 1994b. *The astronomical code of the Rig Veda*. New Delhi: Aditya Prakashan.

———. 1995. *From Vedic science to Vedanta*. Madras: The Adyar Library and Research Center.

———. 1999a. Overview of Indian science. *Pragna Journal*, 3(1) (at *http://www.pragna. org.art17199.html*).

———. 1999b. The Rediff interview/Subhash Kak (at *http://www.rediff.com*), November 18.

———. n.d. A thousand cows standing the one above the other. *India Star: A Literary Art Magazine* (at *http://Indiastar.com/kak3.htm*).

Kakar, Sudhir. 1998. *The colors of violence*. Chicago: Chicago University Press.

Kanan, K. P. 1990. Secularism and people's science movement in India. *Economic and Political Weekly*, February 10: 311–313.

Keddie, Nikki. 1998. The new religious politics: Where, when and why do "fundamentalisms" appear? *Comparative Study of Society and History*, 40(4): 696–723.

Keller, Evelyn F. 1982. Feminism and science. In *Feminist theory: A critique of ideology*, edited by N. Keohane, M. Rosaldo, and B. Gelpi, 113–126. Chicago: Chicago University Press.

———. 1985. *Reflections on gender and science*. New Haven: Yale University Press.

Killingley, Dermot. 1998.Vivekananda's western message from the east. In *Swami Vivekananda and the modernization of Hinduism*, edited by William Radice, 138–157. New Delhi: Oxford University Press.

King, Richard. 1999. *Orientalism and religion: Postcolonial theory, India and the mystic east*. New York: Routledge.

Kishwar, Madhu. 2000. From Manusmriti to Madhusmriti: Flagellating a Mythical Enemy. *Manushi*, Number 117, April–May.

Kitcher, Philip. 1993. *The advancement of science: Science without legend, objectivity without illusions*. New York: Oxford University Press.

———. 1998. A plea for science studies. In *A house built on sand: Exposing postmodernist myths about science*, edited by N. Koertge, 32–56. New York: Oxford University Press.

Klee, Robert. 1997. *Introduction to the philosophy of science: Cutting nature at its seams*. New York: Oxford University Press.

Koertge, Noretta. 1998. Scrutinizing science. In *A house built on sand: Exposing postmodernist myths about science,* edited by N. Koertge, 3–6. New York, Oxford University Press.

Kopf, David. 1969. *British orientalism and the Bengal renaissance*. Berkeley: University of California Press.

Krishna, Venni. 1997. Science, technology and counter-hegemony: Some reflections on the contemporary science movements in India. In *Science and technology in a developing world*, edited by J. B. Spaapen, V. Krishna, and T. Shinn, 375–411. Dordrecht: Kluwer Academic Publishers.

Kuhn, Thomas. 1970. *The structure of scientific revolutions*. 2d edition. New York: New American Library.

———. 1977. *Essential tension: Selected studies in scientific tradition and change*. Chicago: University of Chicago Press.

Lal, Vinay. 1995. Hindu "fundamentalism" revisited. *Contention*, 4(2): 165–173.

Lambert, Yves. 1999. Religion in modernity as a new axial age: Secularization or new religious forms? *Sociology of Religion*, 60(3): 303–333.

Laqueur, Walter. 1996. *Fascism: Past, present and future*. New York: Oxford University Press.

Larson, Gerald. 1995. *India's agony over religion*. Albany: State University of New York Press.

Latour, Bruno. 1993. *We have never been modern*. Cambridge: Harvard University Press.

Laudan, Larry. 1984. *Science and values*. Berkeley: University of California Press.

———. 1996. *Beyond positivism and relativism*. Boulder, CO: Westview Press.

Levi-Strauss, Claude. 1962. *The savage mind*. Chicago: University of Chicago Press.

Levins, Richard. 1986. A science of our own: Marxism and nature. *Monthly Review*, July–August, 3–12.

Lilla, Mark. 2001. *The reckless mind: Intellectuals in politics*. New York: New York Review of Books.

Lloyd, Genevive. 1984. *The man of reason*. Minneapolis: University of Minnesota Press.

Longino, Helen. 1990. *Science as social knowledge: Values and objectivity in scientific inquiry*. Princeton, NJ: Princeton University Press.

———. 1993. Subjects, power and knowledge: Description and prescription in feminist philosophies of science. In *Feminist epistemologies*, edited by L. Alcoff and E. Potter, 101–120. New York: Routledge.

———. 1995. Gender, politics and theoretical virtues. *Synthese*, 104: 383–397.

———. 1996. Cognitive and non-cognitive values in science: Rethinking the dichotomy. In *Feminism, science and the philosophy of science*, edited by L. H. Nelson and J. Nelson, 39–58. Dordrecht: Kluwer.

———. 1997. Interpretation versus explanation in the critique of science. *Science in Context*, 10: 113–128.

———. 1999. Feminist epistemology. In *The Blackwell guide to epistemology*, edited by J. Greco and E. Sosa, 327–353. Malden, MA: Blackwell.

Lutgendorf, Philip. 1997. The monkey in the middle: The status of Hanuman in popular Hinduism. *Religion*, 27: 311–332.

Lyotard, Jean-François. 1984. *The postmodern condition: A report on knowledge*. Minneapolis: University of Minnesota Press.

Mackenzie-Brown, C. 2002. Hindu and Christian creationism: "Transposed passages" in the geological book of life. *Zygon*, 37(1): 95–114.

Madan, Tirloki Nath. 1993. Whither Indian secularism? *Modern Asian Studies*, 27(3): 667–697.

———. 1998. Secularism in its place. In *Secularism and its critics*, edited by Rajeev Bhargava, 297–320. New Delhi: Oxford University Press.

Mahajan, Gurpreet. 1998. *Identities and rights: Aspects of liberal democracy in India*. New Delhi: Oxford University Press.

Mahmood, Cynthia. 1993. Rethinking Indian communalism: Culture and counter-culture. *Asian Survey*, 33: 722–737.

Marglin, Frederique. 1990. Smallpox in two systems of knowledge. In *Dominating knowl-edge: development, culture and resistance*, edited by F. A. Marglin and S. A. Marglin, 102–144. Oxford: Clarendon Press.

———. 1996. Rationality, the body and the world: From production to regeneration. In *De-colonizing knowledge: From development to dialogue*, edited by F. A. Marglin and S. A. Marglin, 142–182. Oxford: Clarendon Press.

Marglin, F. A. and S. A. Marglin. 1990. *Dominating knowledge: development, culture and re-sistance*. Oxford: Clarendon Press.

———. 1996. *Decolonizing knowledge: From development to dialogue*. Oxford: Clarendon Press.

Marriott, McKim. 1990. Constructing an Indian ethno-sociology. In *India through Hindu categories*, edited by M. Marriott, 1–40. New Delhi: Sage Publications.

Martin, Emily. 1998. Anthropology and the cultural study of science. *Science, Technology and Human Values*, 23 (1): 24–44.

Marty, Martin and Appleby, R. Scott 1992. *The glory and the power: The fundamentalist challenge to the modern world*. Boston: Beacon Press.

Marx, Karl. 1978. The eighteenth brumaire of Loius Bonaparte. In *The Marx-Engels Reader*, edited by Robert C. Tucker, 594–617. New York: W. W. Norton & Co.

Mayer, Arno. 1990. *Why did the heavens not darken? The "final solution" in history*. New York: Pantheon Books.

McKean, Lise. 1996. *Divine enterprise: Gurus and the Hindu nationalist movement*. Chicago: Chicago University Press.

Mendelsohn, Oliver and Marika Vicziany. 1998. The untouchables. Cambridge, UK: Cam-bridge University Press.

Merchant, Carolyn. 1980. *The death of nature*. New York: Harper Collins.

Mies, Maria and Vandana Shiva. 1993. *Ecofeminism*. Delhi: Kali for Women.

Moon, Vasant. 2000. *Growing up untouchable in India*. New York: Rowman and Littlefield.

Moore, Barrington, Jr. 1966. *Social origins of dictatorship and democracy: Lord and peasant in the making of the modern world*. Boston: Beacon Press.

Moore, Kelly. 1996. Organizing integrity: American science and the creation of public in-terest organizations, 1955–1975. *American Journal of Sociology*, 101(6): 1592–1627.

Moore-Gilbert, Bart. 1997. *Postcolonial theory: Contexts, practices and politics*. London: Verso.

Mosse, George. 1961. The mystical origins of national socialism. *Journal of History of Ideas*. 22(1): 81–96.

———. 1964. *The crisis of German ideology: Intellectual origins of the Third Reich*. New York: Schocken Press.

———. 1999. *The fascist revolution: Toward a general theory of fascism*. New York: Howard Fertig.

Mukhyananda, Swami. 1997. *Vedanta in the context of modern science: A comparative study*. Mumbai: Bharatiya Vidya Bhavan.

Munson, Henry. 1995. Not all crustaceans are crabs: Reflections on the comparative study of fundamentalism and politics. *Contention*, 4(3): 151–166.

Murphy, Nancey. 1997. *Anglo-American postmodernity: Philosophical perspectives on sci-ence, religion and ethics*. Boulder, CO: Westview Press.

Nader, Laura. 1996. *Naked science: Anthropological inquiry into boundaries, power, knowl-edge*. New York: Routledge.

Nanda, Meera. 1991. Is modern science a western patriarchal myth? A critique of the populist orthodoxy. *South Asia Bulletin*, 11(1&2): 32–60.

———. 1997. Science wars in India. *Dissent*, Winter, 78–83.

———. 1998. Epistemic charity of the social constructivist critics of science and why the
third world should refuse their offer. In *A house built on sand: Exposing postmodernist
myths about science*, edited by N. Koertge, 286–312. New York: Oxford University Press.

———. 2001. A "broken people" defend science: Reconstructing the Deweyan Buddha of
India's dalits. *Social Epistemology*, 15(4): 335–365.

———. 2002. *Breaking the spell of dharma and other essays*. New Delhi: Three Essays
Press.

———. Forthcoming. Science as the standpoint of the oppressed: Dewey meets the Buddha of India's dalits. In *Scrutinizing Feminist Epistemology*, edited by C. Pinnick, R.
Almeder, and N. Koertge. New Brunswick: Rutgers University Press.

Nandy, Ashis. 1980. *Alternative sciences: Creativity and authenticity in two Indian scientists.*
New Delhi: Allied Publishers.

———. 1981. Counter-statement on humanistic temper. *Mainstream*, October 10, 16–18.

———. 1983. *The intimate enemy*. New Delhi: Oxford University Press.

———. 1987. *Traditions, tyranny and utopias: Essays in political awareness*. New Delhi: Oxford University Press.

———, ed. 1988. *Science, hegemony and violence: A requiem for modernity*. Japan: United
Nations University.

———. 1998. The politics of secularism and the recovery of religious tolerance. In *Secularism and its critics*, edited by Rajeev Bhargava, 321–344. New Delhi: Oxford University Press.

———. 2001. A report on the present state of health of the gods and goddesses in south
Asia. *Postcolonial Studies*, 4(2): 125–141.

Nandy, Ashis and Shiv Visvanathan. 1990. Modern medicine and its non-modern critics: A
study of discourse. In *Dominating knowledge*, edited by Frederique A. Marglin and
Stephen A. Marglin, 145–184. Oxford: Clarendon Press.

Narain, A. K. and D. C. Ahir, eds. 1994. *Dr. Ambedkar, Buddhism and social change*. Delhi:
B. R. Publishing Co.

Needham, Joseph. 1969. *The grand titration: Science and society in east and west*. London:
George Allen & Unwin Ltd.

———. 1970. *Clerks and craftsmen in China and the west*. Cambridge: Cambridge University Press.

Nehru, Jawaharlal. 1949. *The discovery of India*. Calcutta: Signet Press.

———. 1988. *Jawaharlal Nehru on science and society: A collection of his writings and
speeches*, edited by Baldev Singh. New Delhi: Nehru Memorial Museum and Library.

Newman, William. 1998. Alchemy, domination and gender. In *A house built on sand: Exposing postmodernist myths about science*, edited by N. Koertge, 216–226. New York: Oxford University Press.

Newton-Smith, William. 1981. *The rationality of science*. Boston: Routledge Kegan & Paul.

Nigam, Aditya. 2000. Secularism, modernity, nation: Epistemology of the dalit critique.
*Economic and Political Weekly*, November 25, 2003–2009.

Noorani, A. G. 1997. The RSS and the Mahatma. *Frontline*, November 28, 93–96.

Nussbaum, Martha. 1995. Human capabilities, female human beings. In *Women, culture
and development: A study of human capabilities*, edited by Martha Nussbaum and
Jonathan Glover, 61–104. Oxford: Clarendon Press.

———. 1996. Patriotism and cosmopolitanism. In *For love of country: Debating the limits of patriotism*, edited by M. Nussbaum, 2–20. Boston: Beacon Press.

O'Flaherty, Wendy Doniger. 1976. *The origin of evil in Hindu mythology*. Berkeley: University of California Press.

O'Hanlon, Rosalind. 1985. *Caste, conflict, and ideology: Mahatma Jyotirao Phule and low caste conflict in 19th century western India*. Cambridge, UK: Cambridge University Press.

Olivelle, Patrick. 1996. *Upanishads*. New York: Oxford University Press.

Omvedt, Gail. 1993. *Reinventing revolution: New social movements and the socialist tradition in India*. New York: M. E. Sharpe.

———. 1994. *Dalits and the democratic revolution: Dr. Ambedkar and the dalit movement in colonial India*. New Delhi: Sage Publications.

———. 1997. Ideology for provincial propertied class? *Economic and Political Weekly*, January 30.

Overing, Joanna. 1985. Introduction. In *Reason and morality*, edited by Joanna Overing, 1–28. London: Tavistock Publications.

Panikkar, K. N. 1995. *Culture, ideology and hegemony: Intellectuals and social consciousness in colonial India*. New Delhi: Tulika.

Papanek, Hanna. 1990. To each less than she needs, form each more than she can do: Allocations, entitlements and value. In *Persistent inequalities: Women and world development*, edited by Irene Tinker, 162–184. New York: Oxford University Press.

Parekh, Bhiku. 1992. The cultural peculiarity of liberal democracy. *Political Studies*, 40: 160–175.

———. 1995. Jawaharlal Nehru and the crisis of modernization. In *Crisis and Change in Contemporary India*, edited by U. Baxi and B. Parekh, 21–56. New Delhi: Sage Publications.

Parikh, Kirit and R. Radhakrishna. 2002. *India development report, 2002*. New Delhi: Oxford University Press.

Parry, Jonathan. 1985. The Brahmanical tradition and the technology of the intellect. In *Reason and morality*, edited by Joanna Overing, 200–225. London: Tavistock Publications.

Patnaik, Utsa and Manjari Dingwaney. 1985. *Chains of servitude: Bondage and slavery in India*. New Delhi: Sangam Books.

Paxton, Robert. 1998. The five stages of fascism. *Journal of Modern History*, 70(1): 1–23.

Peacocke, Arthur. 2000. Science and the future of theology: Critical issues. *Zygon*, 35(1): 119–140.

———. 2001. *Paths from science towards God*. Oxford: Oneworld.

Pois, Robert. 1985. *National socialism and the religion of nature*. London: Croom Helm.

Pollock, Sheldon. 1993. Ramayana and political imagination in India. *Journal of Asian Studies*, 52(2): 261–297.

Potter, Karl. 1980. The karma theory and its interpretations in some Indian philosophical systems. In *Karma and rebirth in classical Indian traditions*, edited by Wendy Doniger O'Flaherty, 241–267. Berkeley: University of California Press.

Prakash, Gyan. 1990. Writing post-Orientalist histories of the third world: Perspectives from Indian historiography. *Comparative Studies in Society and History*, 32(2): 383–408.

Putnam, Hilary. 1992. *Renewing philosophy*. Boston: Harvard University Press.

Putnam, Hilary and Ruth A. Putnam. 1990. Epistemology as hypothesis. *Transactions of Charles S. Pierce Society*, 26: 407–433.

Queen, Christopher. 1996. Dr. Ambedkar and the hermeneutics of Buddhist liberation. In *Engaged Buddhism: Buddhist liberation movements in Asia*, edited by C. Queen and S. B. King, 45–72. Albany: State University of New York Press.

Radhakrishnan, Sarvepalli. 1993. *The Hindu view of life*. New Delhi: Indus. First edition, 1923.

Raghunandan, D. 1989. Walking on Three Legs: Science, Common-sense and Ideology. *Social Scientist*, 25(2): 92–101.

Rahman, Shaikh A. 2002. Indian defense looks to ancient texts. BBC News, May 14 (at *http://www.bbc.co.uk*).

Raina, Dhruv. 1997. Evolving perspectives on science and history: A chronicle of modern India's scientific enchantment and disenchantment (1850–1980). *Social Epistemology*, 11(1): 3–24.

Raina, Dhruv and S. Irfan Habib. 1996. The moral legitimation of modern science: Bhadralok reflections of theories of evolution. *Social Studies of Science*, 26: 9–42.

Rajaram, N. S. 1995. *Secularism: The new mask of fundamentalism*. New Delhi: Voice of India.

———. 1998. *The Hindu view of the world: Essays in the intellectual Kshatriya tradition*. New Delhi: Voice of India.

Ramanujam, A. K. 1990. Is there an Indian way of thinking? An informal essay. In *India through Hindu categories*, edited by M. Marriott, 41–58. New Delhi: Sage Publications.

Rangan, Haripriya. 2000. *Of myths and movements: Rewriting Chipko into Himalayan history*. London: Verso.

Ranganathnanda, Swami. 1991. *Human being in depth: A scientific approach to religion*. Albany: State University of New York Press.

Rao, C. H. Hanumantha. 1994. *Agricultural growth, rural poverty and environmental degradation in India*. New Delhi: Oxford University Press.

Rege, Sharmila. 1998. Dalit women talk differently: A critique of difference and towards a dalit standpoint position. *Economic and Political Weekly*, October 31, WS39–WS46.

Rescher, Nicholas. 2000. *Realistic pragmatism: An introduction to pragmatic philosophy*. Albany: State University of New York Press.

Restivo, Sal. 1988. Modern science as a social problem. *Social Problems*, 35(3): 206–225.

Roland, Alan. 1988. *In search of self in India and Japan*. Princeton, NJ: Princeton University Press.

Rorty, Richard. 1979. *Philosophy and the mirror of nature*. Princeton, N J: Princeton University Press.

Rose, Hilary. 1996. My enemy's enemy is—only perhaps—my friend. *Social Text*, 46–47: 61–80.

Rosenberg, Alfred. 1982. The myth of the twentieth century. Torrance, Calif.: Noontide Press.

Roseneu, P. 1992. *Postmodernism and the social sciences*. Princeton, NJ: Princeton University Press.

Ross, Andrew. 1996. Introduction: Science wars. *Social Text*, 46–47: 1–14.

Rouse, Joseph. 1996. *Engaging science: How to understand its practices philosophically*. Ithaca: Cornell University Press.

Roy, Raja Ram Mohan. 1999. *Vedic physics: Scientific origins of Hinduism*. Toronto: Golden Egg Publishing.

Rudolph, Lloyd and Susanne Rudolph. 1987. *In pursuit of Lakshmi: The political economy of the Indian state*. Chicago: Chicago University Press.

Ruse, Michael. 1996. *But is it science? The philosophical question in the creation/evolution controversy.* New York: Prometheus Books.

Russel, Bertrand. 1935. *Religion and science.* London: Oxford University Press.

Sahlins, Marshall. 1999. What is anthropological enlightenment? Some lessons of the twentieth century. *Annual Review of Anthropology,* 28: 1–23.

Said, Edward. 1978. *Orientalism.* New York: Vintage Books.

Sardar, Ziauddin, ed. 1988. *The revenge of Athena.* London: Mansell.

Sarkar, Sumit. 1975. Rammohun Roy and the break with the past. In *Rammohun Roy and the process of modernization in India,* edited by V. C. Joshi. New Delhi: Vikas.

Sarkar, Tanika. 1996. Educating the children of the Hindu rashtra: Notes on RSS schools. In *Religion, religiosity and communalism,* edited by P. Bidwai, H. Mukhia, and A. Vanaik, 237–248. New Delhi: Manohar.

———. 2001. *Hindu wife, Hindu nation.* New Delhi: Permanent Black.

Savarkar, Vinayak D. 1989. *Hindutva: Who is a Hindu?* New Delhi: Bharti Sahitya Sadan. First edition, 1923.

Sawant, S. D. and Achuthan, C. V. 1995. Agricultural growth across crops and regions: Emerging trends and patterns. *Economic and Political Weekly,* March 25.

Sax, William. 2000. In Karna's realm: An ontology of action. *Journal of Indian Philosophy,* 28: 295–324.

Segerstråle, Ullica. 2000. *Defenders of the truth: The battle for science in the sociobiology debate and beyond.* New York: Oxford University Press.

Sen, Amartya. 1990a. Gender and cooperative conflicts. In *Persistent inequalities: Women and world development,* edited by Irene Tinker, 123–149. New York: Oxford University Press.

———. 1990b. More than 100 Million Women are Missing. *The New York Review of Books,* December 20.

———. 2000. East and west: The reach of reason. *The New York Review of Books,* July 20.

Sen, S. N. 1971. Astronomy. In *A concise history of science in India,* edited by D. M. Bose, S. N. Sen, B. V. Subbarayappa, 58–135. New Delhi: Indian National Academy of Science.

———. 1996. History of science in relation to philosophy and culture in Indian civilization. In *Science, philosophy and culture: Multidisciplinary explorations.* Part I, edited by D. P. Chattopadhyaya and R. Kumar. New Delhi: Indian Council of Philosophical Research.

Setlavad, Teesta. 2001. *Communalism Combat,* November, Rajasthan (at *http://www.sabrang.com/cc/archive/2001/Nov01/cover.html*).

Shapin, Steven. 1995. Here and everywhere: Sociology of scientific knowledge. *Annual Review of Sociology,* 21: 289–321.

Sheehan, T. 1981. Myth and violence: The fascism of Julius Evola and Alain de Benoist. *Social Research,* 48(1): 45–73.

Shermer, Michael. 1997. *Why people believe in weird things.* New York: Freeman & Co.

Sheth, D. L. 1999. Secularisation of caste and making of new middle class. *Economic and Political Weekly,* August 21–28.

Shils, Edward. 1961. *The intellectual between tradition and modernity: The Indian situation.* The Hague: Mouton & Co.

———. 1969. The Intellectuals and the powers: Some perspectives for comparative analysis. In *On intellectuals,* edited by P. Rieff, 25–48. New York: Doubleday and Co.

Shiva, Vandana. 1988. *Staying alive: Women, ecology and survival.* London: Zed Press.

Singer, Milton. 1972. *When a great tradition modernizes.* Chicago: University of Chicago Press.

Smith, Brian. 1996. Re-envisioning Hinduism and evaluating the Hindutva movement. *Religion*, 26: 119–128.

Smith, Donald. 1998. India as a secular state. In *Secularism and its critics*, edited by Rajeev Bhargava, 177–233. New Delhi: Oxford University Press.

Sokal, Alan. 1996. A physicist experiments with cultural studies. *Lingua Franca*, May–June, 62–64.

Sokal, Alan and Jean Bricmont. 1998. *Fashionable nonsense: Postmodern intellectuals' abuse of science*. New York: Picador.

Spengler, Oswald. 1991. *The decline of the west*. Abridged edition, with Introduction by H. Stuart Hughes. New York: Oxford University Press. First edition, 1932.

Spivak, Gayatri, C. 1988a. Can the subaltern speak? In *Marxism and the interpretation of culture*. edited by C. Nelson and L. Grossberg, 271–334. Urbana: University of Illinois Press.

———. 1988b. Subaltern studies: Deconstructing historiography. In *Selected subaltern studies*. edited by R. Guha and Gayatri C. Spivak, 3–34. New York: Oxford University Press.

———. 1993. *Outside in the teaching machine*. New York: Routledge.

———. 2000. The new subaltern: A silent interview. In *mapping subaltern studies and the postcolonial*, edited by V. Chaturvedi, 324–340. London: New Left Review.

Srinivas, M. N. 1966. *Social change in modern India*. Berkeley: University of California.

———, ed. 1996. *Caste: Its twentieth century avatar*. New Delhi: Penguin India.

Staal, Frits. 1999. Greek and Vedic geometry. *Journal of Indian Philosophy*, 27: 105–127.

———. 2001. Squares and oblongs in the Veda. *Journal of Indian Philosophy*, 29: 257–273.

Stenger, Victor. 1995. *The unconscious quantum: Metaphysics in modern physics and cosmology*. New York: Prometheus Books.

———. 1996. Mystical physics: Has science found the path to the ultimate? *Free Inquiry*, 16(3): 1–12.

Sternhell, Zeev. 1998. Fascism. In *International Fascism: Theories, Causes and the New Consensus*, edited by Roger Griffin, 30-34. London: Arnold Publishing.

Sudarshan, K. S. 1998. Hindu Dharma and its scientific spirit. *Pragna Journal* 2(2), April–June (at *http://www.pragna.org*).

Taylor, Charles. 1989. *Sources of the self: The making of the modern identity*. Boston: Harvard University Press.

———. 1993. Explanation and practical reason. In *The quality of life*, edited by M. Nussbaum and A. Sen, 208–231. Oxford: Clarendon Press.

———. 1995. Two theories of modernity. *Hastings Center Report*, 25(2): 24–33.

Teltumbde, Anand. 1997. 'Ambedkar' in and for the post-Ambedkar dalit movement (at *http://www.ambedkar.org*).

Thengadi, Dattopant. 1983. Modernization without westernization (at *http://www.hvk.org/books*).

Thomas, Keith. 1971. *Religion and the decline of magic*. New York: Charles Scribner's Sons.

Thompson, Robert. 1981. *Mechanistic and non-mechanistic science: An investigation into the nature of consciousness and form*. Los Angeles: Bhaktivedanta Book Trust.

Turnbull, David. 1997. Knowledge systems: Local knowledge. In *The encyclopedia of the history of science, technology and medicine in non-western cultures*, edited by H. Selin, 485–490. Dordrecht: Kluwer Academic Publishers.

United Nations Development Program. 1998. *Human Development Report*. New York: Oxford University Press.

Upadhyaya, Deendayal. 1965. Integral humanism (full text available at *http://www.bjp.org*).

Vanaik, Achin. 1990. *The painful transition: Bourgeois democracy in India*. London: Verso.

———. 1997. *Communalism contested: Religion, modernity and secularization*. New Delhi: Vistaar.

Vanita, Ruth. 2002. Whatever Happened to the Hindu Left? *Seminar*, 512, May (at *http://www.india-seminar.com*).

Varshney, Ashutosh. 1993. Self-limited empowerment: Democracy, economic development and rural India. *Journal of Development Studies*, 29(4): 177–215.

———. 1995. *Democracy, development and the countryside: Urban-rural struggles in India*. New York: Cambridge University Press.

Vasavi, A. R. 1999. *Harbingers of rain: Land and life in South India*. New Delhi: Oxford University Press.

Van der veer, Peter. 1994. *Religious nationalism: Hindus and Muslims in India*. Berkeley: University of California Press.

Verma, Roop Rekha. 1995. Femininity, equality and personhood. In *Women, culture and development*, edited by M. Nussbaum and J. Glover, 433–443. Oxford: Clarendon Press.

Viereck, Peter. 1965. *Metapolitics: The roots of the Nazi mind*. New York: Capricorn Books.

Visvanathan, C. 2000. Ayurveda isn't what it used to be: Views of an Indian surgeon. Part I & II (at *http://www.drkoop.com*).

Vivekananda, Swami. 1965. *The complete works of Swami Vivekananda*, Volumes I–VIII. Mayawati Memorial Edition. Calcutta: Advaita Ashram. Original edition, 1907.

———. 1968. *Hinduism*. Mylapore, Madras: Sri Ramakrishna Math.

Vlassoff, Carol. 1994. From rags to riches: The impact of rural development on women's status in an Indian village. *World Development*, 22(5): 707–719.

Walby, Sylvia 1992. Post-post-modernism? Theorizing social complexity. In *Destabilizing theory: Contemporary feminist debates*, edited by M. Barrett and A. Philip, 31–52. Stanford: Stanford University Press.

Warren, Karen, ed. 1997. *Ecofeminism: Women, culture, nature*, with editorial assistance from Nisvan Erkal. Bloomington: Indiana University Press.

Watson-Verran, Helen and David Turnbull. 1995. Science and other: Indigenous knowledge systems. In *Handbook of STS*, edited by S. Jasonoff, G. Markle, J. Petersen, and T. Pinch, 115–139. Thousand Oaks: Sage Publications.

Weber, Max. 1946. Science as vocation. In *From Max Weber: Essays in sociology*, edited by Hans Gerth and C. Wright Mills, 129–158. New York: Oxford University Press. First published, 1922.

Weeratunge, Nireka. 2000. Nature, harmony and the Kaliyuga. *Current Anthropology*, 41(3): 249–268.

Weinberg, Steven. 1992. *Dreams of a final theory*. New York: Pantheon.

Westbrook, R. 1991. *John Dewey and American democracy*. Ithaca: Cornell University Press.

———. 1998. Pragmatism and democracy: Reconstituting the logic of John Dewey's faith. In *The Revival of Pragmatism*, edited by Morris Dickstein, 128–140. Durham: Duke University Press.

Westfall, Richard. 1958. *Science and religion in 17th century England*. Ann Arbor: Michigan University Press.

Whitehead, Anne.1990. Food crisis and gender conflict in the African countryside. In *The food question: Profits vs. people?* edited by H. Bernstein, B. Crow, M. Mackintosh, and C. Martin, 54–68. New York: Monthly Review Press.

Wignaraja, Ponna. 1993. *New social movements in the South*. London: Zed.

Wilson, Bryan. 1982. *Religion in sociological perspective*. Oxford, UK: Oxford University Press.

Witzel, Michael. 2001. Autochthonous Aryans? The evidence from old Indian and Iranian texts. *Electronic Journal of Vedic Studies*, 7(3): 1–93.

Wodak, Jo and David Oldroyd. 1996. "Vedic Creationism:" A further twist to the evolution debate. *Social Studies of Science*, 26: 192–213.

Wood, E. Meiksins. 1988. Capitalism and Human Emancipation. *New Left Review*, 167: 73–100.

Yack, Bernard. 1992. *The longing for total revolution*. Berkeley: University of California Press.

Yandell, Keith. 1986. Protestant theology and natural science in the twentieth century. In *God and nature: Historical essays on the encounter between Christianity and science*, edited by D. Lindberg and R. Numbers, 448–472. Berkeley: University of California Press.

Young, Robert. 1990. *White mythologies: Writing history and the west*. New York: Routledge.

Youzhong, S. 1999. John Dewey in China: Yesterday and today. *Transactions of the Charles S. Peirce Society*, 35: 69–88.

Zachariah, Mathew and R. Sooryamoorthy. 1994. *Science for social revolution? Achievements and dilemmas of a development movement*. London: Zed.

Zaehner, R. C. 1962. *Hinduism*. Oxford, UK: Oxford University Press.

Zelliot, Eleanor. 1992 *From untouchable to dalit: Essays on the Ambedkar movement*. New Delhi: Manohar.

———. 1994. New voices of the Buddhists in India. In *Dr. Ambedkar, Buddhism and social change*, edited by A. K. Narain and D. C. Ahir, 195–208. Delhi: B. R. Publishing Co.

Zimmer, Heinrich. 1951. *Philosophies of India*. Princeton, NJ: Princeton University Press.

Zimmerman, Michael. 1994. *Contesting the earth's future: Radical ecology and postmodernity*. Berkeley: University of California Press.

# Index

actor-networks, 23; inscriptions, 275n. 3; ontological monism of, 143–144. *See also* Latour, Bruno

Agarwal, Anil, 225

Agarwal, Bina, 234, 235, 237, 239, 255; on women's right to property, 241–242

Ahmad, Aijaz, on postcolonial intellectuals, 27, 229

al-Azm, Sadiq, 65

Alberuni, on Hindu astronomers, 102–103

alternative modernity, concept of, 6–7, 263; Gandhi as icon of, 151; multiculturalism and, 7. *See also* critical traditionalism; Hindu modernity

alternative science, concept of, 95; enchanted science, 6; multiculturalism and, 162–187. *See also* Vedic science

alternative science movement (India), context of, 210–221; de-secularization and, 220–222; Gandhian thought and, 216–217; Hindutva and, 214–216; other new social movements and, 217–222. *See also* people's science movements; scientific temper debate

Alvares, Claude, 25, 156, 213

Ambedkar, Bhimrao, Ramji, xiii, 261; against standpoint epistemology, 192; American experience of, 182–183, 189; *Annihilation of Caste*, 190, 193; *Buddha and His Dhamma*, 183, 189, 194, 197; Buddhist critics of, 196; call for Indian Enlightenment, 191–192, 197; conversion to Buddhism, 189, 190; debates with Gandhi, 191; Dewey's influence on, 182, 183–184, 188–189, 202; on Hinduism and caste, 190–191; people's science movements and, 219, 221; relevance to feminism, 181–182,

197–199; significance of 199–202, 266–267; struggle for dalit rights, 190. *See also* Deweyan Buddha

*Annihilation of Caste* (Ambedkar), 190, 193

anti-caste movements, 184, 192; Indian feminism and, 198–199

anti-Enlightenment, feminism and, 146–147; postcolonial intellectuals and, 76, 152, 254; postmodern intellectuals and, 1, 19; reactionary modernism and, 7–9; religious-political movements and, 10–11. *See also* Enlightenment; Enlightenment in India

anti-essentialism, alternative science and, 169; hybridity, and, 169; postmodern dogma as, 67, 156, 263

Appleby, Scott, 11, 43

Armstrong, Karen, 30, 42, 273n. 8

Aryan, "Aryan Hindus," the, 14, 15; Aryanization of science and technology, 8, 111–112; India as the homeland of, 89, 272n. 11

astrology. *See* Vedic astrology

associative logic, decline of, in the West, 116; illustration of, 112–113; macrocosm-microcosm, 114–115; Vedic science defense of, 110, 116–117

authenticity, 6; as a political force, 42–48; authentic Hindu secularism, 52–61; reactionary modernism and, 7–9. *See also* Indian intellectuals

Ayureveda, 65, 213

Babb, Lawrence, 87

Balgopal, K., 252

Banaji, Jairus, 252, 253, 258

Barnes, Barry, 23, 128, 129, 131, 133

*About the Author*

---

Meera Nanda received her Ph.D. in science studies from Rensselaer Polytechnic Institute, Troy, New York. She was trained as a microbiologist, and has worked as a science writer in India and in the United States. She received a fellowship from American Council of Learned Societies and has been a visiting scholar at Columbia University

Her other books include *Breaking the Spell of Dharma and Other* Essays (2002) and *Planting the Future: A Resource Guide to Sustainable Agriculture in the Third World* (1990).